Visions of Paradise

Visions of Paradise

Glimpses of Our Landscape's Legacy

John Warfield Simpson

UNIVERSITY OF CALIFORNIA PRESS
Berkeley   Los Angeles   London

University of California Press
Berkeley and Los Angeles, California

University of California Press, Ltd.
London, England

© 1999 by John Warfield Simpson

Illustrations © 1998 Abigail Rorer

Library of Congress Cataloging-in-Publication Data

Simpson, John W. (John Warfield)
    Visions of paradise : glimpses of our landscape's
legacy / John Warfield Simpson.
        p.   cm.   Includes bibliographic
    references (p.   ) and index.
      ISBN 0-520-21364-5 (alk. paper)

      1. Landscape changes—West (U.S.)—History.
    2. Landscape changes—Middle West—History.
    3. Human ecology—United States—History.
    4. Environmental policy—United States.   I. Title.
    GF504.W35S56   1999
    333.73'13'0978—dc21   98-3620

Printed in the United States of America
9 8 7 6 5 4 3 2 1

To
Katherine Anne
and
Edward Warfield

Contents

Acknowledgments

Like many Americans I've lived virtually my entire life in suburbia, safe, comfortable, and generally content, yet basically blind to the underlying landscape. I was oblivious to its fascinating physical and cultural heritage. Fortunately I'm beginning to look beneath the surface. I'm learning, at least in part, to *see* the landscape, not just to *be* in the landscape. As my vision sharpens, I'm linked to the land more and more with inextricable bonds that comfort me and provide a pleasing sense of belonging. Each new glimpse shows me a landscape alive with history and rich with meaning. The landscape is my heritage made visible.

Many guides have aided this search for insight. My university home has proven an inexhaustible wellspring of gracious people who have generously contributed their time and wisdom. I'm indebted to Professors Ralph Boerner, George Hall, Charles King, Garry McKenzie, and Bill Mitsch for their cheerful willingness to assist a colleague in need. Each shared his special knowledge on fascinating field trips into the central Ohio landscape. Professors Hazel Morrow-Jones, Ken Pearlman, Lewis Ulman, and Rick Yerkes provided other assistance along my intellectual journey.

Thanks, too, to Mac Albin, chief naturalist at Battelle–Darby Creek Metropolitan Park, for many wondrous afternoons, despite frequent cold and damp conditions, exploring the landscape's many treasures hidden from the view of those who don't know how or where to find them. Mac and his colleagues truly understand ceremonial time. Thanks as well

to Harold Gibbs at the National Park Service. He introduced me to the
American West during a dozen summer expeditions to the high country.
I've learned much about the land from him. And I'm grateful to colleagues at the Tennessee Valley Authority who taught me about the Appalachian landscape and gave me a greater appreciation of differing values than I possessed before.

My most important guides, though, were those who shaped my basic
landscape point of view. To John Brinckerhoff Jackson and Charles
King I owe my greatest debt. While many others helped point the way,
they were my principal guides. Each showed me how to see the landscape and how to interpret its meaning in different, yet fundamental,
ways.

J. B. Jackson helped me, as he helped scores of others, to see the wonders of the vernacular landscape. He showed me how to interpret our
everyday world, both the obvious and profound as well as the subtle and
sublime. Unlike scholars who examine the minutiae of the extraordinary and monumental, he focused on the mundane elements that combine to make our world unique and personal. Those are the truly important elements. They better reflect who we are, how we came to be,
and what we aspire to be.

Charlie King helped me see another side of our landscape. He was
my guide into the inner workings of the biophysical world. To wander
about an old field or forest with Charlie was to enter a magical world
of wonders unseen by all but the very fortunate few who share his rare
knowledge and keen vision.

Special recognition goes as well to those who reviewed parts of my
manuscript, including Ralph Boerner, Jot Carpenter, Mike Hansen, Henry
Hunker, Ed Lentz, Bonnie Loyd, Garry McKenzie, Jack Nasar, Ken Pearlman, Bernie Rosen, Randy Roth, Hal Rothman, Paul Starrs, John Stilgoe, Lewis Ulman, Rick Yerkes, and several anonymous readers. Their
collective expertise humbled me in my dabbles in their disciplines, and
their sharp editorial eyes spotted flaws that I time and again failed to
find. The resulting comments sharpened my message and made me keenly
aware of the many dangers one inevitably encounters when attempt-
ing to cross so many disciplinary bounds. But important insights can be
gained by crossing them. Oversimplification is the greatest risk, balancing breadth and depth the greatest challenge.

Howard Boyer, my editor, suggested the manuscript's final form. Bo-
nita Hurd's skillful copyediting polished its prose, and Abigail Rorer's
wonderful illustrations lent it a special sparkle.

I remain a neophyte learning to see the landscape as clearly as my many gifted guides, and I struggle to express that insight as eloquently as they do. The shortcomings and errors in my work are entirely the result of my own limitations.

Finally, much love to my family for their patience during my preoccupation with this project.

J.W.S.
Columbus, Ohio
Summer 1998

Visions of Paradise

Prelude

Landscape Vision

If we want to understand ourselves, we would do well to take
a searching look at our landscapes.

Peirce Lewis

We all share a rich legacy: our landscape. It is our heritage made visible, for it records the forces, both physical and human, that shaped it. Each landscape is unique, for those forces vary endlessly in a wonderful form of life-giving alchemy. The landscape is often our autobiography as we imprint it with our actions, values, policies, and programs.[1] Some imprints last for centuries; others are erased and rewritten more rapidly, leaving behind a complex layering of messages—past and present, physical and cultural, public and private, local and national, legal and moral, rational and emotional, aesthetic and economic, Euro-American and Native American. The landscape tells a story of many meanings, written by many hands, that reveals the past, explains the present, and foreshadows the future. That story is my subject. It is a partly factual and partly anecdotal story, because the landscape is both objective and subjective.

"Landscape is a slippery word," according to Harvard historian John Stilgoe.[2] By landscape I mean not just the appearance of our surroundings, whether urban, suburban, rural, or wild. The landscape is far more than scenery. To see it as simply scenery implies a separation between the viewer and the view. In this sense the landscape is merely a two-dimensional object of our gaze, little different from an Albert Bierstadt or Thomas Cole painting. This perspective strikes me as distant, lifeless, and limited.

Another common definition focuses on physical character—topography, vegetation, and human-built forms. In this sense the landscape is three-dimensional space, an architectural volume defined by attributes

I

like scale, proportion, color, and texture. I find this definition too analytical, too cold, too impersonal. The landscape is a living element, not
just an abstraction.

Landscape is about people, too. Our activity animates and enlivens
it, transforming space into place known through personal experience.[3]
Yet landscape is more than place. Our sense of landscape is informed
by our knowledge of the historical events and forces that created the
place, together with our actions and experiences. Myth, legend, fact, and
folklore communicate its physical and cultural legacy.

Connection with the landscape begins when we are children and deep-
ens as a result of mundane experiences throughout our lives. Two sets
of experiences proved significant for me because they opened my eyes
to the vast difference in landscapes. The first set was family visits as a
child to my maternal grandparents in Ridgeville, Ohio; the second was
visits to my paternal grandparents in Margate City, New Jersey. These
places are etched in my memory.

Ridgeville was, and is, a small crossroads community in southwestern Ohio. Without the town sign, you would not know for sure that
Ridgeville was a real place—a place with a name, a place on the map. I
can still feel the rise and fall of the car, and my stomach, as we approached it on Route 48. The beautiful rolling hills and valleys of the
rural land around Ridgeville differed just enough from my suburban
home to make it an alien world.

Although surrounded by farms, Ridgeville was not an agricultural
community. Many residents were employed in other occupations. My
grandfather was a master machinist who owned a small machine shop
that he operated out of a nondescript concrete-block building behind
the house. The shop was a smörgåsbord for the senses of a small boy. It
was a world of oil, metal, and machines: the cool dampness of the concrete floor stained by spilled oil and scarred by the heavy machines; the
splintery metallic filings sprinkled about like dust; and the dull luster of
the lathes and presses coated in a film of fine machine oil. The shop's
scent lingers in my memory.

Margate City was, and is, a different type of wonderland. It was the
shore, sand, surf, and salt marsh. I recall the fecund smell of the tidal
marshes intervening between the mainland and Absecon Island, the
slender barrier island on which Margate, Atlantic City, Ventnor City,
and Longport sit. This was the landscape of Monopoly—Boardwalk,
Park Place, and Marvin Gardens; ironically, it's a landscape of opulent
casinos now.

My memories of our summers in Margate are vivid too: the neat homes adorned with monogrammed window awnings; the miniature, immaculately maintained lawns; the bitter, oily smell of the creosote pilings and bulkheads; and the sand and salt spray. Sand and salt defined this world: the feel of searing sand on white beaches cluttered with broken bits of shell; the muddy, drippy texture of wet sand at the surf's edge, perfect for making magical sand castles; the hot, pebbly pavement on the streets and sidewalks covered everywhere by a dusting of sand; the gritty texture of the breeze. At the shore, the wind is not just an ephemeral vapor; you feel a palpable force on your face. Sights, sounds, smells, experiences, these are the things that connect people to place.

Knowledge of the many forces that shape a landscape also connects us to it. To visit Ridgeville or Margate only as a passerby unfamiliar with its history would be to acquire only place memories. My memories of them go beyond those of place. My mother's family settled in southwestern Ohio along the banks of the Little Miami River in the mid-1800s. Our family lore is filled with stories of this land and its history. So too for my father's family, which has lived around Margate and Cape May since before the Revolutionary War.

This book, then, is about our landscape vision—our ability to not just be in a place for which we have little interest or understanding, but to see the landscape, to know and be captivated by it. Such knowledge enriches us in the same subtle and subconscious way as our knowing our lineage. It delights and inspires us as we learn to appreciate the landscape's minute workings. It provides harmony as we align the patterns of our life with the subtle rhythms of the land. The resulting connection creates a sense of social and environmental continuity and a feeling of warmth, comfort, and security.

An enchanting short story by Jean Giono, first published in *Vogue* in 1954, illustrates these beliefs.[4] Giono, a popular French author, wrote about an unforgettable acquaintance named Elzéard Bouffier. Giono had met Bouffier nearly forty years earlier during a short visit to Bouffier's remote home deep in the Provence region in southeastern France.

Bouffier had died peacefully in 1947 at the hospice in Banon, France. Few noticed his passing; he was eighty-nine and had lived a solitary life in a small cabin high in the mountains after the death of his wife and child thirty-seven years before. There he tended a small flock of sheep in the company of his dog. He had no friends, only a handful of acquaintances. Nor was he famous or rich. Yet those around him unknowingly owed much of their happiness to him.

When Giono first met Bouffier during a trek through the region in 1913, Giono was a young man; Bouffier was fifty-five. The landscape was desolate, harsh, and windswept, a virtual desert of dry streambeds and wells, abandoned farms, decaying villages, and barren hillsides of scrub clinging to the impoverished soil. The few people who remained in the valleys lived meager, desperate lives. Where, unbeknownst to Giono, a gentle, happy people once flourished in a fertile landscape, the harsh conditions of the landscape he found made the few remaining residents equally harsh and inhospitable.

As Giono was wandering over the vacant mountains in search of water, after several days without, he spotted the silhouette of a still figure leaning on a staff high on a nearby mountainside. It was Bouffier tending his sheep. The strange, gentle man generously opened his cabin to Giono that night, offering him a simple meal, shelter, and silent company. Giono watched his host spend the evening carefully sorting a hundred acorns from a bag of thousands he had collected. Early the next morning Bouffier left, staff in hand, to tend his flock. He carried the hundred acorns in a small satchel. Curious, Giono followed.

From a polite distance away, Giono watched as Bouffier poked at the dormant soil with his staff, dropped an acorn in each hole, and covered it with soil. By day's end, he had planted one hundred oak trees. Bouffier said he had planted nearly one hundred thousand trees over other nearby hillsides in the past three years. He thought about twenty thousand had sprouted and hoped half would survive to maturity. That had become his purpose—to restore the landscape to the health it possessed before human ignorance and mismanagement damaged it. Deliberately, day by day, he toiled toward that goal. Giono left the next day, never expecting to see the results of Bouffier's quiet quest.

Seven years later, in 1920, Giono again trekked through the area to allow its solitude to purge the horrors of World War I from his soul. Bouffier was still at work with his staff. He had replaced his sheep with bees, which posed less of a threat to the young trees that now blanketed the mountains with a delicate lace of green. The hillsides felt just a bit less dry to Giono as the saplings seemed shrouded by a vague mist each morning. Bouffier had begun to plant other species as well. Beech and birch would soon cover other valleys and hillsides, each species in its proper place. Ever so patiently, he was transforming the landscape with a hundred seeds a day, week after week, year after year. The miracle gradually unfolding enticed Giono to return several more times over the next twenty-five years to see it more fully revealed.

When he last visited Bouffier in 1945, the landscape was scarcely recognizable as the one he had first traversed in 1913. It was again a prosperous, fertile place. The once desolate landscape bloomed as new life incrementally transformed the valleys and hillsides. Streams flowed through valleys of willows and rushes. Hillsides were cool in the dappled shade. Hunters stalked the forests for hare and wild boar. And the villages were again full of people optimistic about the future. All the result of one man's lifelong work, one man's landscape vision.

The story of Elzéard Bouffier is fiction. Yet it is so convincing and compelling that people all over the world take the beloved little tale as true. Giono's story is of landscape healing and human well-being, a story of connection and reconnection to the land. Perhaps the beautiful story answers a deep-rooted hope many share to once again live in harmony with the land.

Giono's story also suggests the close relationship between what we know about the land and how we care for it. A typical residential landscape in America makes the point. About fifteen miles north and a bit west of Columbus, Ohio, is a community I will call Concord Estates.[5] A bedroom suburb of Columbus built around two golf courses, it is growing as fast as the corn that grew there a generation earlier. Homes, Stilgoe noted, are the last crop grown on many farms.[6] Concord Estates is a pleasant residential community of subdivisions that includes mostly single-family, detached homes. Because concern for the landscape was of primary importance in its design, much of the land in the core was set aside as open space with bikeways, wildflowers, ponds with flocks of geese, quiet woodlands, and small, tree-lined streams trickling down shallow ravines.[7]

In many ways, Concord Estates is an ideal landscape, a garden in which nature's raw splendor has been polished like a diamond. Homes sit amid an idyllic, miniature world of manicured shrubbery and neatly mulched planting beds enforced by strict landscaping standards. And the design of each home meets strict standards, giving the community a striking uniformity of color, material, and style. Its boundaries are marked by elaborate stone or wooden walls bearing its name. Some of the smaller subdivisions have elaborate gatehouses standing sentinel on the islands dividing their entry drives.

The development is little different from scores of suburban communities in the Sonoran desert of the Southwest, the prairie grasslands of the Great Plains, the Pine Barrens of the mid-Atlantic Coast, or the till plains of the Midwest. Its homes proudly stake their claim to the land, as did the pioneer farmsteads in the wilderness nearly two hundred years ago. But the homes are intruders in the surrounding farmland, as the farms they displaced were once intruders in the wilderness. I do not say this as the next voice in the chorus of romanticized calls for a neo-Jeffersonian world of yeoman farmers. Instead, I sense that many developments like Concord Estates respond little to the landscape's legacy. And I fear that few of their residents understand that legacy, missing the benefit of its knowledge.

As one drives along Concord Estate's curving roads through romanticized *tapis vert* with scattered coppice, past pastoral ponds and streams, one sees how little these elements reflect the site's legacy. They reflect mostly recent imprints. The "green carpet" is composed of grasses bioengineered from species imported from Europe, grasses that require mowing, fertilizing, weeding, and watering. The sensuous shapes of the roads and paths bear little resemblance to the angular or traditional gridlike layout of local roads, and the form and function of the constructed streams and ponds differ from those found before. Overall, the community's appearance, as well as that of individual homes and yards, derives from an eighteenth- and nineteenth-century romantic style of English garden and estate design. The beautiful landscape is an illusion, created and maintained at considerable economic, ecological, and cultural cost.

This design style prevails across diverse landscapes and sociocultural settings. Many factors contribute to its popularity. Concord Estates is a compromise landscape, a middle landscape that appeals to our basic landscape values. There we enjoy the benefits of a home near a city yet in close contact with "nature." There we avoid the threats associated with life in the city, or on the farm, or in the wilderness. There we feel in control of nature while still enjoying its splendor. There we feel simultaneously a part of and apart from nature.

This style appeals to our traditional concept of land as property and a commodity primarily for functional use. It appeals to our historical sense of separation from and superiority over a land of endless abundance and resilience, a sense that has blinded us to the environmental consequences of our actions. And it appeals to our traditional hostility to the environment, partly practical and partly moral and philosophical, that has colored our actions. Landscapes like Concord Estates reflect our reliance on rationalism and science, a reliance that further deadens our perception of the land as a living element of which we are an

integral part. And they reflect how we have time and again used technology to intervene between the land's inherent characteristics and our
wants and wishes.

Our mobility is another contributing factor. Americans move to new
homes on average about every seven years. Movement has been at the
core of our culture since the settlement of the New World. The Conestoga wagon and the mobile home are much more than simply two
forms of transient shelter. This mobility has left indelible marks on our
landscape, lending it a temporary feel quite different from that of foreign landscapes. Frequent movement has also discouraged our desire to
understand each successive place and simultaneously fostered a preference for the familiar.

As we move about and resettle, like pioneers we often assume occupancy of a place—whether rural, suburban, or urban—with little knowledge of its history. Unlike our predecessors, though, we cannot easily
acquire such understanding. Often, the physical imprints of its legacy
are lost when we replace the existing building styles and landscape
forms. Local folktales, myths, and stories that carry a landscape's history from one generation to the next are also often lost, or few longtime
residents remain to tell the landscape's complete story. As a result, many
of us content ourselves with a shallow knowledge of the readily apparent landscape, overlooking the land's more subtle history. Sadder still,
few of us are skilled at seeking that story. These problems create a general paralysis: even if we want to learn about our new home, circumstances inhibit us.

Fortunately, there are ways around it. Consider time as cumulative,
like water flowing into a bucket, rather than as a linear, ephemeral phenomenon that passes like the wind. Henry David Thoreau sensed this
viewpoint, writing in Walden, "Time is but the stream I go a-fishing in.
I drink at it; but while I drink I see the sandy bottom and detect how
shallow it is. Its thin current slides away, but eternity remains."[8] From
this perspective, the physical and cultural events that shaped a landscape still exist as physical marks on the land, or as the intangible values, policies, or histories attached to the land. These imprints reflect both
idiosyncratic local forces and generic forces that shaped the country.
Remnants of the past, they lie all around us, usually unnoticed and un-
appreciated. Reading them makes the landscape more meaningful to
us. It becomes a home in which we reside rather than merely occupy: a
real *home* alive with history.

Reading landscape is not easy. As geographer Peirce Lewis noted, landscapes are messy, like a book with missing pages or a manuscript that has been written and rewritten by people with bad handwriting. But landscapes were not meant to be read like books. Their lack of an editorial intent makes most an unbiased record of our tastes, our values, our aspirations, even our fears. The landscape records without judgment our glories and failures and our commonplace qualities.[9]

This book pursues those imprints in an attempt to better reveal our contemporary landscape, especially our suburban landscape. I look at people, philosophies, policies, and programs that have most affected its development. I also consider the evolution of our landscape attitudes and values. These shape how the landscape is perceived, how it is interpreted, and how it is manipulated. Subsequent actions are better understood in this context.

The literature is full of wonderful books that tell the detailed story of a specific landscape, or landscape element, whether from a scholarly or literary perspective. Engaging, warm, and personal, these intimate stories convey a love of the land and offer great insight to its cultural and physical richness. The essence of the land is readily captured at the local scale. Excellent regional and national overviews also abound. These describe the broad forces that shape our landscapes, forces often overlooked at the site-specific scale. This book seeks the middle ground. It explains broad forces within the context of local landscapes to illustrate their interplay.

The book's title tells the central themes. The word *vision* reflects the belief that through better understanding of a landscape, through better landscape vision, we become better connected to that landscape and so derive greater fulfillment and satisfaction. One's values are integral. To have vision—to see—means one must interpret the view. The word *paradise* expresses the belief that our landscape truly was, and is, a paradise; a world of magic and majesty far beyond any we might imagine. And *glimpses* indicates that this is not the complete story of the landscape. Instead, it is a set of snapshots of formative forces over the past two hundred years that I believe most shaped our contemporary setting. None of these snapshots presents previously unseen material. Historians, geographers, and other academics provide the original scholarship. These snapshots synthesize that scholarship across many disciplinary boundaries to clarify and find general meaning in the landscape story, as a giant lens might focus light to form a clearer image.

With these as guideposts, I look at our landscape and try to see it clearly, as free as possible from the many tints that commonly color

our vision. To do so I search for the imprints of those forces. Donald
Meinig wrote that such study "is an immense realm which needs many
kinds of explorers. Any landscape is so dense with evidence and so complex and cryptic that we can never be assured that we have read it all or
read it aright. The landscape lies all around us, ever accessible and inexhaustible. Anyone can look, but we all need help to see that it is at
once a panorama, a composition, a palimpsest, a microcosm; that in every prospect there can be more and more that meets the eye."[10]

Most of all, I pursue a love of the land and the continuous wonder
its infinite variety can provide each of us when we learn how to *see* the
landscape.[11]

CHAPTER I

Paradise Lost, Paradise Found

Sell a country! Why not sell the air, the clouds and the great
sea, as well as the earth? Did not the Great Spirit make them
all for the use of his children?

Tecumseh

One pleasant May morning in 1781, Hannah Alder sent her eight-year-old son, Jonathan, and his teenage brother, David, into the woods surrounding the family's log cabin in Wythe County, Virginia, to round up a mare and colt that had strayed from the homestead. Living in the relative isolation of the wilderness frontier, the Alders often let the animals loose to feed in the shelter of the surrounding forest.[1] On this day, however, a small band of Mingo Indians from central Ohio also searched the forest for horses, or for human captives.

The boys were jumped by the Indians. David was speared, scalped, and killed, and young Jonathan was taken captive nearby. The attackers then went in search of the boy's cabin but found, instead, the Martin homestead several miles from the Alders. There they kidnapped Mrs. Martin and her older child and murdered Mr. Martin and the couple's infant.

The band set off for its village, promising its terrorized captives it would take them to "a fine country, and take good care of them, and they could live easy, and need never work."[2] True to that promise, young Alder was adopted by the chief of the small community and began a new life. His new father, Succopanus, was a Mingo Indian, and his new mother, Winecheo, a Shawnee. Young Alder adapted well to Indian life and became a cherished member of the village. For twenty years he lived happily as an Indian around the headwaters of the Big Darby, Scioto, Maumee, and Mad Rivers, farming the rich till soils and

alluvial bottomland and hunting in the surrounding forest, bur oak savanna, and prairie openings. Even though he had the opportunity to return to the white world, he chose to remain with his adopted family;
such behavior was not uncommon among white captives, for life with
the Indians was to some an attractive alternative to the hardships of
frontier life. There he also witnessed firsthand the growing conflict over
land between the natives and settlers. As Alder approached manhood,
he grieved Winecheo's, and later Succopanus's, death as if each were his
biological parent.

By the 1790s, as Alder entered his twenties, he was "a little over six
fet [sic] in height, and straight as an arrow ever was. His hair and eye-
brows were as 'black as coal,' his complexion dark and swarthy, his face
large and well formed, denoting strength of character and firmness of
purpose; his eyes were bright and piercing, while his whole appearance,
gait and actions were characteristic of the Indian." In 1795 he married
Barshaw, an Indian woman, and "got us a brood mare apiece and packed
our goods and safely landed on Big Darby or Crawfish Creek, as it was
then called. . . . " The couple built the first cabin in the county and
"commenced life in good earnest" because the "Darby . . . was the
greatest and best hunting ground of the whole Indian territory."[3]

Other whites soon began to settle along the Big Darby as hostilities
were temporarily halted by the Treaty of Greenville: first Benjamin
Springer, then Usual Osborn. Richard Taylor, Joshua Ewing, John Story,
and others arrived the next spring, bringing horses, hogs, and cattle that
thrived on the luxuriant prairie grasses and plentiful mast from the
oaks, hickories, and beech. Alder went into the livestock business, selling horses and hogs to the whites, milk and butter to the Indians, and
furs and skins to the traders. He wrote, "I was now in a manner happy.
I could lie down at night without fear, a condition that had been rare
with us[,] and I could rise up in the morning and shake hands with the
white man and the Indians all in perfect peace and safety. Here I had
my own white race for neighbors and the red-man that I loved all min-
gling together. Upon the whole I felt proud over it."[4]

Alder's knowledge of the land, its native people and wildlife, made
him a favorite of those first white settlers. He acted as an interpreter between the two cultures and assisted whites during their initial seasons
of hardship as they struggled to adapt the wilderness to their vision of
permanent homesteads, fields, and pastures. Often that vision overlooked the simple means Alder and the Indians used to harvest the land's
riches and survive with relative ease in paradise. Alder wrote, "And as

for work, I did but very little for I did not know how to work. I pretty much hired all my work done and I was forced to hire white men[,] for the Indians were like myself and didn't know how or couldn't work. If they had known how they wouldn't have worked, for it is not natural for an Indian to work."[5]

Barshaw did not adapt well to the couple's lifestyle that mixed white ways with Indians ways, and they soon separated. Afterward Alder lived alone for several years, content to be an intermediary between the two peoples. In 1802, though, all that changed quite by accident on a crisp summer Sunday. On that day, Alder met John Moore lounging in the shade under a tree near Alder's cabin. The conversation soon centered on Alder's captivity and his faint childhood memories of Virginia. He vaguely remembered having heard the word *Wyth,* but did not know what it meant; and he recalled that the family had lived near a lead mine in a place called Green Brier. The name of the only neighbors he could recollect were the Gullions. Amazingly, Moore responded that he was familiar with Wythe County, Greenbrier Township, Virginia, and had even spent a night at the home of a man named Gullion. And, co-incidentally, Moore was heading back to the area the coming fall; he agreed to make inquiries for Alder regarding the fate of his family.

Unfortunately, he was unable to locate the Alder family. Yet while in Virginia, Moore had spread the word about Alder's interest in tracing his family in the hope the message might eventually reach the Alders or at least someone who knew their fate. Even that appeared to have been in vain as 1803 passed with no news.

As fate would have it, though, Moore and Alder stumbled upon one another in 1804, nearly a year after Moore's return from the South. Both were in nearby Franklinton on personal business when they met by accident. As they talked, Alder was told by another that he had a letter waiting at the post office. Together Alder and Moore went to pick it up. The letter was from his long-lost family in Virginia.

Events happened quickly for Alder. In 1805, he returned to Virginia and met his mother and other remaining family members, whom he then moved to his home in Madison County as soon as affairs were settled the following year. There he and his biological family lived near his former Indian family. He tried his hand at marriage a second time too, this time to Mary Blont, a white woman. The couple prospered and raised a large family, and Alder became a leader of the fledgling white community.

Jonathan Alder, the first long-term white resident of the county, died in 1849. Before he died, he dictated his memoirs to his son, giving readers today a fascinating firsthand account of central Ohio as it was transformed by Euro-American contact. Through such accounts of captivity, as well as early accounts of pioneer life, we better sense Native American cultures, their significant variation, and their fundamental differences with European cultures. We also sense how European contact inflicted catastrophic change on those native cultures and the landscape, contact that resulted in a range of interactions between natives and aliens, at times cooperative, conciliatory, and sensitive, but mostly competitive, suspicious, and hostile.

That contact triggered a cascade of change in the Native American societies: European diseases decimated Indian populations; the exchange of commodities and values disrupted their lifestyles; and competition for land displaced populations, leading to territorial conflicts and a domino-like effect of dislocation that spread westward. Initial contact was often economic-based for the exchange of food and other goods. During the first century of colonization along the Atlantic coast, the two peoples struggled to coexist based on tenuous cooperation and trade. But coexistence became impossible as the Euro-American population swelled and spread across more and more of the landscape. Settlement meant clearing the forest and removing wildlife. It meant parceling and partitioning the land into private property for production agriculture and pasture. Such landscape changes were incompatible with the Native American lifestyle and exposed a core conflict between the two peoples—land. Its ownership, its use, its philosophical and spiritual significance, even how it was perceived aesthetically, as much as any other difference, separated the Euro-American and Native American cultures. Alder wrote, "Such is Indian life. It is either a feast or famine, as the whites sometimes say. They (the whites) live off the fruit of farms; but sometimes their crops fail. Yet, if ever a people live on the game of the land, when it is plenty and fat, that people are the Indians. What more delicious eating could a man desire than fat deer, bear, buffalo, elk or wild turkeys, all of which the Indians frequently had in abundance? Then they were happy, and for all this prosperity, gave thanks to the Great Spirit."[6]

To Alder and most Indians, humans were an integral part of nature, little different from other species. Nature was revered as a benevolent force that sustained life. Its appreciation was central to their religion and general philosophy. Alder believed the landscape was a nurturing home shared with spirits, alive with the history of his adopted people.

Although the eastern woodland Indians acted as agents of environmental change by burning the forest, hunting, and farming, at times
drastically disturbing the landscape, their impacts were localized and
short-lived as a result of their small numbers and mobile, subsistence
lifestyle. They accumulated little material wealth and consumed few resources, living, to a large extent, hand-to-mouth, season to season. Nature always provided for them, even though at times better than others.
Compared to Euro-Americans, it might seem Indian agriculture and the
Indian lifestyle made little provision for the future. Yet in many ways,
both were more compatible with the indigenous landscape.

The Indian concepts of property and ownership also differed from
those of the pioneers. While each community typically claimed a territory that it defended, doing so was intended to protect access to the
basic resources it needed to survive. Rarely did a community claim an
area greater than that necessary to supply its fundamental needs. And
individuals had few personal possessions since the group was mobile,
at times moving on a seasonal basis. The landscape provided all they
needed or wanted free for the taking, so personal property, from a Euro-
American perspective, was unnecessary. One "owned" only what one
made or applied some labor to in order to obtain. Even those possessions were held loosely: if no longer useful, or if needed by someone else,
they were easily given away.[7] As William Cronon noted in *Changes in
the Land,* "What the Indians owned—or, more precisely, what their
villages gave them claim to—was not the land but the things that were
on the land during the various seasons of the year. It was a concept
of property shared by many of the hunter-gatherer and agricultural
peoples of the world, but radically different from that of the invading
Europeans."[8]

To James Kilbourne the landscape was much different. On September 22, 1786, Josiah Kilbourne called in his fifteen-year-old son from
working on the remaining portion of the family's small, run-down
farm near Farmington, just west of Hartford, Connecticut. The farm
was failing and the family barely able to survive in the economic depression resulting from the Revolutionary War. Josiah and his wife,
Anna, had concluded their beloved James should strike out on his own
in search of better opportunities.

The next day young James set out to seek his fortune. Walking into
an unknown future, leaving the care and protection of his family behind on the farmstead, he wandered for several days northward along

the Farmington River until he reached the town of Granby, twenty-some miles upriver. There began a truly remarkable life that still shapes the central Ohio landscape today.

By October 1, James found a four-year apprenticeship in a clothing mill where he received room and board while he trained seven months each year; the remaining five months were his to use as he wished. Only able to return home for short stays, what James wanted to do during those long breaks was work. Fortunately the mill was adjacent to the prosperous five hundred–acre farm of Elisha Griswold. The farm needed help and James was hired. By season's end, James's industriousness and ambition were rewarded. The Griswolds invited him back the next season, and his master at the mill allowed him to continue work at the farm when production at the mill was slack.

His relations with the Griswolds quickly expanded as the family took young Kilbourne under its care. Over the next decade, James was tutored at night by his "adopted" brother Alexander Griswold, four years older than James and the second of Elisha's ten children. Unfortunately a farm accident had nearly killed Alexander at age ten, an accident that left him sickly for several years. While he recuperated, Alexander turned his energy to reading and independent study under the guidance of his favorite uncle, the Reverend Roger Viets, with whom he lived for several years after the accident. Young Alexander devoured the parish library and his uncle's extensive private collection.

Alexander was a brilliant student, and his uncle an excellent tutor. Whereas under other circumstances the young man would have attended Yale College when he reached his late teens, he was forced instead to return to work on his father's farm in the mid-1780s. The farm suffered from the same economic hardships caused by the Revolutionary War that affected the much smaller Kilbourne family farm. Besides, Alexander's elders judged his present education to be far superior to the one offered by the college.

So in 1786, when James was first hired by the Griswolds, Alexander worked all day in the fields, then read late into the night; and as the Reverend Viets had tutored Alexander, Alexander now tutored his inquisitive new friend. The same year, Alexander joined the Episcopal church of his uncle. He eventually was ordained and rose to be bishop of the four New England states.

In 1788, after two years with the Griswolds, James Kilbourne also converted to the Episcopal faith of his adopted family. Although raised like most in Connecticut as a Congregationalist, he found the Calvinistic doctrine of the state-supported church unacceptable as he matured and discussed religion with Alexander. He was a willing convert to the Anglican beliefs.

Kilbourne soon married Lucy Fitch, the beautiful daughter of John Fitch, inventor of the steamboat. As a young man in his early twenties, Kilbourne was a prosperous entrepreneur in the clothing business. Success enabled him to pay off his father's debts and expand the family farm. By 1800, when not yet thirty, James Kilbourne was a wealthy businessman and civic leader noted for his contributions of time and money to literary societies and other voluntary associations. He also served as a lay minister in the church. Good fortune, goodwill, diligence, and devotion had served him well.

In just fifteen years from the day fifteen-year-old Kilbourne walked out of his parents' cabin, he had worked his way from rags to riches, from anonymity to a community leader. Yet this success was merely a prelude, an apprenticeship of sorts, for another adventure in which he would again leave behind the safety and comfort of home for unknown opportunities in an unknown place. Despite economic success, life in Granby was not easy for Kilbourne and the Griswolds. As Episcopalians, a branch of the Church of England, they were persecuted in the postwar period. Moreover, decades of exploitation and mismanagement had left much of the once fertile Connecticut landscape eroded and stripped bare of natural resources. Perhaps frustrated with those limitations, Kilbourne formed a speculative land company with seven other wealthy Episcopalian friends in 1802. Their intent was to buy land on which to start a new community somewhere beyond the Appalachian Mountains in the Northwest Territory, recently acquired by the United States from England in the treaty ending the war.

That vast territory, all land north and west of the Ohio River and east of the Mississippi, was then opening to white settlement based on three key provisions. The first was the Land Ordinance of 1785, which established the basis for land survey and disbursement. The Northwest Ordinance followed in 1787, establishing the basis for governance. The third key provision was the Treaty of Greenville in 1795. Signed following General Anthony Wayne's victory over the Indians at the Battle of Fallen Timbers, the treaty temporarily settled the conflict with the Indians by separating the territory into an Indian zone and a white zone.

The effect of the three provisions was profound. Like a hand opening a dam's floodgates, they opened the Ohio territory—the easternmost portion of the Northwest Territory, thus the first to be reached—

to a surge of settlers. In 1790, whites in Ohio numbered only 3,000, compared to about 5,000 Indians. By 1800, the first wave of settlement brought another 40,000 pioneers. By 1810, the number surged to 230,000 persons. Many in the human tide drifted down the Ohio River from Wheeling by flatboat; others entered overland, on Zane's Trace or, later, the National Road.

Where should the Kilbourne group go? The newly opened territory was extremely remote and still very dangerous despite the treaty. It was the forefront, the farthest point west in the advancing frontier boundary. Other, less remote land was available for the many westward-moving New Englanders; in particular, Oliver Phelps and Nathaniel Gorham's huge tract in the Genessee Valley of western New York was popular. But Kilbourne visited the Phelps-Gorham site and found it wanting. Instead, his group targeted the Ohio territory between the Muskingum and the Great Miami Rivers. Why there? Why risk the many threats in the western wilderness?

John Fitch, Kilbourne's father-in-law, was a surveyor in the territory during the war and was briefly imprisoned by the British at Fort Detroit. While there, he drew a detailed map of the Northwest Territory on which Kilbourne and many other pioneers relied (Fitch used the proceeds from the sale of the map to finance his invention of the steamboat). The Ohio Territory, Fitch felt, offered the greatest opportunity because the land was still cheap, uninhabited, and extremely fertile. Kilbourne was convinced, even though only Fitch was familiar with the region. Kilbourne's group subsequently named itself the Scioto Company, after the river that flowed through the heart of the region.

Over the summer of 1802, Kilbourne and a young assistant, Nathaniel Little, traveled to Ohio in search of land to purchase for the company. Following Zane's Trace from Wheeling, they arrived in Chillicothe in late August. There they met Colonel Thomas Worthington, the federal land agent, who befriended the explorers and guided their survey of the available land along the Scioto River between Chillicothe and Franklinton (now Columbus), twenty-five miles upriver. Kilbourne and Little's attitude toward the landscape is seen in Little's personal journal kept during the trip. He wrote:

Tuesday, August 24, 1802

We rode on to Peirsols 21 miles and took breakfast, thence on to Pickaway plains. The prairies here are extensive[,] in many places twenty miles long and from two to four wide, with scarcely a tree to be seen on them, but they are covered with an abundant crop of wild grass. It is a grand sight to us who were raised among the hills of New England. We crossed over the Scioto river and on to Chillicothe and put up at Wm. Keys. The lands from about 12 miles southwest of the Muskingum are in general of good soil and covered with timber[:] on the high lands white oak generally, and on the bottoms with a great variety as can be found in almost any country; such as walnut, butternut, sugar maple, elm, white, red, blue and black ash, white and yellow poplar, white, black, red, and swamp oak, hickory, honeylocust, gumtree, cherry and sycamore with a great variety of shrubbery.

The Scioto is a fine clear stream, about 100 yards wide, but very low now. About 5 miles from Chillicothe we passed one of the ancient fortifications that are found in this part of the country. It is of a circular form, enclosing a few acres of land. On the embankments and inside, the trees growing, are as large as any in the surrounding forest. In the town of Chillicothe there is a regular mound of earth, about 30′ high, on which is now growing five sugar maples of good size. Some old stumps yet remain on the mound, one that I noticed particularly, must have been a very large tree, either white oak or black walnut. The owner of the lot has an ice house inside the mound. By what information I could gain there has never been any particular examination to ascertain the contents of this mount.

Saturday, August 28, 1802

On the Franklinton road we passed an Indian camp today. Their huts are built of bark and there is a considerable number of them. As we approached the camp, which is a short distance from the road, we were met by a large number of dogs, which annoyed our horses very much. These Indians appear to live a miserable life.

Sunday, August 29, 1802

This afternoon we walked about two miles up the Scioto, where there is one of those ancient fortifications. This is of circular form, containing 30 or 40 acres. The earth that is thrown up is now about 10′ high. There are 12 gateways through the wall of earth. This grand circle is situated on a plane and is covered with heavy timber on the walls and insides. In the center of this enclosure is a mound of earth, similar to the one in Chillicothe. There are several of these fortifications or earth works, on or near the banks of the river. One in particular near the Pickaway plain, impressed us, as being very magnificant [sic]. These works are to my mind strong indications that many centuries ago this extent of country was inhabited by a people much farther advanced in civilization and intelligence than the present race of Indians. In many places also are found where they had wells but no traces of regular built towns are discovered.

We have seen but few farmers in this country that appear to be industrious. Generally their cabins are poor and dirty. Their improvements look slovenly, and many live on lands belonging to the public. They clear a little patch so that they can raise a little corn and the rest of their time is spent hunting and lounging around. The inhabitants of the small towns appear to be rather indolent. The tavern keepers seem to be doing the best of any of them.[9]

Little and Kilbourne, like most Euro-Americans, could not understand why the Indians and other "indolent" whites squatting preemptively in the public domain lived so "miserably" in the midst of such
potential. The attitude of the two men reflected the landscape values of
the day, values rooted in their Judeo-Christian, European heritage.

As Europeans left their "civilized" landscape to come to what they
considered to be the vacant, virgin wilderness of the New World, they
brought along their landscape values and land use practices, as well as a
set of preconceptions about the New World. Those values and practices, like their general social values and religious beliefs, arose during
the Enlightenment and the Reformation as their homeland cast off the
manorial and feudal systems.

The New World was at once familiar and alien. Because of its familiarity, many of the traditional European settlement practices transferred
intact, as did the underlying landscape values regarding the relation of
people to the land and to nature. Those values shaped the new society.
If the New World had been totally alien to the Europeans, if they had
encountered instead a desert, tundra, or rain forest, they would have
been forced to quickly abandon their values and practices as impractical. Differences between the New World and the Old did force some
adaptation and change, but since Kilbourne's day American landscape
values have remained remarkably constant at their core.[10]

Here Europeans found a temperate climate similar to that of the Old
World, though seasonal variation in New England was much greater; in
particular, the winters were harsher. Those extremes offered phenomenal seasonal abundance. The New World was a treasure trove of forest
resources and game that were scarce in Europe as a result of centuries
of settlement.

Here they encountered many familiar plant and animal species intermixed among the exotic. They were surprised, though, at the absence of
the domesticated animals that underpinned European agriculture, such
as cattle, horses, sheep, and swine. The new landscape was also devoid
of many nuisances commonly associated with dense, dirty European
cities. Lacking species of fly, mouse, rat, and roach, and many microorganisms that carried or caused diseases common in Europe, the New
World was a pristine paradise.

Euro-Americans like Kilbourne considered the landscape a virtually
limitless source of potentially valuable natural resources. It was primarily a collection of commodities viewed mostly in functional, utili-

tarian terms. The landscape was a resource for economic exploitation
rather than the living element perceived by the Native Americans. Land
was property that could be bought and sold. No single concept has so
shaped the American landscape, nor more distinguishes it from others
around the world. Private ownership of land rests at the very core of
America. The famous French immigrant Hector St. John de Crèvecœur
noted this in *Letters from an American Farmer* (1782):

> The instant I enter on my own land, the bright idea of property, of exclusive
> right, of independence exalt my mind. Precious soil, I say to myself, by what
> singular custom of law is it that thou wast made to constitute the riches of
> the freeholder? What should we American farmers be without the distinct
> possession of that soil? It feeds, it clothes us, from it we draw even a great
> exuberancy [*sic*], our best meat, our richest drink, the very honey of our bees
> comes from this privileged spot. No wonder we should thus cherish its pos-
> session, no wonder that so many Europeans who have never been able to say
> that such portion of land was theirs, cross the Atlantic to realize that happi-
> ness. This formerly rude soil has been converted by my father into a pleasant
> farm, and in return it has established all our rights; on it is founded our rank,
> our freedom, our power as citizens, our importance as inhabitants of such a
> district. These images I must confess I always behold with pleasure, and ex-
> tend them as far as my imagination can reach: for this is what may be called
> the true and the only philosophy of an American farmer.[11]

While our European heritage recognized several fundamental bases
of land ownership, including royal grant, or inheritance, the most important, for my purpose, was fee simple purchase and, underlying it, the
basic right of ownership derived from occupancy and improvement of
the land.[12] Occupancy meant the construction of a permanent farmstead or settlement, and improvement meant the clearing of the forest
to create an agricultural landscape of permanent fields and pastures.
Since the Indians did neither of those things, people like Kilbourne did
not recognize their title to the land. Indians were merely temporary occupants who could be legally displaced. Colonists often considered trea-
ties and land purchase agreements with Indians as ceremonial rather than
as legally or morally binding. The failure to recognize Indian property
rights trivialized the Indian way of life and opened the door to its destruction.[13]

Colonial theorist John Winthrop posited two fundamental ways of
owning land, one natural and the other civil. The natural right, Winthrop argued, existed at some past time "when men held the earth in
common[,] every man sowing and feeding where he pleased." That

was the basis of Indian rights. However, he felt that primitive right was superseded by another, superior civil right when people began to raise crops, keep cattle, and improve the land by enclosing it: "As for the natives in New England, they inclose noe Land, neither have any setled habytation, nor any tame Cattle to improve the Land by, and soe have noe other but a Naturall Right to those Countries[;] . . . the rest of the country lay open to any that could and would improve it."[14]

Similarly, the colonial minister John Cotton wrote, "In a vacant soyle, hee that taketh possession of it, and bestoweth culture and husbandry upon it, his Right it is."[15] In 1831, Timothy Flint rationalized the displacement of New World aborigines on similar bases in his influential description of the Mississippi Valley:

> It is no crime of the present civilized races [Euro-Americans], that inhabit these regions, that their forefathers came over the sea, and enclosed lands, and cut down trees, where the Indians had hunted and fought. If they will not, and cannot labor, and cultivate the land, and lead a municipal life, they are in the same predicament with a much greater number of drunkards, idlers and disturbers of society, who are a charge and a burden upon it, in all civilized communities. Like them, they ought to be treated with tenderness; to be enlightened and reclaimed, if possible; and, as far as may be, to be restrained from hurting us, and each other. But it is surely as unjust, as it is preposterous, to speak of the prevalence of our race over theirs, as an evil; and, from a misjudging tenderness to them, do injustice to our own country, and the cause of human nature.[16]

Native Americans held much different beliefs. Wilderness was a familiar, nurturing home whose geographical and seasonal diversity, whether manipulated or not, meant stability and an abundance of the necessities and luxuries of life. For many, mobility was the key to both the exploitation and preservation of that diversity. Villages were not fixed in size or location, rather they were moved and reassembled to match changing social needs and the ecological patchwork of the land. As a result, their subsistent, mobile lifestyle limited their population as well as the degree of their long-term landscape disturbance.[17]

In contrast, wilderness posed a direct threat to the pioneers' survival and their preferred form of society and landscape management. Wilderness was an alien place, not a nurturing provider, even though initially European colonists often used a subsistence form of agriculture not dissimilar to that of the Indians. That technique soon evolved into the familiar row crop and livestock bases. By Kilbourne's day, Euro-America was based on the accumulation and consumption of far more resource-derived commodities than was native America. The resulting form of settlement was incompatible with the wilderness. The magnificent forest, prairies, wetlands, and floodplains of the new land were not hospitable to their row crops and livestock, so the wilderness had to be modified. That malevolent attitude was typified in a section titled "War on the Woods" in W. H. Venable's *Footprints of the Pioneers in the Ohio Valley* (1888):

> The trees are the backwoodman's [*sic*] enemy, for they occupy his ground. They will not run away, like the buffalo and the Indian, so they must be hewn down and cremated.
>
> The labor of clearing, like that of building, was lightened for each by the union of all in the war upon the woods. "Choppings," and "log-rollings," were among the toilsome pleasures of the settlers.
>
> A small army of stalwart men, with strong muscles and sharp axes, soon cut away a regiment of trees, and let daylight upon a plot of ground large enough for a "patch" for planting corn, beans, and pumpkins. The trees felled, their branches were lopped off, their trunks were cut into lengths of from twelve to twenty feet. Then came the log-rolling. Ox teams and handspikes dragged and rolled the slain giants of the forest into high piles, which, when dry, were burnt to ashes. When the task of the day was ended, such games as racing, wrestling, and boxing were in order.[18]

The needs of the native wildlife and the native peoples also conflicted with the pioneers' row crops and livestock. They too had to be removed and other major environmental modifications made before one could prosper. It was not a time or place for the conservation of resources or preservation of nature: the landscape was to be altered and exploited as quickly as possible.

People of Kilbourne's day believed there was an inexhaustible supply of fertile land and saw little necessity to husband natural resources. If sloppy land management depleted one area, one need only move to another where the land was untouched. The old axiom "The grass is always greener on the other side of the hill" had elements of truth early in our landscape's history. By the mid-1800s, however, it became mostly myth, one we still struggle to overturn.

Similarly, most settlers failed to fully realize the relationship between their environmental actions and the resulting outcome, despite experience that people often deplete the land. They thought the land had unbounded fertility and resiliency. Whatever wounds they caused would quickly heal. Common beliefs that people were fundamentally separate from nature and that nature was merely a collection of commodities further blinded them to the true consequences of their land use practices.

Those beliefs led them to underestimate the full effect of Native American land use practices, so they saw the New World as a virgin wilderness rather than a settled landscape shaped by its human occupants.

Euro-Americans like Kilbourne considered the landscape mostly in rational terms: reason applied by the enlightened mind, not the folklore and mysticism practiced by the natives, was the primary means used to understand the New World. Their rationalism reinforced another key difference between native and invader. While the Indians lived in a landscape full of spirits and rich in the myth and folklore of their ancestors, to the Euro-Americans the New World was a place devoid of mystical or ancestral significance, an alien land lacking cultural heritage and meaning separate from certain moral or religious connotations. The ghosts and gnomes, fairies and folktales of their European heritage were left behind in the Old World. The Euro-Americans had no emotional, psychological, or historical linkage to the new land. Their ancestors had not lived here. The resulting disconnection from the new land made it easier for them to alter the landscape with little guilt or self-restraint.

Euro-Americans also believed that people existed separate from and were superior to nature in a moral sense; they did not see themselves as integral components of nature as commonly believed in Eastern and Native American philosophies. That sense of separation, together with the perception of land as property, sits at the very core of American landscape values. Drawing on their Judeo-Christian heritage, pioneers like Kilbourne believed God created people in his image, and that this separated humans from all other forms of life. They believed humans were given dominion over the earth and all its creatures and so were ethically free, even empowered, to use the land to fulfill God's will—in part, to go forth and multiply, to prosper in the paradise they sought to create.[19]

They also considered the wilderness a moral threat in the context of that heritage. Roderick Nash detailed those attitudes in his classic study of American environmental values, *Wilderness and the American Mind*. He wrote, "Wilderness as fact and symbol permeated Judeo-Christian tradition. Anyone with a Bible had available an extended lesson in the meaning of wild land. Subsequent Christian history added new dimensions. As a result, the first immigrants approached North America with a cluster of preconceived ideas about wilderness. This intellectual legacy of the Old World to the New not only helped determine initial responses but left a lasting imprint on American thought." He continued:

> Wilderness not only frustrated the pioneers physically but also acquired significance as a dark and sinister symbol. They shared the long Western tra-

dition of imagining wild country as a moral vacuum, a cursed and chaotic wasteland. As a consequence, frontiersmen acutely sensed that they battled wild country not only for personal survival but in the name of the nation, race, and God. Civilizing the New World meant enlightening darkness, ordering chaos, and changing evil into good. In the morality play of westward expansion, wilderness was the villain, and the pioneer, as hero, relished its destruction. The transformation of a wilderness into civilization was the reward for his sacrifices, the definition of his achievement, and the source of his pride.[20]

The wilderness was lawless. It was the darkness beyond God's light. Whether forbidding, primeval forest or desolate desert, wilderness was the devil's domain. And even though wilderness as a harsh, alien landscape had traditionally been sought in Judaism and Christianity as a place of spiritual atonement or cleansing, it remained one's moral duty, the pioneers believed, to subdue wilderness, to civilize it, to bring the light of his word and his law to it, and in its place to create a true paradise more reflective of his will.

That paradise, they believed, was a peaceful, pastoral landscape, a gardenlike place common to many ancient cultures and echoed in the Bible's description of the Garden of Eden—a fertile landscape of rolling fields, a well-watered land of milk and honey. Virgil's pastoral poetry and the Arcadian landscapes painted by Claude Lorrain further popularized this imagery, the antipode of wilderness, and profoundly affected the European and American landscapes. Today it still directs our landscape thoughts and actions.

The dark, desolate New World wilderness also triggered a contradictory set of perceptions in the Euro-American mind. Its virgin purity and fertile promise offered the European immigrant the opportunity to escape the moral, social, and environmental degradation of the Old World and begin anew. Here they could achieve a utopian life in an idyllic land free from the bonds of history.[21] Hence the American wilderness beckoned and beguiled the immigrant with opportunity while it simultaneously physically and morally threatened.

Like most Euro-Americans, James Kilbourne set out to transform his piece of the New World wilderness into a Garden of Eden where his diligent labors would yield a life of plenty and foster harmony with the land and his Lord. Central Ohio was a paradise in the rough, like an uncut diamond. His righteous labor would polish it into a true earthly garden and make him a handsome profit in the process.

Armed with those values, he and his young assistant surveyed central Ohio in search of the perfect land to buy for their new community.

As the survey concluded, Kilbourne reserved with Thomas Worthington a twelve thousand–acre tract about ten miles south of Franklinton, along the eastern bank of the Scioto River. The survey complete, Kilbourne and Little then began the difficult trip back to Connecticut as the wondrous Ohio forests blazed with the colors of autumn. On return to Granby, Kilbourne reported the abundance of the newfound paradise; the description reads like an inventory of resources awaiting exploitation.

But Kilbourne worried from the beginning about the quality of the land he had reserved, sensing it might be too wet and conducive to malaria, a prevalent malady thought to emanate from stagnant water. He was also concerned about the political stability of the region and its stance on slavery. Ohio was just becoming a state, and serious questions remained regarding its boundaries and governance. Kilbourne's boundary concern centered on the location of the state's western edge. One proposal by the Federalists sought to make the Scioto River the boundary. This would place Kilbourne's property at the edge of the frontier and make it more susceptible to Indian attack. The Federalists feared another popular proposal to locate the boundary more than one hundred miles to the west. That they thought would create too large a state, one that might become so politically and economically powerful it would overwhelm their eastern power base. Therefore Kilbourne made the purchase contingent on three conditions: the adoption of a state constitution that prohibited slavery; the establishment of the state's boundary well to the west of the Scioto River; and the approval of his investors.

By the time of his return to Connecticut late that autumn, word had spread of their plan. To satisfy a growing interest in the plan, the Scioto Company soon expanded to forty members. The enlarged company needed more land and it wanted better land. Fortunately it learned of another promising tract for sale in the same area. Dr. Jonas Stanbury, a New York City land speculator, and his partner, Jonathan Dayton of Elizabethtown, New Jersey, were offering sixteen thousand acres on higher ground about twenty miles north of the twelve thousand–acre tract Kilbourne had optioned. When word reached the Scioto Company that the Ohio constitution was passed and the boundary issue resolved, meeting its approval on both accounts, the company dispatched Kilbourne to New York to negotiate the purchase of the Stanbury-Dayton property.

In 1803, James Kilbourne and thirty-nine other investors, including Ezra and Roger Griswold, Alexander's brothers, purchased the Stanbury-

Dayton tract in central Ohio sight unseen for twenty thousand dollars. That spring, members of the company, including Kilbourne and Ezra Griswold, liquidated their holdings in Connecticut and emigrated to the wilderness home of Jonathan Alder in search of religious freedom and economic opportunity. In a letter Kilbourne wrote to his wife, Lucy, sent from Chillicothe on May 13, 1803, shortly after he first surveyed the purchase, he concluded, "I have been carefully over the land and find it full as good as I expected, and I am fully persuaded that the half township lying on the [Olentangy] River is worth all the money we have given for the whole and that the Situation will be healthy[.] I can see nothing to prevent it[;] . . . every body speaks highly of the . . . Country both for goodness of land and healthyness of Situation,—& I must say that if my affairs were well settled, & I had my Lucy in my Arms & the children around us I could very well dispense with ever going to Connecticut again."[22]

The Scioto Company named the new town Worthington, in honor of Thomas Worthington, the "father of Ohio statehood." Before leaving civilization behind in Granby, the members signed a compact detailing the company's basic operating rules, including the terms of land purchase, repayment, and resale, as well as the design and character of their new frontier town. Worthington would be a New England–style village in layout and social structure, a tight-knit, agriculture-based community devoted to family, education, work, and worship. The village formed a 160-acre rectangle situated on high ground along the eastern bank of the Olentangy River. A 4-acre village green, or public square, formed the town's physical and psychological focus. There, the local militia drilled and celebrations were held. The village was then divided into 160 lots of about an acre each. Double-sized lots facing the green were set aside for the school and the Episcopal church; each was also allocated an outlying glebe of 100 acres, from which supporting revenues were drawn. Land outside the village was divided into farm lots of about 100 acres, so most farmers lived in the village and commuted to their outlying fields. This kept the community socially and physically compact, as intended by the rules governing its creation. Town and farm lots were auctioned to company members, and lots not purchased were then sold to others. Receipts were distributed to the original forty investors permitting them to recoup their initial investment and make a profit.

Worthington still reflects that design and those values. Today, the street names and layout remain little changed in "old" Worthington, and the city still prides itself on its schools and close-knit community structure. The school glebe is now the site of the Thomas Worthington High School complex. The church glebe was eventually sold and subdivided for homes, with the receipts placed in a trust fund to support the church in perpetuity.

Did Kilbourne know Alder? Probably. As the first settler in Madison County, Alder was likely well known. And Kilbourne was a prominent leader of central Ohio's burgeoning white community. He held a variety of local, state, and federal offices during the early 1800s. In addition, the town of Worthington was only twenty miles or so from Alder's Madison County cabin. It seems unlikely the two did not cross paths as white settlement converted central Ohio from wilderness to an agricultural landscape in the first half of the 1800s.

Their lives and philosophies illustrate wonderfully the differences between the way the Native Americans and Euro-Americans perceived and acted toward the landscape. Similar stories of white settlement are easily found. Many tell the same story of hard work, self-reliance, idealism, optimism, and community spirit. The landscape Alder knew as paradise and his landscape values were quickly overwhelmed by the landscape Kilbourne sought to create and the values he represented. Those values—of property and permanence, of separation and superiority, of accumulation and consumption—still dominate today.

Euro-American settlement of Ohio, and the New World in general, wrought profound change in the wilderness and in the native peoples as settlement raced westward in the early 1800s. In 1810 roughly one-seventh of the nation's population lived west of the Appalachian Mountains; in 1840 it was one-third. During that period Ohio grew from 230,000 people to more than 1.5 million people. Illinois grew from a territory with only 12,000 people to a state with more than 450,000 people, and the Michigan population grew from fewer than 5,000 people to more than 200,000. Driving the wave of settlement was the phenomenal growth in the nation's population. Between 1800 and 1840 it rose from 3.9 million persons to more than 17 million. New states were created virtually overnight as the wilderness was settled and converted to an agrarian landscape within a single generation.

"What is there better here in the western states than in those of the East?" wondered Swedish novelist Fredrika Bremer as she traveled throughout the country west of the Appalachian Mountains in the mid-1850s. "More freedom and less prejudice [and] more regard to the man than to his dress and his external circumstances: a freer scope for thought and enterprise," she was answered.[23] As John Jakle noted in *Images of the Ohio Valley,* the West was a way of thinking, a state of mind quite different from that in the East. The West was the future, the East the past. Those differences shaped the landscape as well as politics, economics, and culture.

So too did differences between North and South, free and slave state. Ohio was also the breakpoint between these, as Pulitzer Prize–winning author Louis Bromfield wrote in his novels extolling the Jeffersonian value of rural life on his Malabar Farm outside Mansfield. New Englanders, like James Kilbourne, dominated settlement of the state's mid and northern sectors as they entered through eastern portals such as Wheeling and Zanesville. Settlement in some of the state's southern sectors was dominated by people from Virginia and Carolina who passed through southern portals such as the Cumberland Gap.[24] Bromfield also saw Ohio as the boundary between industrial and agrarian America, since it sat on the eastern edge of the vast midwestern till plain that formed the corn belt. The Ohio landscape documented those differences too.

Political integration of the Northwest Territory into the Union occurred simultaneously with statehood. Economic and social integration followed more slowly. Yet the unprecedented speed of that expansion and integration differed from that of the nation's previous, more gradual growth. The rate of change left little time for reflection on or civility toward the landscape and its native people.

CHAPTER 2

Actions and Outcomes

Nature has already done her part for this region, and man
has done, is doing, and will continue to do his, to make it all
that man can ever desire it to be, forever.

> Caleb Atwater, *History of the State of Ohio,
> Natural and Civil*

Transformation of the New World wilderness into the Euro-American
vision of paradise radically altered the indigenous landscape. The preferred form of pioneer settlement based on row crops and livestock inherently conflicted with the landscape's physical character and its native
people. Those people, as well as the predominant forest cover and wildlife, had to be cleared. Little in the New World escaped change.

The starting point in that transformation was removal of the vast
forest that covered much of the eastern United States. Most trees in a
forest that covered 95 percent of Ohio at the time of white settlement
were cut during the 1800s. In 1832, John James Audubon wrote of the
Ohio Valley, "All this grand portion of our Union, instead of being in a
state of nature, is now more or less covered with villages, farms, and
towns, where the din of hammers and machinery is constantly heard . . .
[and] the woods . . . [are] fast disappearing under the axe. Whether
these changes are for the better or for the worse, I shall not pretend to
say." Yet he concluded, "The greedy mills, told the sad tale, that in a
century the noble forests . . . shall exist no more."[1]

Settlers armed only with hand tools and a strong team of horses or
oxen cleared the state for "the plow and the cow." Their herculean effort quickly transformed Ohio from forest to field, farm by farm. The
forest fell to the ax, saw, and wedge. Other stands were girdled to die
slowly and decay. Snags, stumps, and slash were burned and new crops
planted in their ash. After the forest was felled, many fields still had to

be cleared of glacial erratics forced to the surface each spring by soil heave from the freeze-thaw cycle. Many fields were also too wet in the spring for productive use, requiring ditches to be dug to drain them.

The rapid removal of the forest suggested the need to replant to provide for future needs. Few paid heed. Historian Caleb Atwater noted in 1838, "Most of our timber trees, will soon be gone, and no means are yet to restore the forests which we are destroying." Still, he thought their removal was a mark of progress: "We do not regret the disappearance of the native forests, because by that means, more human beings can be supported in the state, but in the older parts of Ohio, means should even now begin to be used to restore trees enough for fences, fuel and timber, for the house builder and joiner."[2]

The maximum amount of open space in the state was likely reached in the second decade of the twentieth century. Since then, land of marginal agricultural quality has been allowed to reforest, particularly on the unglaciated Appalachian Plateau in the southeastern quadrant of the state. Today, the Ohio landscape nears 30 percent forest. Yet where once nearly 25 million acres of forest blanketed Ohio, today fewer than two thousand acres remain uncut.

Even prior to the Euro-American clearing of the forest, Native Americans routinely manipulated plant communities to better suit their hunting and agricultural needs. Fire, though, was their principal tool. In 1848, historian Samuel P. Hildreth noted the Indian's use of fire around Marietta, the first permanent white settlement in the state: "The yearly autumnal fires of the Indians, during a long period of time, had destroyed all the shrubs and under growth of woody plants, affording the finest hunting grounds."[3] The Indian practice of burning is illustrated by the following story recorded in the mid-1750s by James Smith when Ohio was still a wilderness inhabited by Indians.

In 1755, as an able-bodied eighteen-year-old, Smith volunteered to join a three hundred–man work crew mustered from Pennsylvania to construct Braddock's Road. The British ordered the road constructed to move the men and material of General Edward Braddock's military campaign against the French at Fort Duquesne (now Pittsburgh) during the initial stages of the French and Indian War. The road passed near the Smith family's small Franklin County farm (near present-day Mercersburg) at the frontier boundary between wilderness and settlement in south-central Pennsylvania.

Shortly after he began work, Smith and a companion were jumped by Indians while returning to a supply station. His companion was killed and Smith was taken captive. Like Jonathan Alder, Smith was adopted into a small group of Ohio Indians (mostly Caughnewago). Renamed Scoouwa, he spent five generally happy years with the group until homesickness triggered his return to the white world. In 1757, Scoouwa and several "brothers" left Sunyendeand, a summer Indian gathering place on Sandusky Bay, to hunt and explore central Ohio to the south. Indians from throughout the Great Lakes region "summered" at Sunyendeand, playing games, socializing, and generally lounging about, since food was easily obtained and they had little else to do. As they walked south across the Sandusky Plains, they set the grassland ablaze in their common practice of fire-hunting. Smith later wrote:

> When we came to this place we met with some Ottawa hunters, and agreed with them to take, what they call a ring hunt, in partnership. We waited until we expected rain was near falling to extinguish the fire, and then we kindled a large circle in the prairie. At this time, or before the bucks began to run a great number of deer lay concealed in the grass, in the day, and moved about in the night; but as the fire burned in towards the centre of the circle, the deer fled before the fire: the Indians were scattered also at some distance before the fire, and shot them down every opportunity, which was very frequent, especially as the circle became small. When we came to divide the deer, there were above ten to each hunter, which were all killed in a few hours. The rain did not come on that night to put out the outside circle of the fire, and as the wind arose, it extended thro [sic] the whole prairie which was about fifty miles in length, and in some places nearly twenty in breadth. This put an end to our ring hunting this season, and was in other respects an injury to us in the hunting business; so that upon the whole, we received more harm than benefit by our rapid hunting frolic.[4]

Burning, cutting, and clearing were only part of the landscape changes caused by Euro-Americans and, to a lesser extent, Native Americans. Both peoples dispersed a host of plant species across the land as they sought more functionally or aesthetically pleasing plant material. Many of the grasses that welcomed the first pioneers to the Ohio Valley, and now comprise our lush lawns, were not native, although frequently mistaken as such. The first Englishman to extensively explore Ohio and record his observations noted their presence. In 1750–51, Christopher Gist crossed much of the state, surveying its resources and native peoples on behalf of the Ohio Company, a Virginia-based English fur-trading and land development company. On February 17, 1751, he wrote that the land around the Miami River was "full of beautiful natural Meadows, covered with wild Rye, blue Grass and Clover, and abounds with Turkeys, Deer, Elks, and most Sorts of Game[,] particularly Buffaloes, thirty or forty of which are frequently seen feeding in one Meadow: In short it wants Nothing but Cultivation to make it a most delightfull Country."[5]

Because much of the eastern seaboard was forested during initial colonization, few grass species were native to the Atlantic Coast. Those that were, offered little use for pasturing. By the mid-1600s a regular trade had developed in "English" grass seed such as white clover and bluegrass. The first seeds were brought to the New World inadvertently in animal dung and shipboard fodder, and began to spread in advance of settlement. With systematic cultivation by the colonists, such species were so common as to be considered native within several generations. European "weeds" came along in the process as well. In 1672, one colonist listed no fewer than twenty-two European species common around Massachusetts Bay, including dandelions, chickweeds, bloodworts, mulleins, mallows, nightshades, and stinging nettles.[6] The importation of plants also inadvertently spread pests and plant diseases, such as Dutch elm disease, which virtually eliminated the majestic American elm.

The climatic effect as the forest was cleared was surely significant. In 1838, Atwater noted the effects on soil moisture and general humidity as much of the forest had by then been cut. A decade later, Hildreth followed suit, describing the dramatic change in climate as paradise was cleared and cultivated:

> While the earth was defended from the rays of the summer sun, and protected from the cold blasts of winter by an impenetrable covering of fallen leaves, and a thick growth of forest trees, there can be no doubt of the winters being milder, and the summers more temperate, than at present. It was especially noticed in the summer nights, which were so cool as to render a blanket both a pleasant and desirable covering to the sleeper. On the alluvions the earth, when protected by the forest from the influence of cold winds, and covered with a thick coat of fallen leaves, never froze; while in an adjacent cleared field it froze to the depth of several inches. The warm vapour constantly rising from the earth, served to temper the atmosphere and render it more mild than at present. . . . It is true there were some very cold winters and deep snows, but they were not so changeable as now; nevertheless it yet remains certain, that the winters are much more uniform, while a country is covered with forest, and not subject to such sudden changes of temperatures as they are in an open region. . . . The summers are as much changed as the winters; fifty years since, they were more humid, and there was more generally that condition of the atmosphere which we call *sultry,* and now experience in warm weather after a heavy rain. This constant humidity of the air was occasioned by the regular evaporation of moisture from the leaves of the trees, shrubs, and plants, that clothed the face of the earth, and shut out the drying influence of the sun and air.
>
> The same causes kept the surface constantly moist, and afforded a regular

supply of water to the springs, during the summer as well as the winter, protecting the tender roots of the grasses and other plants from the cold, caused them to vegetate early in the spring, and bring forth a plentiful supply of herbage for the wild animals of the forest, and the domestic cattle of the new settler.[7]

With rare foresight for the time, Atwater also speculated about possible climatic effects of further deforestation across the whole Mississippi Valley. He questioned whether the seasons would continue to be warm and dry as the remainder of the eastern forest was removed, and expressed fear for other unforeseen effects. He even thought it possible that soil productivity would be diminished by the process.

That suspicion was well founded, although for reasons he may not have considered. Early settlement practices profoundly affected the soil. Soil erosion was extensive, as was the depletion of soil fertility, as a result of the clearing and subsequent farming practices. Few farmers used manures, compost, or lime to replenish the natural soil fertility. Few even rotated crops. Most thought those practices a waste of time since land was cheap and labor dear. Instead, they simply cultivated a field until it was exhausted, then cleared another.

The state's extraordinary wildlife diversity posed another problem for the settlers. Many viewed that abundance as a threat and frequently sought to systematically eliminate unwanted plant and animal species. Whether predators, competitors, or merely nuisances, the early pioneers hunted some species, such as the buffalo and bear, to local extinction, while others were inadvertently eliminated by landscape disturbances such as the clearing of the forest.

The passenger pigeon, perhaps the most abundant bird in the world at the time of initial settlement in the Ohio Territory, was driven to extinction in a single generation by hunting and the loss of habitat in the late 1800s. Once, not long ago, the skies over Ohio were clouded with this migratory relative of the mourning dove. Alexander Wilson, a famous American ornithologist, described in 1806 the migration of a single flock a mile wide and 240 miles long. Traveling at nearly a mile per minute, the flock of several billion birds, he estimated, passed overhead continuously for four hours. "On several occasions," wrote Dr. J. M. Wheaton in his 1882 *Report on the Birds of Ohio,* "we have been favored with a general migration of these birds, when they have appeared in congregrated [sic] millions. This was the case in 1854, when the light of the sun was perceptibly obscured by the immense, unbroken, and apparently limitless flock which for several hours passed over the City [Columbus]."[8]

The passenger pigeon's great numbers made it a nuisance since it fed partly on farm crops. It was also easily hunted. Organized hunts attacked the bird's roosts, such as one at Buckeye Lake outside Columbus, where the unwary pigeons were slaughtered by the millions. By March 24, 1900, the carnage in Ohio was complete as the last kill in the wild was recorded. "Martha," the very last of the species, died in captivity at the Cincinnati Zoological Gardens on September 1, 1914. This passing also affected one of the native trees in the Ohio forest. Since the passenger pigeon was the primary dispersal agent for the beech tree, the tree is now less likely to (re)colonize local landscapes.

The passenger pigeon was not the only species native to Ohio that people may have helped extinguish. Megafauna like the mammoths and mastodons no longer roam Ohio either, their extinction likely affected by Native American hunting as climatic change triggered a sudden change from coniferous to deciduous forest, beginning about ten thousand years ago. Still, our culture and our form of settlement, although not the only ones to have contributed to the extinction of a species, may be triggering more extinctions than any other.

It was more common, however, for human disturbance to extirpate a species from an area. Extirpation was usually a by-product of changes that settlement triggered in the landscape matrix that reduced the habitat mostly as a result of deforestation. It also resulted from purposeful effort. Organized hunts designed to eradicate unwanted species were common in the 1800s, as were bounties for killing unwanted species, including the puma and timber wolf. Hildreth reported that the timber wolf "for thirty years was a great hindrance to the raising of sheep, and for a long period the state paid a bounty for the scalps. Neighboring farmers often associated and paid an additional bounty of ten or fifteen dollars, so as to make it an object of profit for certain hunters to employ their whole time and skill in entrapping them."[9] Squirrels were so abundant and bothersome that the Ohio General Assembly in 1807 required that each person, already subject to county tax, "shall in addition thereto, produce to the clerk of the township in which he may reside such number of squirrel scalps, as the trustees shall, at their annual meeting, apportion, in proportion to their county levies, provided that it does not exceed 100 or be less than ten."[10] The penalty for noncompliance was a fine of three cents per pelt.

As settlers removed native animal species, they also introduced selected exotic species to the landscape, especially domesticated animals. Christopher Columbus began the process when he disgorged a panoply of exotic plants and animals in the New World on his second voyage in 1493. Other species of plants, animals, and microorganisms were inadvertently transported in the cargo holds of ships, on animals, or on people. Among the diseases Europeans infected the New World with were amoebic dysentery, bubonic plague, diphtheria, influenza, malaria, measles, smallpox, whooping cough, and yellow fever. The effects on the immunologically naive Native Americans were catastrophic. Smallpox, first introduced to the New World in Hispaniola by the Spanish around 1518, spread like wildfire throughout the Caribbean, then to the mainland, decimating every native population it infected. While the Europeans were little affected by the plague they inadvertently triggered, it constituted virtual genocide to the immunologically defenseless natives. Wave after wave of disease swept through the Indian population of the New World over the ensuing centuries, often well in advance of direct physical contact with Europeans. Cultures were devastated as populations were reduced by as much as 90 percent.

As we recall the Euro-American transformation of the New World paradise, it's easy to make moral judgments, cast blame, and assign guilt based on our current values and environmental understanding. Too frequently, however, we rush to judgment and explicitly or implicitly assign blame for that landscape disturbance without fully understanding the context in which it occurred. To better understand and interpret our landscape, to place it in proper context, we must understand its indigenous condition and the way people, whether Native American or Euro-American, affected it. We must also understand the way people in general affect every landscape.

In fact, the extent of landscape disturbance in the New World is not unusual. Humans have always been agents of profound landscape change—we are a disturbance species, especially those societies based on agriculture and industry. We modify our surroundings to better suit us in our quest to acquire the food and shelter necessary for survival. We also change the world around us to create the comforts and communities we desire. People have always manipulated the landscape to satisfy these common physical and psychological requirements and, in so doing, have created fundamental environmental disturbances that have exhibited striking similarities throughout history and across cultures. In the

New World, that change was based on Judeo-Christian, Euro-American values as pioneers like James Kilbourne moved from the slender Atlantic foothold of initial colonization inland over the Appalachian Mountains. Those values have driven settlement since, westward across the continent and forward through time.

· Forest Cathedral ·

I once walked through a magnificent seven hundred-year-old Douglas fir forest in the Oregon Cascades, my first trip into a true old-growth forest. From ridge top openings, I could see the snow-capped summits of the Three Sisters in the distance. The towering trees, like massive columns, carried the canopy far above, forming a forest cathedral, cool and damp, bathed in twilight. My walk was nearly a spiritual experience, one for which my previous concept of "forest" was poor preparation. I left changed, my sense of time and change and landscape altered.

Until then, to my midwestern sensibilities an old landscape was one little changed for a century. The landscapes of my experience were dynamic places, changing dramatically over a lifetime, whether from human action or natural process. An old forest was the seventy-five- to one hundred and fifty-year-old second-growth stands common in our national forests, or the young eastern woodlands scattered over abandoned fields once logged in favor of agriculture. A big tree stood more than fifty feet tall, and a giant was more than a hundred feet tall. Few old-growth forests remain, particularly in the corn belt of the midwestern till plains.

To walk among immense trees towering several hundred feet tall, trees with nearly twice the girth of the largest I had even seen, trees whose almost incomprehensible massiveness dwarfed my senses, had a profound effect on me. Yet what defined this forest was more than just the monumental size of the Douglas firs. The forest canopy, nearly ten stories above, filtered a soft, diffused light to the forest floor, a floor littered with the fallen behemoths laying askew like mammoth matchsticks randomly scattered. These snags, shrouded with moss and ferns, remain recognizable as Douglas fir for up to three hundred years and may require several more centuries to fully decompose. The massive, standing trunks tower one against another, crowded tightly together like commuters jostling, pushing, straining to get aboard the day's last train. The extraordinary size of the Doug- las fir, combined with this density, make the forest one of the most productive ecosystems in the world.

It was hard to comprehend: to grasp that the standing monarchs had sprouted long before Gutenberg invented the printing press, and that the decaying snags, nurseries for the next generation of infant rulers, had lain silently in state since Columbus set sail. I was merely an ephemeral insect crawling about a landscape of giants formed far back in time.

Despite its age, the forest hadn't reached "climax," in which the plant community is in dynamic equilibrium with the energy budget and site conditions of the location. Perhaps it never will. A forest dominated by Western hemlock and Western red cedar will likely fol- low the Douglas fir if conditions remain stable. These shade-tolerant species persist in the understory beneath the Douglas fir awaiting a break in the dense forest canopy to mature. But the Douglas fir, a pioneer species that grows best in sunny conditions and so establishes itself first, can live more than a thousand years, a span so long that a forest fire or other landscape disturbance often disrupts the sequence before its successors have their day in the sun. After a disturbance, the

Douglas fir often reestablishes itself first, forcing the hemlock and red cedar to again wait their turn for perhaps another millennium. Old-growth Western hemlock–Western red cedar forest is very rare in the Pacific Northwest, not because we've logged these forests, rather, because the successional sequence seldom reaches that point.

As I left these forested mountains, I found that they stand in stark contrast to the Willamette Valley floor they border. The flat fields of row crops felt familiar to me, comfortable and compatible with my midwestern landscape experience. That sense of familiarity attracted scores of settlers 150 years ago, pioneers who had bypassed the alien landscape of the Great Plains to find paradise at the end of the Oregon Trail.

Logging now shapes the forest. Farming and residential subdivision now shape the valley. Each is based on the same fundamental form of landscape disturbance—the manipulation of plant succession. The chain saw in the forest, the giant John Deere in the farm field, and the miniature John Deere in the yard do this, as do most forms of landscape disturbance. To understand a landscape, we must understand plant succession there and the effect of disturbance regimes, whether human-caused or not.

Where scores of species once comprised forest and prairie plant communities across the continent, we now find vast expanses of monocultures mixed in a much different matrix of open and closed canopy, islandlike patch, or linear corridor than existed before. Many agricultural crops are annuals imported from distant lands that we selectively cultivate while suppressing all other species trying to germinate in the fertile soil. We hold succession in its disturbance stage with herbicides and the plow. By economic measures, these commercial fields are extremely productive. Ecologically, they are relatively inefficient, unstable, biological deserts.

In suburbia, lush lawn replaces the forests and prairies. We decorate the green carpet with flower beds, shrubbery, and stately specimen trees, most of them exotic, in both connotations of the term. These landscapes are ecological curiosities, schizophrenic jumbles of plants from various successional stages and site conditions that coexist as a result of our coercion. The injection of energy in the form of our labor, a mower, a good rake, a bag of fertilizer, and a gallon of herbicide compensates for our plant preferences. The management of most farm fields and commercial forests follows a similar approach. Defiant plants pushing through cracks in city sidewalks remind me that the "control" of succession, like the "control" of nature, is an endless effort and mostly an illusion.

When I left the cathedral of the old-growth Douglas fir forest, I passed an abandoned farmstead cut in the forest at the edge of the valley floor. Remnants of the previous use remained. Ghostly tire tracks led off into fields, still discernible though overgrown with herbaceous perennials and young shrubs and trees. Without the farmer's efforts, a new forest was reclaiming the fields. The previous forest was lost forever. Plants were also reclaiming the old house, weathered and dilapidated, just the latest in the landscape's succession of forms.

CHAPTER 3

Designs for a National Landscape

It is impossible not to look forward to distant times, when
our rapid multiplication will . . . cover the whole northern, if
not the southern continent, with a people speaking the same
language, governed in similar forms, and by similar laws; nor
can we contemplate with satisfaction either blot or mixture
on that surface.

Thomas Jefferson

Nation-building is a messy process. What grand design should shape a
new nation's political, economic, or social structure? What design should
shape its landscape? Designs evolve incrementally, so the resulting landscape is a messy, often chaotic combination of many layers. Yet like layers of artifacts uncovered by an archaeologist, layers in a landscape
provide a wonderful record of a nation's history. For those skilled in
their excavation, the layers reveal the legacy of our landscape. In many
ways, the story of America is told in its landscape.

Construction of America from the New World wilderness began
when Europeans first stood on the shores of the Caribbean. But intense
work did not begin until the Revolutionary War period. Then the designs
of our nation, and our national landscape, were first drawn.[1] Those designs still shape our landscape.

When Jefferson expressed his vision for the young nation, he did not
speak alone. To some a generation before, it was clear the population of
the colonies was growing far faster than that of its European parents.
That growth, they suspected, would propel settlement westward across
the Appalachians into a vast fertile land of tremendous economic opportunity. In addition, the continent as a whole occupied an advantageous geographical position between the markets of Europe and Asia.
By the time of the Revolutionary War, many leaders knew Americans
would inevitably spread across the continent and perhaps form a single

nation extending from sea to sea. The future direction of the infant nation was both apparent and problematic to Benjamin Franklin, Jefferson, and others. That direction inexorably pointed west even though knowledge of the western lands extended little beyond the Mississippi River, and even though the relationship between the new lands and the new states posed many vexing problems. Land represented tangible wealth, economic and political power, and individual freedom and opportunity. The beckoning call of western expansion tempted and tantalized, yet threatened the very future of the fledgling nation. Westward expansion was a dangerous and uncertain future. It was also an inescapable one. How was that destiny to be accomplished? Acquisition and disposition of land, both public and private, lay at the heart of the answer.

Destiny took a giant step forward with the formal separation of the colonies from England. The terms of the Treaty of Paris (1783), negotiated by Benjamin Franklin, John Adams, and John Jay, were remarkably favorable for the new Union in light of the tangled nature of European politics and the extensive English, French, and Spanish interests in the New World. The American delegation was instructed to negotiate with the English only for formal recognition of the new nation, then defer to the French for other concessions. France, however, had made a military alliance with Spain and chose not to consider the details of the Anglo-American settlement until its new ally had recaptured Gibraltar from England. Chances of that were remote, for England was far superior to Spain militarily. In addition, Parisian subcurrents hinted that the three European powers might finally strike a mutual agreement over the remainder of the North American continent without involving or perhaps even recognizing the United States. The centuries-old land grab in the New World appeared far from settled by the war and its creation of a new nation with whom the Europeans had to share the continent.

Deftly the American delegation privately negotiated a tentative agreement with the British while publicly placating the French. The agreement called for full political recognition plus transfer of all lands between the Mississippi River and the Appalachian Mountains that England had received from France just twenty years earlier in the settlement of the French and Indian War. The English, in return, received a promise from the new American government to help British merchants recoup prewar debts from the colonies and to compensate loyalists for lands lost to the states. While the Americans negotiated the treaty with England, the English negotiated a separate agreement with the French, enabling the full treaty to be signed on September 3.

With the stroke of a pen America more than doubled in size as it received claim to more than four hundred thousand square miles of land outside the original colonies; land largely unoccupied by Europeans; land zealously coveted by many in the new nation and many abroad. America was now nearly nine hundred thousand square miles, one of the world's largest nations. But it was a nation divided physically and psychologically into two halves by the Appalachian Mountains—old and new, east and west, coastal and interior.

Acquisition of the new land created extraordinary opportunities and posed extraordinary problems. While the treaty transferred claim, it left unanswered how the land was to be stitched into the existing national fabric of the Atlantic Coast. Practical questions included: Who would hold initial claim to the land, the states or the federal government? For what purpose would the land be used? Would it be sold, given away, or leased to individuals or companies, or would it be held in common? How would this massive undertaking be accomplished? And what would be the predominant form of settlement and development? Politically, how would the new land be governed and what would be its relationship to the original states? How was the lengthy and remote border to be secured from foreign encroachment? And how were the existing inhabitants to be treated, especially the Indians? The answers to those momentous questions laid the foundation of our national landscape.

Few issues received more intense debate during the nation's birth. The corridors of power were cluttered with private interests jockeying for position as leaders struggled to balance conflicting land claims from the colonies, land speculators, and common citizens, a balance complicated by larger geopolitical, economic, philosophical, and practical concerns. Pioneers, though, did not wait for the answers. By the 1780s Americans poured over the Appalachian Mountains even though a formal settlement process had yet to be established. In 1775 only a hundred Euro-Americans lived in the Kentucky territory. By 1784, in the aftermath of the war, the population surged to thirty thousand. The nation's western land rush had begun.

Settlers were often left to fend for themselves as the states and federal government struggled to resolve the larger land issues. Infighting was frequently as common and unruly as the fighting with the Indians with whom eager entrepreneurs and land speculators were hastily striking private deals. The British, still smarting from their defeat, and as yet unable to collect on their debts, were slow to withdraw from the region.

They often fueled Indian resistance to the wave of pioneers. Frontier settlement remained risky business despite being within the United States. And Native American claims to the land were yet to be resolved.

Important precedents that shaped the settlement process across the continent were established in those first unofficial years of westward expansion. Although pioneers moved west for many reasons, reasons that often varied even between husband and wife, westward expansion was driven by wave upon wave of optimistic, idealistic, independent people such as James Kilbourne—people in search of freedom, opportunity, and profit. As the surge of people into the Kentucky territory illustrated, initial settlement also often preceded official sanction, organization, and rule. And rampant land speculation, often unethical or illegal, accompanied the unruly rush for land. The planning historian Christopher Tunnard wrote, "The whole history of American land, even of urban land, until not long ago, is a record of the wildest speculation. It is a part of our inheritance, and although the continent could have been developed by other means, the speculator's role is of the greatest importance in the founding and developing of the American community."[2]

The colonial government was politically and practically incapable of controlling the speculation, preemptive settlement, fraud, and other abuses that plagued the dispersal of the new public domain. Many leaders were participants in the profiteering. Indecision and political deadlock stemming from the immense practical and philosophical issues related to the land paralyzed others. Those problems persisted over the next century.

Charles Dickens satirized the national land frenzy in *Martin Chuzzlewit* (1844). In his story the prodigal son, a young English architect named Martin, set off to America to make his fortune but instead was swindled of his savings when, for $150, he bought sight unseen a fifty-acre lot of supposedly prime land in a thriving western community called "Eden." The land agents represented Eden as a "flourishing," "architectural city" with "banks, churches, cathedrals, market-places, factories, hotels, stores, mansions, wharves; an exchange, a theatre; public buildings of all kinds, down to the office of the Eden Stringer, a daily journal."[3] There, Martin dreamed, his architectural training would bring him fame and fortune. Unfortunately the naive foreigner fell easy prey to the extravagant claims of unscrupulous New York land agents, as did so many people in real life. The Eden that Martin found was worthless swamp land in an anemic, backwater collection of shacks and shanties somewhere along the Mississippi River:

There were not above a score of cabins in the whole; half of these appeared untenanted; all were rotten and decayed. The most tottering, abject and forlorn among them, was called, with great propriety, the Bank, and National Credit Office. It had some feeble props about it, but was settling deep down in the mud, past all recovery.

Here and there, an effort had been made to clear the land; and something like a field had been marked out, where, among the stumps and ashes of burnt trees, a scanty crop of Indian corn was growing. In some quarters, a snake or zigzag fence had been begun, but in no instance had it been completed; and the fallen logs, half hidden in the soil, lay mouldering away. Three or four meagre dogs, wasted and vexed with hunger; some long-legged pigs, wandering away into the woods in search of food; some children, nearly naked, gazed at him from the huts; were all the living things he saw. A fetid vapour, hot and sickening as the breath of an oven, rose up from the earth, and hung on everything around; and as his foot-prints sunk into the marshy ground, a black ooze started forth to blot them out.[4]

Dickens based his jab partly on his personal perception of "a people who have shewn themselves so shamelessly dishonest," acquired during a lecture tour in 1842, as well as on observations made during a related trip westward to St. Louis, via Cairo, Kentucky;[5] Cairo served loosely as the model for Eden. He subsequently published a travelogue of his unflattering glimpse at the nation's sparsely settled interior and its scarcely civilized people, entitled *American Notes* (1842).

Even before the treaty officially concluded the Revolutionary War, the colonies began political maneuvering in preparation for the western land grab they assumed would be one of the spoils of victory. The bankrupt colonies hoped to acquire huge tracts as a means of repaying war debts and promoting general economic growth and development. New territory, the colonies thought, also meant greater political power and autonomy in the infant confederacy. Six colonies sought to reestablish old claims to land west of the Appalachians based on their original royal charters, despite little geographical knowledge of the continent's interior. In the Northwest Territory, the colonies of Connecticut, Massachusetts, New York, and Virginia reasserted extensive overlapping claims.

Other colonies lacked the opportunity for expansion. Delaware, Maryland, New Hampshire, New Jersey, Pennsylvania, and Rhode Island had well-defined boundaries and no legitimate claims in the anticipated new territory even though some, such as Pennsylvania, proudly proclaimed intentions to simply appropriate large tracts since they too had contributed life and limb to the war effort. The landlocked colonies stubbornly resisted the impending land grab, insisting on a more equitable distribution of the wealth to all members of the Union in the immediate aftermath of the war. In protest, Maryland, the group's leader, held Virginia and the other colonies hostage by refusing to ratify the Articles of Confederation.

Virginia was the linchpin for resolution of the land issue, since its claims to the eagerly awaited new territory predated the others. As the debate raged, more and more Virginia leaders like Jefferson feared the state would be unable to govern its vast claim across the Appalachians from its coastal capital. They envisioned the same potential problems between the state's Atlantic core and its remote "colony" as occurred between England and its remote colonies. In February 1780, New York's decision to establish its boundary and cede extensive land that it claimed beyond to the new federal government provided further precedent. So in 1781, prompted as much by self-interest as a sense of republican responsibility, Virginia agreed to cede most of its land claims to the new federal government, even though a treaty ending hostilities with England was yet to be negotiated and, hence, the disposition of English claims west of the Appalachians were yet to be determined. The other colonies soon, albeit in some cases begrudgingly, followed suit, enabling the Articles to be ratified. Although the details of the compromise took three more years to resolve, the landmark precedent was established—the West belonged to the United States, not the separate states.

At the same time, the new Continental Congress was besieged from all sides by special interests seeking preferential treatment in the disposition of the spoils. With the end of the war and the removal of colonial constraints to western settlement, pressure mounted from individuals and giant land companies for a national land policy to resolve basic questions of land ownership, sovereignty, survey, and distribution. Speculative land companies, like Christopher Gist's Ohio Company, that had made huge claims before the war without crown approval petitioned Congress for official recognition. Congress declined as a condition of Virginia's acceptance of the 1781 compromise (most of those claims, based on purchases from the Indians, lay on land in the Northwest Territory claimed by Virginia). Pressure also came from thousands of poor individuals who hoped the government would either give them enough land in the West for a farm or would at least sell it at a low price with liberal credit. Members of Congress joined in the clamor, seeking a system that would ensure their personal profit or replenish the national coffers.

The fate of any new land was further complicated by genuine fears that the English, French, and Spanish might again contest claims to it. A treaty was merely a piece of paper, while land was wealth and power, a tangible asset not readily ignored. Jefferson and others recognized the urgent need to settle those lands as a means of securing the borders. But, again, how was the settlement to be managed? What claims were to be recognized, and how were others to be compensated? The colonies bitterly debated the form of governance, even the method of survey and distribution.

By that time, too, the colonies had long developed profound differences in general land use and land management practices. While there were differences within each region, the southern plantation system in Virginia and the Carolinas typically used haphazard boundary surveys, the dispersion of population, and large holdings utilizing slave labor. In contrast, the New England village system relied on more precise surveys and often organized the landscape into townships composed of small villages where homes were clustered neatly around the commons and meeting hall to form a compact community encircled by farm fields. Land settlement systems were at the heart of a region's culture and social structure as well, so each colony naturally sought to pattern new settlement after its own system.

Resolution of the many issues emerged incrementally in the early 1780s, with the Land Ordinance of 1785 as the landmark compromise that laid the foundation for national expansion westward. The ordinance stamped the American landscape west of the Appalachian Mountains with its most distinctive characteristic—the national grid. Ultimately the grid covered 69 percent of continental America (excluding Alaska), the most extensive area in the world surveyed with a uniform system.

Officially titled "An Ordinance for Ascertaining the Mode of Disposing of Lands in the Western Territory," the Land Ordinance derived from a report drafted a year earlier by Jefferson as the chairperson of a committee composed of two representatives from the New England states and two from the Southern states. Congress adjourned in the summer of 1784 with the issue unresolved. Before it reconvened, Jefferson reluctantly left the committee for Paris as the new ambassador. When Congress did reconvene, it rejected many of Jefferson's pet proposals but accepted the general principles. The version that emerged was a political compromise of northern and southern settlement conventions and customs, as well as a compromise between traditional survey practices and those of the Enlightenment. Enacted on May 20, 1785, the ordinance provided for the survey and distribution of the public domain but was initially applied only to a portion of eastern and southeastern Ohio abutting the Ohio River, known as the Seven Ranges. It was later revised and applied to the Northwest Territory (220,000 square miles) and virtually the entire remainder of continental America. From the moment of its promulgation, no other land law has been as contentious or as significant in the shaping of the American landscape.

The Land Ordinance seemed to establish a practical land survey and distribution system with straightforward rules and procedures. It began the systematic division of the landscape into the checkerboard of one-square-mile parcels that today blankets the country. Each square-mile section contained 640 acres, a size later found convenient to subdivide into smaller parcels of 320 acres (a half section), 160 acres (a quarter section), or 80 acres (a half-quarter section). A township, as traditionally sized in New England, was an aggregate of thirty-six sections, amounting to six miles square. All section and township boundaries were oriented north-to-south/east-to-west along lines of longitude and latitude. Prior to the opening of an area, base survey lines would be set and the region surveyed into townships and sections. Half of the townships would then be sold to large land companies and speculators; the other half would be sold by sections to individual white males over twenty-one years of age. One section in each township would be set aside to support public education, and four others for the exchange of land warrants given in payment for military service during the Revolution. Sale would be by auction at a minimum price of one dollar per acre plus survey costs of one dollar per section.

Settlement south of the Ohio River in Kentucky and Tennessee received far less congressional attention for political reasons, even though Daniel Boone and others had passed through the Cumberland Gap since the war. The Land Ordinance never applied there, as Virginia and North Carolina retained claims to the land. Parceling followed the southern model, resulting in a free-for-all. By 1790, the region was a tangled mess of claims and counterclaims that kept lawyers busy for decades.

The use of a grid for land parceling did not originate in the Land Ordinance. Grids were used for the layout of cities dating back to Roman times, and were used in several colonial cities, such as Philadelphia, Savannah, and Charleston. Colonial governments had also specified the use of square parcels, to little avail. Farmers had traditionally resisted use of the grid on agricultural landscapes, as they preferred a less rigidly structured system that enabled them to adjust field boundaries to the general quality of the land. Still, in 1784 the new states of Massachusetts and New York specified the use of square townships in the survey of lands opening for settlement. Convenience and procedural practicality overwhelmed traditional hesitancy.

The ordinance's use of the square-mile section also had earlier origins. In *Common Landscape of America,* John Stilgoe described how Edmund Gunter, an English surveyor, invented a surveying chain of one hundred links, each .66 feet, in the early 1600s. Gunter's sixty-six-foot-long chain became the standard in England, and gradually gained acceptance and widespread use in the colonies during the 1700s as it simplified and standardized the arcane art of surveying. The attractiveness of a grid system based on the chain and the square mile rested in its combination of the most commonly used units of land measure, the acre (originally defined as the amount of land one could plow in a day) and the mile, with the precision and rationality of the decimal system— with an acre defined as ten square chains, then 640 acres fit precisely into one square mile. And the system was easily expandable across vast areas.

Yet the grid was fundamentally flawed. Some of its flaws still haunt us today. The most obvious was its underlying assumption that each section of land was the same. The grid did not respond to locational or environmental differences as it was draped over rivers, ravines, and ridge tops. It ignored topography, soil fertility, vegetation, and surface water, just as it ignored proximity to roads and cities. That foolishness meant many impractical parcels were inadvertently created. Certainly, people selected sections in response to the land's character and individual needs. And as the minimum parcel size was later decreased, more opportunities arose to configure properties responsive to the land, especially after an 1832 law allowed the purchase of four 40-acre blocks to constitute the 160-acre minimum. Still, those were jerry-rigged reactions to the fundamental problem, and they left the less suitable parcels for later purchase.

Nor were the western lands vacant as the grid presupposed. Thousands of squatters preemptively set up homesteads in advance of official government survey and land sales, as the grid sequentially spread across the continent. Native residents remained too. And both groups were manipulating the land to satisfy their wants and needs. Congress knew the western lands were not a tabula rasa. Few colonial leaders were innocent idealists, and few failed to see the folly of artificial, square property boundaries. But as John Opie noted in *The Law of the Land,* the alternatives were worse. The utility of the grid and the order it brought to chaotic land survey and distribution cannot be overestimated.

Another nagging problem was more subtle. Lines of longitude converge as they approach the poles. The only place where a square is created by equal distances north-to-south/east-to-west as measured by longitude-latitude coordinates is at the equator. The parallax in the meridians makes all other longitude-latitude "squares" actually trapezoids wider at the bottom than top. While the parallax is unnoticeable for surveys over small areas, boundary problems arise between tracts when applied over hundreds and thousands of miles. Congress was aware of this too, but found no better alternatives.

The grid also made no provision for the layout of roads. Over whose land would they cut? As a result, most roads were routed along section and township lines, instead of routed across the grid diagonally or routed parallel to natural features such as rivers and ridge tops. Unfortunately that doomed many townships and counties to build and maintain far more road mileage per unit area (and landowner) than other alignments. It also meant roads crisscrossed the landscape without regard for topography or other natural features, creating many of the same practical problems incurred by property boundaries.

As the ordinance took effect, some states still retained individual claims to large tracts of the new land, such as the Virginia Military District and the Connecticut Western Reserve in Ohio. Reasons varied. Some states were slow to officially transfer claim to the federal government as agreed in the 1781 compromise. Other states like Virginia and Connecticut retained claim as a special exemption, or as a political payoff for support of the compromise. Leaders like Jefferson knew land was surely key to the nation's future. State governments knew it too, as did many private land speculators and entrepreneurs. The compromise and the Land Ordinance were only words whose real meaning and authority were yet to be established, so the unruly land grab continued despite the agreed-upon rules.

The first auction under the ordinance ran from September 21 to October 9, 1787, in New York City, despite the fact that the survey was completed for only the first four of the Seven Ranges (a range was formed by one column of townships). Congress pressed ahead with the sale because of mounting financial problems, public pressure, and uncertainty over when the survey would be completed for the remaining three ranges. Public response was tepid, for several reasons. Few people had the $640 plus surveying costs necessary to purchase a section at the specified terms—payment of one-third down and the remainder due in three months. And travel to an eastern city for the auction was difficult for most. A full section in Ohio was also far more than most could farm or needed in order to farm profitably, and much of the land was hilly. Scores of squatters simply ignored the law and preemptively claimed property prior to survey; some purchased it directly from the Indians.

In addition, many prospective buyers and speculators felt they could obtain better terms from Congress by waiting (this soon proved true). The financial windfall anticipated by the federal government did not materialize, nor did the many other benefits it anticipated: the borders were not secured from British interference and harassment; economic growth and development were slow; and settlement in the nation's interior lagged. Further survey and sale were halted; however, a potential solution to the urgent problems soon appeared from a most unlikely source, a New England minister.

Manasseh Cutler was an idealistic minister turned land speculator and congressional lobbyist. In several incremental deals, he and his associates persuaded Congress to sell in total 6.5 million acres of unsurveyed land in southeastern Ohio to two groups of land speculators they represented: one group was the Ohio Company (not the same as Gist's Ohio Company); the other was the Scioto Company (not the same as Kilbourne's group).

Cutler was simply the latest in a long line of large-scale American and European land speculators searching for property and profit, and social reform, in the New World. The deal he negotiated with a bankrupt Congress, desperate to distribute its land windfall, permitted the companies to buy land using warrants purchased from soldiers who had received them from the government as compensation for military service. The glut of land had driven down the value of the warrants, so land speculators bought these nearly worthless IOUs from veterans on the open market for far less than face value. Cutler also persuaded Congress to sell the land at a one-third discount, rather than at the prevailing one dollar per acre minimum set by the Land Ordinance. This enabled the Ohio Company to purchase nearly a million acres from the government for $66\frac{2}{3}$ cents per acre using warrants purchased from veterans for about 12 cents on the dollar. To finance the deal, the company raised one thousand dollars cash from each of its 250 investors. Consequently the company actually paid only about 9 cents per acre. To finance the deal, the company sold shares to subscribers purchased with gold or continental currency.

The Ohio Company hoped to settle most of its vast holdings and resell the small remainder at market rate. The more speculative Scioto Company hoped to settle only some of its much larger holdings and resell most. The government hoped to rid itself of land and watch settlement surge—everyone would benefit.

Cutler's deal appealed to Congress for many reasons. Some members supported such speculative ventures out of greed, since they hoped to personally profit from the deals. More significant, however, were fundamental philosophical and political reasons. While Congress assumed that the public domain would be transferred to freehold private ownership as quickly as possible, it was deeply divided on how that was to be accomplished. Jefferson championed dispersal of small parcels to individuals either for free or for a nominal fee. Alexander Hamilton argued that few pioneers had cash to purchase land. He urged the sale of large tracts to companies and wealthy speculators at discounted prices or with liberal credit, if necessary. This he felt would move the most property from public to private ownership. Fee simple purchase, he believed, would raise badly needed revenue for the federal government, so it mattered little whether the land was sold to wealthy individuals or to companies that would then survey it into small parcels for resale to pioneers. The result was the same. And by then, fee simple purchase was deeply engrained philosophically as the preferred basis for ownership.

Few shared Jefferson's infatuation with the yeoman farmer; most agreed with Hamilton. And with Jefferson away in Paris, philosophical concerns for the humble American citizen were secondary to more practical considerations. By that time, too, the distribution of western land had become a bargaining chip in the debate on the Constitution, as the Confederate government was about to be replaced. Economic concerns about land value and revenues, and political concerns about the size and power of new western states, shaped congressional thinking. The Land Ordinance's emphasis on government-controlled sale of land to individual citizens was temporarily abandoned (although soon reaffirmed) as Congress gave greater autonomy to the private sector.

In 1788, amid much fanfare and optimism, the Ohio Company founded Marietta, the first permanent white settlement in the Northwest Territory, at the confluence of the Muskingum and Ohio Rivers. Despite grandiose plans, the settlement struggled to survive as conflict with Indians, the general hardships of frontier life, and the ready availability of illegal land for squatters resulted in the settling of only several hundred residents. By 1793 the Ohio Company was bankrupt. Marietta survived but the company's unsold land reverted to the government.

The Scioto Company fared even worse. Outright fraud and deceit quickly led the company to bankruptcy; it left behind as its only legacy the shattered lives of five hundred French immigrants that it duped and the small Ohio River town of Gallipolis they founded in 1791 (those who could afford to, later left for New Orleans).

Congressional hopes that dispersal of the public domain based primarily on large-scale land speculation run by the private sector would prove more efficient dimmed as most of the ventures failed. Settlement still languished in the Ohio Territory and generated few funds for the national coffers. Young Martin Chuzzlewit's sobering experience in Eden became all too common. After several years, Congress reasserted federal control by pressing forward with the survey and sale of land in the Seven Ranges under the terms of the Land Ordinance, which it revised in 1796. It soon opened several other portions of the territory to federally supervised survey and dispersal as well. With each, application of the grid was refined. Settlement also began in the Connecticut Western Reserve and the Virginia Military District sectors of the territory at this time. Connecticut quickly sold most of its reserve to a consortium of private speculators, while Virginia retained control of its district to repay war debts to veterans holding land warrants. Parceling in it relied on the southern mode of land survey, not the grid. Yet everywhere, rampant speculation and abuse continued.

Neither price nor the speculative process was the major stumbling block to land sale and settlement in the Northwest Territory. Instead, the greatest hindrance was the unresolved Indian claim to the land and the ensuing fight as Indians resisted Euro-American intrusion. The Treaty of Paris gave America only the English claim to the land, but the English had not resolved the Native American claims. As settlers purchased lands in Ohio from the government or speculators like Cutler, or as they cashed in warrants, or as they simply squatted on vacant land, they had to fight off the Indians who, like Jonathan Alder's adopted people, had hunted and farmed the same land the Euro-Americans sought to settle.

The government initiated a military response to the conflict, mounting campaigns to squelch the Indian resistance. Results were disastrous until General ("Mad") Anthony Wayne defeated a coalition of Miami warriors on August 20, 1794, at the Battle of Fallen Timbers along the Maumee River, upstream from what is now Toledo. In the aftermath of his victory, Wayne intimidated the Indians into signing the Treaty of Greenville in 1795. For just twenty thousand dollars in merchandise and an annual annuity of nine thousand dollars to be divided among the tribes, the Indians were forced to forfeit their ancestral lands.[6] The surge of white settlement then began.

The lives of Jonathan Alder and James Kilbourne, like those of most settlers, were dramatically affected by the government effort to distribute the public domain. Jonathan Alder's paradise home in Madison

County became part of the Virginia Military District. When Alder and his family returned from Virginia in the early 1800s, white settlers were acquiring property all around his cabin, and the Indians with whom he had lived for nearly twenty years were being evicted. The land on which they and their ancestors lived for millennia was no longer open to them. Shortly after his return, Alder learned the land he had occupied had been purchased by another. Like many whites settling the area, Alder had to purchase land on which to live, land he had before hunted and farmed freely as an Indian. He eventually bought a tract adjacent to his previous plot from Lucas Sullivant, the land speculator who had founded Franklinton several years before.

Unlike Cutler's company, James Kilbourne's Scioto Company found that land speculation and settlement worked just as the government and speculators planned. When Kilbourne bought the Stanbury-Dayton property in 1803, the dispersal of the public domain remained dominated by speculators despite the reaffirmation of the general process defined by the Land Ordinance. The direct sale of small parcels to individuals simply was not replenishing the national coffers, nor were warrant holders attracted inland to settle, to the extent expected. Consequently speculators continued to buy warrants from cash-starved individuals at a fraction of face value and exchange them with the government for large tracts of land. They then sold the land in small plots to individuals for a profit at the prevailing market price, which was often much less than the official government price. The government also continued to sell large tracts at a discount directly to speculators, who then settled some and sold the rest at market price for a profit. Stanbury and Dayton held the property they sold to Kilbourne's group for only several months and made a four thousand dollar profit in the transaction. They likely purchased the land from the government at one-half the minimum two-dollar-per-acre prevailing price and held it until the market price rose after the Ohio constitution was resolved. This enabled Kilbourne's investors to buy large, contiguous tracts of prime land at a price well below the government minimum, settle some, and sell the unused portions. Certainly speculation made many of the wealthy much wealthier and widely abused Jefferson's noble intentions reflected in the principles underlying the Land Ordinance. Yet as Hamilton realized, speculation also made settlement possible for most pioneers while still earning income for the treasury.

Kilbourne's land speculation did not end with the settlement of Worthington. As a federal land surveyor, local agent for eastern speculators including Jonathan Dayton (a signatory of the Constitution who was one of the region's largest land speculators), and private speculator, he was responsible for the sale and settlement of tens of thousands of acres throughout the state. Hundreds of families owed their relocation to Ohio to his efforts, for which he was frequently, but not always, paid. His years of speculation ended around 1811, as his business interests returned to manufacturing. With that shift, his influence on the landscape followed a different path, one that again illustrated fundamental national forces shaping the land.

So the ordinance did not instantly resolve the many land distribution issues that confronted and confounded the young nation. Land distribution remained complicated by technical snags for decades. Preemption, special exemptions, and preferential treatment tangled the process. Trial and error in its implementation gradually refined its operation. Over the next century Congress revised the grid again and again in response to procedural problems, politics, and economics. The minimum parcel size, sale price, terms of payment, and response to preemption were repeatedly adjusted. But the basic approach remained steadfast across the continent and across time.

Despite its many faults, a number of landmark precedents underpinned that approach. The Land Ordinance established that the new western lands, the public domain, belonged to the American people, not to the states nor to the native peoples. It established that the public domain was not a permanent or an irrevocable possession of the federal government. Instead, prompt land disposal into private ownership would be the national policy and priority. The ordinance stated, "Be it ordained . . . that the territory ceded by the individual states to the United States . . . *be disposed of*. . . . " Similarly, Article IV, Section 3 of the Constitution stated, "The Congress shall have power *to dispose of* and make all needful rules and regulations respecting the territory or other property belonging to the United States . . . " (emphasis added).

Disposal meant fee simple sale, not the long-term lease nor the holding of land in a national commons. The ordinance affirmed and reinforced the national preference for autonomous private property rights and universal ownership. It rejected alternative land-ownership patterns based on tradition and geography and broke any ties to the nation's European feudal roots. Land would be bought and sold. Ownership would be decentralized and dispersed to the common citizen under government rather than private control, although historically the ordinance aided initial acquisition by wealthy speculators and big business more than the yeoman. From the beginning, this disposal tolerated widespread abuse of policies and procedures and granted endless exemptions, preferential priorities, and special privileges. However, at its core, the Land Ordinance reflected the traditional Euro-American, Judeo-Christian landscape values of Kilbourne: values that considered land as an abstract commodity, as property; values that considered people as separate from the land they sought to reshape into an earthly paradise.

Like Jefferson's ideals, the national grid system was rational, efficient, and systematic, very much a reflection of the seventeenth- and eighteenth-century Enlightenment that glorified democracy, capitalism, classicism, Protestantism, and science. Each section of land was treated the same as the next, and each citizen (as then defined) had the opportunity to acquire one in concept, if not practice. Egalitarian. Orderly. Pragmatic. The grid dispersed people across the landscape to promote Jefferson's cherished rural society. It made a national landscape and obfuscated local and regional differences. As J. B. Jackson, a pioneering commentator on the American landscape, observed, adoption of the grid marked a profound shift in social philosophy: its free-title land ownership promoted a sense of individualism and isolation and, as the land was reordered to fit the grid's abstract geometry, it deprived people of community support, sacred spaces, and custom. People believed in the grid. They identified with it. The grid indelibly etched the land and the American psyche as it governed the dispensing of 1.3 billion acres of the public domain. "Squareness" was good. A "square deal" was a good deal; to be a "four-square man" was to be solid, honest, and hardworking. The grid defined America physically and psychologically.

Ohio was the testing ground because it was the easternmost area in the Northwest Territory and, hence, the first to be officially opened for settlement west of the Appalachians. Today it is a mosaic of nine different land distribution systems, the 1785 version of the Land Ordinance having been applied only to the Seven Ranges in extreme eastern Ohio (Monroe, Belmont, Harrison, and Jefferson Counties). Visible remnants of the state's varied land survey, speculation, and settlement process are seen in the township and county political boundaries; the layout of roads, fields, and property boundaries; and the cultural heritage of the communities. These are among the most telling imprints left on the land and can be seen most everywhere. They offer important clues to the landscape's legacy, linking us with the past and foreshadowing the future.

The Land Ordinance was one means Jefferson used to shape the young nation that he understandably sought to make reflect his fundamental belief in the Constitution, the common sense of the American people, and the desirability of a rural society that would best encourage civic responsibility. Jefferson deeply distrusted big government and the selfish interests of the wealthy. Instead, he envisioned a truly republican nation of small family farms scattered across the landscape. While he realized the importance of international trade, as well as commercial and industrial development, perhaps even their inevitability, he did not want them to overshadow the nation's rural roots. America, he believed, should remain agrarian at its heart; it should be a nation that would, to the extent possible, ship raw materials and resources to Europe in exchange for finished goods. In *Notes on the State of Virginia* (1787) he wrote,

> The political œconomists of Europe have established it as a principle that every state should endeavour to manufacture for itself: and this principle, like many others we transfer to America, without calculating the difference of circumstance which should often produce a difference of result. In Europe the lands are either cultivated, or locked up against the cultivator. Manufacture must therefore be resorted to of necessity not of choice, to support the surplus of their people. But we have an immensity of land courting the industry of husbandman. Is it best then that all our citizens should be employed in its improvement. . . . Carpenters, masons, smiths, are wanting in husbandry: but, for the general operations of manufacture, let our workshops remain in Europe. It is better to carry provisions and materials to workmen there, than bring them to the provisions and materials, and with them their manners and principles.[7]

He thought that "those who labour in the earth are the chosen people of God, if ever he had a chosen people, whose breasts he has made his peculiar deposite [*sic*] for substantial and genuine virtue. It is the focus in which he keeps alive that sacred fire, which otherwise might escape from the face of the earth. Corruption of morals in the mass of cultivators is a phænomenon of which no age nor nation has furnished an example."[8] He believed people, power, and wealth should be decentralized and dependent on the land. And he believed in the grid because it forced most people to live on the farm in relative isolation rather than in cities or even New England–style villages that required a daily commute to the surrounding fields.

His travels in Europe convinced him that cities were corrupting and unhealthful morally, politically, and physically. He wrote, "The mobs of great cities add just so much to the support of pure government, as

sores do to the strength of the human body."[9] Jefferson believed even
Williamsburg, a city of only eighteen hundred people when he attended
the College of William and Mary there, was too big and centralized. He
supported his antiurban fervor with scholarly studies by doctors and
public health officials that purported European cities were less healthful
than the rural landscape, and with the admonitions of the clergy who
preached Christian values were better maintained in rural settings, based
on biblical references to pastoral landscapes and agrarian life. Country
life made country people more virtuous than urbanites.

Yet Jefferson's vision was not the only one shaping the new nation.
Alexander Hamilton's political philosophy and personality made him the
mirror opposite of his contemporary. We can only imagine the brilliant
ideological struggle between those two rivals as they fought passionately for their principles in President Washington's first cabinet—Jefferson as secretary of state and Hamilton as secretary of the treasury.

The differences in their views became the foundation for the first
major political parties. Jefferson's political philosophy of "pure republicanism," tempered by James Madison, James Monroe, and his other
supporters, became the cornerstone of the Republicans (not the same as
our current GOP). Hamilton's economic policies laid the foundation
for the Federalists. Philosophically, the Republicans favored a small,
weak, almost invisible federal government in which the House of Representatives was the dominant force and France our chief European ally.
The Federalists, led by Hamilton and John Adams, favored a proactive
federal government in which the executive and judicial branches used
consolidated power to shape markets and set both financial and political agendas. England was its primary role model and European ally. The
Federalists felt national stability required an energetic federal government to forge coherence and stop disparate factions and interest groups
from separating into regional units. They believed that the nation's economic development was best directed by the social elite who better understood the complexities of big business and international trade; thus
they could better direct such critical affairs on behalf of the nation, com-
moner and wealthy alike.

Economic policy, as a manifestation of fundamental philosophical differences regarding the role of government, most differentiated the two
parties. Republicans believed big government and powerful financial
institutions with central authority would corrupt the new nation just as
they had corrupted the English government. Instead they placed their
faith in the public's practical wisdom to equitably and effectively move
the country toward prosperity and economic independence. They be-

lieved that the best way to grow and develop, while promoting positive civic values and avoiding corruption, crass greed, and materialism, was to free the citizen and the individual entrepreneur from government interference. John L. O'Sullivan's statement in 1837, "the best government is that which governs least," encapsulated the philosophy.

Hamilton believed America's future was better served by a diversified economy and the prosperity of the wealthy. Private greed and success, he believed, were desirable, for they promoted general public prosperity. A staunch Anglophile, he revered English culture and promoted a closer relationship with the mother country. Like Jefferson, Hamilton was passionately committed to the new republic, but one for which he saw a different future that he fought vigorously to direct. Hence, when asked by Congress and the president to design the nation's basic financial structure, he capitalized on the opportunity to advance his national vision, just as Jefferson used the Land Ordinance to further his.

When Washington assumed the presidency, the government was a financial mess. He and the House turned to thirty-two-year-old Hamilton in August 1789, only days after Washington took office, and asked him to chart a path toward economic independence, stability, and security. By December 1791, Hamilton had completed the immense task by outlining an economic structure for the nation as well as a physical structure for its landscape. That he accomplished with three visionary reports he prepared for Congress.

His first report, *Report on the Public Credit,* submitted to Congress on January 14, 1790, assessed the government's financial status ($54 million in debt) and the states' collective status ($25 million in debt). Nearly 75 percent of the national debt was owed to twenty thousand private investors and speculators who held the government certificates sold to raise money during the war. Hamilton committed the government to repaying those certificates (or to replacing them with new interest-bearing bonds) at full face value regardless of who owned them; many were originally purchased by ordinary citizens who had subsequently sold them at deep discounts to wealthy speculators. Similarly he proposed that the government honor all foreign debt at full face value too. He also recommended the federal government assume the remaining state debts even though some states had already exhibited the financial responsibility, and willingness to sacrifice, necessary to settle them. Such actions, Hamilton believed, were critical in order to transfer central economic power from the states to the federal government and to establish sound credit and trust, particularly with wealthy domestic and foreign investors. Critics believed Hamilton was simply promoting the self-interests of the wealthy over the interests of the country, but he saw the two as inextricably linked. The controversial report was signed by Washington the following August after intense political maneuvering, including the creation of a new capital on the banks of the Potomac River in economically depressed northern Virginia. Jefferson never forgave himself for begrudgingly supporting the plan, support bought by a "dinner table bargain" brokered by Madison between the two rivals the preceding June. Jefferson later considered the federal assumption of state debt, and the resulting creation of a national debt, to be the nation's original sin. He pledged to never again deal with the devil.

Hamilton further advanced his vision with the submission of his second report in January 1791. It proposed the creation of a national bank and mint like those in England. The central bank, although privately owned, would be funded in part by the government and would hold significant assets in government bonds; therefore, the fate of the two would be interdependent. The bank would be the government's main depository and would issue legal tender.

Uproar over the plan began as soon as the report was issued, much of it again critical of the plan's perceived preferential benefit to big business and the wealthy. Others questioned the constitutionality of the government chartering a bank. Congress approved the plan on February 8; but a cautious President Washington delayed signing until he obtained the advice of his cabinet, especially on the constitutionality question. Jefferson's predictable attack was far outweighed by Hamilton's eloquent defense when the latter asserted the idea that the Constitution granted government certain *implied* powers, including the power to carry out the provisions of his second plan. His persuasive "loose construction" for constitutional interpretation won the support of Washington, who approved the plan on February 25 (loose construction later became a guiding principle used by the Supreme Court).

Hamilton's third plan met even greater resistance. Submitted to Congress in December 1791, the *Report on Manufactures* proposed direct means to stimulate manufacturing and make the nation economically independent from Europe. He called for government intervention in the private sector by imposing protective tariffs and special industrial bounties to stimulate American business. He believed such protections would enable America to soon rival England and France economically.

Critics attacked the sophisticated plan on moral and political grounds more than on economic grounds. Madison, for example, championed state's rights and claimed the plan centralized too much power in the federal government. Jefferson railed at the plan's promotion of cities at the expense of the nation's rural roots. Congress soundly defeated it. Washington's first term ended with his cabinet torn by contentious debate, fractured by irreconcilable philosophical differences. Although the plan was never formally adopted, many of its proposals gradually went into effect. Today, protective tariffs and taxes of the sort Hamilton proposed are common.

Hamilton's vision of a diversified national economy composed of agricultural, manufacturing, and commercial centers laced together by an interconnecting network of transportation corridors spread slowly, initially taking root only in New England, where his programs reinforced existing social forces and economic trends. Together they hastened the development of a landscape (both urban and rural) different from that created by initial colonization. His emphasis on internal improvements and the establishment of transportation ribbons to move raw materials and finished products set precedents later drawn upon by government grants to wagon roads, canals, railroads, and highways. His reports set the financial basis for a business-oriented nation of entrepreneurs, innovation, and ingenuity—as Calvin Coolidge later recognized, "The chief business of the American people is business." However, the full effect of his vision would not become visible until the Industrial Revolution after the Civil War.

Today most of America blends Jefferson's and Hamilton's visions, reflecting the struggle begun in the 1780s to establish a national identity as the new republic organized after the Revolution. Jefferson's rural view still forms the nation's basic fabric, a patchwork quilt of fields, fencerows, and family farms stitched within the national grid. As he wished, most of America remains a decentralized rural landscape.

But embedded within are rapidly growing commercial and manufacturing centers linked by transportation corridors, just as Hamilton envisioned. From far above, the cities resemble cell-like structures spreading along tentacles that reach into the surrounding agrarian tissue to seemingly sap the necessary nutrients for continued growth. Perhaps to Hamilton, cities today might function like nerve cells that process the nation's basic business, and the transportation tentacles like neuroganglia along which our vital information is passed. Perhaps to Jefferson, they might better resemble cancerous growths.

By the start of the nineteenth century, three hundred years after the Spanish began the exploration and exploitation of the New World, the

struggle for continental supremacy among the European powers gave
way to a new nation that controlled much of the eastern third, a nation
based primarily on Anglo-European landscape values. The designs for
the national landscape were drawn. But the struggle for the remainder
of America, its heartland, its center physically and psychologically, was
far from complete.

That heartland remained shrouded in myth and mystery. For two and
a half centuries, the nature of the nation's interior was known mostly
by Coronado's report from his extraordinary 1540–42 survey of the
American Southwest in search of Cíbola, the fabled "Seven Cities of
Gold." Certainly other explorers, hunters, and trappers traversed the region, yet reliable knowledge of the interior was scant. What was known
was perhaps more fiction than fact. Was it a hostile, uninhabitable desert wasteland? How far was it from the Mississippi River to the Pacific
Ocean? What else did that vast expanse include? How extensive were
the mountains (the Rockies) reported by the Indians and trappers? Was
there a navigable water route, the long-sought Northwest Passage, across
the land mass? No one really knew.

CHAPTER 4

Westward the Course of Empire

There is no line straight or crooked, suitable for a national
boundary upon which to divide. . . . Our national strife
springs not from our permanent part; not from the land we
inhabit; not from the national homestead. There is no possible severing of this. . . . In all its adaptations and aptitudes
it demands union and abhors separation.

Abraham Lincoln

The blueprints for our national landscape were drawn as early as 1800
by Jefferson and Hamilton. But at the time the fate of the young nation
was far from certain. President Jefferson and the Republicans sought
to return the country to what they considered the true path set by the
Revolution. They felt that path had been lost as a result of the policies
of the "imperial" Federalist presidents, George Washington and John
Adams. Fear of political retaliation for the last-minute court appoint-
ments by Adams, the Jay Treaty, the Alien and Sedition Acts, and most
important, Hamilton's federal fiscal policy gripped the nation. The mis-
print of a line in Jefferson's first inaugural address that made it read
like a statement of reconciliation did little to ease those fears. Jefferson said, "We are all republicans—we are all federalists," but the paper capitalized republican and federalist, leading people to interpret the
two as direct references to the political parties. Amid this political turmoil and instability, Jefferson, Madison, and Albert Gallatin set out to
free the citizen from federal intrusion and restore a minimalist government by slashing its size, including the military, eliminating taxes, and
eliminating the greatest evil—the national debt. Hamilton had used the
debt to concentrate federal power, as the debt created the need for federal taxes, banks, and fiscal policies. Only by making these profound
changes, the Republicans believed, could the promise of the Revolution
be achieved.

Nor was America's future secured internationally. After two hundred years of interaction in the Americas, England, France, and Spain had pervasive political, economic, and cultural linkages to the continent, linkages not easily or quickly broken despite the Revolution. American affairs remained tightly tangled with those of Europe. Most Americans were European in origin, so popular opinion toward European nations varied from person to person. Jefferson, remember, favored the French, Hamilton the English. Even the borders, except those defined by water bodies, were disputed and the right to free international commerce and neutrality challenged as the English and French sought to threaten and bully the United States. The fate of the new nation was still very much in doubt. War was likely with either or both of these European competitors. And the internal ability to maintain the Union was not certain, nor was it clear that a larger continental nation could be or would be achieved. Many leaders had serious doubts about whether such a vast union could be governed politically or held together socially, as a result of its geographic size and physical diversity.

In 1800 England controlled the Canadian territory and its strategic Atlantic gateway through the St. Lawrence River. English settlement was well under way along the river valley and around the Great Lakes. Yet England still coveted the lands lost to the Americans in the Revolutionary War, in particular the Northwest Territory as a source of furs. Without American forces to police the borders, the English continued to tap that wealth during the decade after the war. They also fueled Indian resistance to American settlement. By 1800, however, the recalcitrant English had finally withdrawn from the Territory, begrudgingly leaving the riches behind. Still, two giant English trading companies, the Hudson Bay Company and rival North West Company, dominated the immensely profitable fur trade around the Great Lakes and upper Mississippi valley.

Those interests were propelled toward the Rocky Mountains and the Pacific Coast in 1793, after Alexander Mackenzie found a viable route to ship the fabulous wealth of western pelts eastward across the continent. The two fur-trading giants expanded westward even though Mackenzie failed to find an easy trade route, and England soon established a menacing presence in the Pacific Northwest to capitalize on the awaiting treasures. Many Americans feared that its presence on the Pacific Coast might potentially leave the United States landlocked and stymie its expansion and economic growth just as French, then English, possession of the land west of the Appalachian Mountains had before the

Revolutionary War. Others feared England might expand eastward into the Louisiana Territory and directly threaten the American Midwest.

The English still posed a real threat to the United States overseas as well. The 1790s was a decade of continual confrontation between the British and Americans (and French). The powerful English navy continually harassed American interests internationally. England might not be able to win a land war on continental America, but it could still inflict catastrophic damage to the young nation's vulnerable economy via an embargo, blockade, and sea war, perhaps damage sufficient to fracture the shaky Union and extract significant concessions or regain some form of control. The Jay Treaty (1795) had lessened tensions, but fears remained just below the surface.

At the same time, Spain controlled the Louisiana Territory and Florida and retained a minimal presence in Texas, New Mexico, and California, although Spanish interests remained focused below the Rio Grande. With its political and economic power now greatly diminished in comparison to its rivals, Spain harbored little hope of contesting for the continental heartland. Yet Spain and its possessions were critical pawns in the European powers' larger geopolitical chess game for North America.

Many Americans coveted the Spanish holdings above the Rio Grande, especially New Orleans. The city was pivotal to American economic development as a result of its strategic location at the mouth of the Mississippi River. It was the river's choke point, the gateway to commercial development of the Midwest, for it was far more expensive to transport goods over land to the eastern markets than to float them down the river and ship them from the Gulf of Mexico to the Atlantic coastal ports. Jefferson wrote, "There is on the globe one single spot, the possessor of which is our natural and habitual enemy. It is New Orleans, through which the produce of three-eighths of our territory must pass to market, and from its [the West's] fertility it will ere long yield more than half of our whole produce and contain more than half our inhabitants."[1] The French and English eyed the prize too.

As American leaders planned the nation's future, they sought to ensure free access to the Mississippi River and to consolidate the nation's land area by acquiring New Orleans and Florida, two key parcels adjacent the national borders. The Louisiana Territory, they felt, would follow. So long as Spain acted alone, the Americans were confident the three parcels could be easily acquired since Spain was a shell of its former self. It struggled just to maintain control over Mexico, the heart and soul of its continental interests; consequently, it could not seriously resist

American coercion for the coveted areas, since the areas were far from the Spanish power base. In addition, the surge in the American population adjacent to the Spanish possessions seemed unstoppable. The areas would be American soon; only the timing and conditions were uncertain. Failing negotiation, many American leaders were prepared to take them militarily.

The international political maneuvering over the areas twisted and turned at the end of the century. In the decade after the Revolution, a dispute over the border between Spanish Florida and the United States, resulting from the vague boundaries established by the Treaty of Paris, simmered between the two nations. Pinckney's Treaty finally resolved that argument peacefully in 1795. The treaty set the border at the top of the Florida panhandle, and it established American rights of navigation on the Mississippi and duty-free transport of goods through New Orleans.

The apparent calm created by the treaty lasted less than six years, shattered by Napoleon's global ambitions. By 1801 rumors suggested that Spain had secretly ceded Louisiana back to France. If true, the consequences could be catastrophic. Spain was subject to American manipulation; it posed no real threat to America. Napoleon did. French control of Louisiana was intolerable.

Jefferson wrote that French control of Louisiana would completely reverse all political relations of the United States and would form a new epoch in our political course.[2] In short, it would "change the face of the world" because Americans would be forced to "marry ourselves to the British fleet and nation" as the only means of combating the French.[3] In response, the president instructed Robert Livingston, the newly appointed ambassador in Paris, to see if the rumor was true and, if so, to begin negotiations to acquire West Florida to lessen tensions (possession would give the United States Baton Rouge and several alternative waterway routes around New Orleans). The rumor was confirmed in May. Tensions tightened. By April 1802, Jefferson pressed Livingston to negotiate for New Orleans and all of Florida.

Tensions tightened further when word spread that Napoleon had dispatched a large army in early 1802 to regain control of Santo Domingo, the sugar-rich Caribbean colony ruled by the charismatic mulatto Toussaint L'Ouverture. The provocative move implied that the emperor was planning to militarily occupy New Orleans and the Louisiana Territory, enabling old glories and grand designs for New France to be realized.

The crisis reached a climax in October when Spain, which had not yet transferred control of New Orleans to the French, closed the port to unobstructed American commerce. Jefferson assumed, incorrectly, that the Spanish action was at France's behest. While he saw ways around even a complete blockage using other nearby waterways or canals, none of the alternatives was as desirable as acquisition of the city. He also knew the growing number of American settlers surrounding the city would not tolerate the alien presence in their midst much longer. War loomed directly ahead. And were it to occur, the United States might be forced to ally itself with England. For some, the fear of war was overshadowed by the nauseating thought of seeking British help and the potential political repercussions this might trigger; for others, that possible realignment meant opportunity for reconciliation.

Still, Jefferson sought a peaceful resolution by dispatching James Monroe to Paris in January 1803 to join Ambassador Livingston in the complex negotiations. He told his trusted lieutenant, "On the event of this mission, depends the future destinies of this republic."[4] At the same time, he strategically leaked through his old French friend du Pont de Nemours carefully crafted threats of alignment with England and the assured annihilation of the French army at sea should Napoleon press his plans. Perhaps it was merely bluff and puffery, realpolitik disinformation to dissuade the emperor. Regardless, it changed the political equation. So too did other events.

Jefferson likely knew by this time that Napoleon's imperial dreams for North America were dissipating in the Caribbean heat and humidity. The savage fighting of the slave insurrectionists and the biting of malarial mosquitoes dispatched tens of thousands of his veteran troops within scarcely a year. "Damn sugar, damn coffee, damn colonies . . . I renounce Louisiana," Napoleon shouted in a fit of rage.[5] Quickly he recharted the course of his empire through Germany and across the English Channel. Yet without a French presence in New Orleans, he knew his historical antagonist would likely use its naval power to control the Mississippi portal. That could tilt the Anglo-Franco balance of power unfavorably. Napoleon wrote, "To emancipate nations from the commercial tyranny of England, it is necessary to balance her influence by a maritime power that may one day become her rival; that power is the United States."[6]

Knowing little of those events, Monroe and Livingston were pleasantly surprised in April when the Spanish reopened New Orleans. American fears softened. Perhaps, they hoped, some progress might be possible in the frustrating negotiations, now over a year long. As Bernard DeVoto described in *The Course of Empire,* Charles-Maurice de Talleyrand, the French foreign minister, had toyed with, ignored, scorned, misled, and lied to Livingston month after month. So on April 11, as Livingston prepared to begin another round of routine discussions, he was shocked when Talleyrand abruptly asked, what would the United States pay for all of Louisiana? The immediate crisis for America was averted; however, the tangled nature of international geopolitics left others dangling.

Once again the stroke of a pen doubled the size of the young Union, adding another nine hundred thousand square miles to the infant nation. Unlike the land transferred by the Treaty of Paris in 1783, though, the new American land was shrouded in mystery. America had quadrupled in size in just a generation. The political and social pressures that unprecedented expansion created were matched only by the economic opportunities it also created. Despite constitutional questions about whether the government was empowered to purchase land, Congress and President Jefferson gleefully approved the purchase (Jefferson's democratic idealism had been tempered by the practical nuances of geopolitics). The public announcement was made, appropriately, on the fourth of July. Among those unceremoniously signing the formal transfer documents several months later were two little-known American army officers who would soon shape the fate of the land received from the French: one a young captain named Meriwether Lewis, who served as the president's personal secretary; the other General James Wilkinson, commander of the U.S. western army.

Florida remained the last foreign property east of the Mississippi. With negotiations for its purchase stymied, the United States used a thinly veiled pretense to launch a series of quasi-military skirmishes across the Florida border to intimidate the Spanish. The plan worked. Spain sold the territory for $5 million on February 22, 1819, accepting the settlement of private American claims against the Spanish as payment. The sale was one part of the Adams-Onís Treaty that also gave to the United States the Spanish claims to the Pacific Northwest established by their settlement of California (English claims to the region still remained).

The effect of the Louisiana Purchase and the Adams-Onís Treaty on the future of America and its landscape was manifold. These transfers of land marked the last gasp of the French as a powerful presence in the New World and further weakened the Spanish as a serious contender for control of the continent. They propelled American dreams of a continental nation dramatically forward. They also assured the new nation's continued growth and economic development, and they secured the borders of its Midwest heartland. As many perceptive leaders at the time foresaw, the Louisiana Purchase would inevitably thrust the new nation into the ranks of major world powers. Napoleon knew it: "The sale assures forever the power of the United States, and I have given England a rival who, sooner or later, will humble her pride."[7] He and Jefferson had manipulated the Western world's geopolitical landscape like deft puppeteers whose fingers made the figures dangling below dance an intricate ballet. Three centuries after Columbus, America stood at the brink of becoming a continental nation from sea to sea, a single nation dominated by a single set of landscape values, perhaps as Franklin, Jefferson, and Hamilton might have envisioned.

Thus as Kilbourne and his Connecticut partners left the impoverished land of New England in 1803 for economic opportunity and religious freedom in central Ohio, as Alder returned with his Virginia family to his paradise home nearby in Madison County, as the native Indians who inhabited the landscape for generations were displaced to northwest Ohio and westward to Indiana, as the Ohio Territory opened to rampant land speculation by groups like the Ohio Company, as settlement divided the landscape into parcels based on Jefferson's grid, as Hamilton's economic plans for the nation gradually began to take effect, as all of those things happened, the country expanded westward once again, never content with its existing borders, always looking westward as if driven or propelled by dreams and opportunity. Into that unknown future Jefferson quietly dispatched Meriwether Lewis and William Clark to explore its promise.

Jefferson's interest in the immense area lying between the Mississippi River and the Pacific Ocean was deep-rooted, reaching even to adolescence. As an adult, he was, like many people of the day, intensely curious about the nature of the vast landscape known primarily by myth, rumor, and speculation. Not since Coronado's expedition in 1540 had a significant European group ventured into the continent's interior. For nearly 260 years it remained little explored beyond the principal tributaries of the Mississippi River, and even those were investigated only a relatively short distance upstream from their confluence with the mighty river.

In 1783, Jefferson had asked General George Rodgers Clark to lead an expedition across the "terra incognita." Clark declined. Three years later, while serving as the U.S. Ambassador in Paris, Jefferson saw another opportunity amid swirling suspicions of European interests in the

Pacific Northwest, especially French interests. There he chanced upon John Ledyard, an adventurous Connecticut Yankee, part genius, part moon-gazer, who had accompanied the English Captain James Cook on his third voyage to the Pacific (1776 to 1779). Cook had traced the American coast in search of the Northwest Passage—a navigable water route across the continent—as far north as Alaska and the Bering Strait. While the *Resolution* anchored off Nootka Sound on the Pacific Coast side of Vancouver Island, Cook and the twenty-six-year-old American marine corporal noted the trade opportunities for the fabulous sea otter fur worn by the natives. But both underestimated their full abundance and incredible value. When the ship later stopped in Canton, China, the small supply of otter pelts they acquired for mere trinkets at Nootka easily sold for two thousand pounds sterling.

British trading vessels returned to the Sound by 1785, and the British fur trading companies moved westward toward the Pacific Coast. Ledyard tried feverishly after the war to organize American trade in the Pacific Northwest. It would also enable him to pursue his related dream of crossing the continent eastward from the Pacific Coast. Close many times, he finally gave up as economic and practical considerations focused American maritime interests on international trade.

Ledyard then tried to realize his dreams in France, where Cook's discovery had prompted furious activity. In Paris he formed a business partnership in the mid-1780s with Admiral John Paul Jones, the retired naval hero of the Revolutionary War. It too failed. Another opportunity followed with the retirement of the American minister to France and the arrival of his replacement. Ledyard used his contact with the outgoing minister, Benjamin Franklin, to meet the new minister, Thomas Jefferson. Their dreams of exploration were more compatible.

The two quickly formulated a plan for Ledyard to cross Russia and reach the Pacific port of Kamchatka. From there he would take a Russian vessel working the Aleutian trade to the North American coast and set off on foot to cross the (Rocky) mountains to the headwaters of the Missouri River, float downriver to the Mississippi, and eventually make his way to New York. Jefferson used diplomatic contacts to approach the Russian empress Catherine the Great for the requisite papers and passports. She refused. Jefferson tried again, but patience was not one of Ledyard's virtues. While both awaited a response from St. Petersburg, Ledyard was off to London in 1786 on the chance of catching a ship bound directly for Nootka Sound. British customs officials seized the ship as it set sail. The next year he began the trans-Siberian journey in the hope he could obtain permission in the Russian capital even though the empress had refused Jefferson's second request. Fortunately, permission was received, although not from the empress.

Ledyard traversed virtually all of Russia, reaching the town of Yakutsk, several hundred miles from Kamchatka, when representatives of Russian trading and political interests, who were following him, forced him to hold up for the winter. In February 1788, he was arrested ostensively for espionage on orders from St. Petersburg (Catherine was furious on learning of his passage), spirited nonstop back across the continent in a closed carriage, and rudely dumped at the Polish border. Ledyard planned one more attempt to begin an expedition westward from Kentucky but died in Egypt before it started.

Jefferson's interest was steadfast. In 1793 he championed a plan to explore the Pacific Northwest sponsored by the American Philosophical Society of Philadelphia. The members of the Society, including Jefferson, Hamilton, and Washington, would finance the venture as private investors. Jefferson's neighbor from Albemarle County, Virginia, eighteen-year-old Meriwether Lewis, applied to lead it. Jefferson felt the militiaman was too young and denied his request (his chance would come later). Instead, the noted French botanist André Michaux was selected. Michaux had lived in the United States for about nine years, studying the Smoky Mountains, Florida swamps, and the Hudson Bay region of northeastern Canada. Now he proposed to go up the Missouri River to its headwaters, portage across the mountains, and float down the nearest river to the coast. Again fate worked against Jefferson. Michaux turned out to be tangled in a complex web of international intrigue involving the highly inflammatory new French ambassador, Edmond Genêt. The expedition strangled in that web.

Finally in 1803 President Jefferson had success. In January he sent a secret letter to Congress requesting twenty-five hundred dollars to fund an expedition (the request was secret because the Louisiana Territory, through which they would pass, was still in French control). The expedition was to follow the same route Michaux had proposed. Congress quickly and quietly appropriated the money. Jefferson placed his personal secretary, Meriwether Lewis, in command with instructions "to explore the Missouri river, & such principal stream of it as, by it's [sic] course & communication with the waters of the Pacific Ocean, may offer the most direct & practicable communication across this continent, for the purposes of commerce."[8] Over the spring, as Lewis organized his party, Jefferson tutored his protégé in various scientific fields and the latest scientific techniques for use during the expedition.

On June 19, Lewis invited his friend and former commanding officer Lieutenant William Clark to join the expedition as coleader. Lewis had served in Clark's rifle company in the Ohio Territory shortly after the Battle of Fallen Timbers, a battle in which Clark had fought. When the news of the Louisiana Purchase stunned Washington on July 4, 1803, Lewis was only a day from his departure on the first leg. Clark, the younger brother of General George Rodgers Clark, solicited his brother's advice and, on July 24, responded to Lewis, "My friend, I join you with heart and hand."[9] Jefferson was thrilled with Lewis's choice— both Clark and Lewis had been his neighbors, and both families were longtime friends of the president's family.

Jefferson's motivations for the expedition were many. A man of science, he was intensely curious about the distant landscape—the plants, animals, topography, and climate of the unknown land. He was equally curious about the region's inhabitants, with whom he hoped to establish friendly relations to enable ethnographic study and to facilitate friendly cooperation in the fur trade. The enhancement of the American fur trade was of particular interest as a means of countering the emerging British interests and activities in the Pacific Northwest, and as a means to simply enlarge the lucrative business for American entrepreneurs. Jefferson hoped the expedition would establish a better route for transporting pelts to the Atlantic markets than that used by the British, enabling American traders to undercut British prices. He also hoped friendly relations with the Indians would align them with American fur companies instead of the dominant British companies. Such relations might also set the stage for the conversion of the Indians to his preferred agrarian lifestyle. Jefferson's fears that the British had already made significant inroads with the Indians in the Montana, Minnesota, and Dakota Territories later proved well-founded.

Jefferson, like many people of the day, also sought a way, whether overland or by water, to link the Pacific Coast with the nation's eastern core for more direct trade with China, India, and Asia. Dim hopes of a Northwest Passage were still harbored by some. Shipping costs on such a route would be far less expensive than on overland routes or the much longer sea route around Cape Horn. Most educated people like Jefferson, however, realized such a route was unlikely. Instead, they hoped that no more than a short portage from the headwaters of Missouri to those of the Pacific drainages would be required, which would diminish the overall economic benefits only slightly.

Jefferson also used the exploration of the region as a form of geopolitical and economic imperialism. Even though it was still in foreign possession, its exploration was the necessary precursor to American settlement and economic exploitation. Only with American settlement of the western region would American borders and American claims be secured. Only with American settlement would the nation realize the unlimited economic opportunities offered by the region's natural resources. And only with settlement could he extend his "pure republicanism" across the continent and achieve the full promise of the Revolution. Even though the Louisiana Territory was part of France as the expedition was planned, Jefferson and others presumed it would be American soon. Perhaps the Pacific Northwest would be as well. Such internal and international intrigue would continue to shape the exploration, acquisition, and settlement of the Louisiana Territory, and much of the American West, for decades.

The expedition would proceed up the Missouri River to its headwaters, then portage across the mountains, intercept a tributary of the Columbia River, and float downriver to the Pacific. There the party hoped to gain return passage, probably via China, on a Yankee trading ship working the Pacific Coast. If that failed, they would retrace their overland route.

On May 14, 1804, the expedition of approximately forty-five people began its amazing adventure as it launched a small flotilla into the Missouri River just above the river's confluence with the Mississippi River, several miles upstream from St. Louis. Lewis, moody and solitary, was twenty-nine years old. Clark, thirty-three, was an extrovert, genial, socially comfortable, and more sensitive to the Indians. Both were hearty and accustomed to wilderness. Rather than cause friction, the opposite personality of each blended with and balanced the other. Their close relationship proved instrumental in the group's stunning success.

The expedition spent the winter of 1804–05 in a fort they built among the Mandan Indians just upriver of Bismarck, North Dakota's location. There the party was reduced to around thirty, including its Shoshone Indian guide and translator, Sacagawea, her French husband, Toussaint Charbonneau, and their baby. Lewis and Clark used the relatively easy phase of the journey to test the expedition members. Two were found wanting and were sent back downriver to St. Louis with the *engagés* (professional rivermen) enlisted to serve as porters. When the river ice finally broke in early April 1805, the expedition continued.

As the group progressed farther up the Missouri River, across the northern Great Plains in what is now Montana, the landscape it traversed was a wonderland of wildlife, an endless sea of grass across which flowed unimpeded great currents of buffalo and antelope. Perhaps in the same way we might marvel at the plenitude of life on an African savanna, Lewis recorded this view of the American Great Plains on April 22, 1805: "I ascended to the top of the cut bluff this morning, from whence I had a most delightful view of the country, the whole of which, except the valley formed by the Missouri, is void of timber or underbrush, exposing to the first glance of the spectator immense herds of buffalo, elk, deer, and antelopes feeding in one common and boundless pasture." Again and again, the life on the Plains impressed him:

> Game is still very abundant. We can scarcely cast our eyes in any direction without perceiving deer, elk, buffalo, or antelopes. The quantity of wolves appears to increase in the same proportion. . . .
> As usual, saw a great quantity of game today: buffalo, elk, and goats or antelopes feeding in every direction. We kill whatever we wish. The buffalo furnish us with fine veal and fat beef. We also have venison and beaver tails when we wish them. The flesh of the elk and goat is less esteemed, and certainly is inferior. . . . The country is, as yesterday, beautiful in the extreme.[10]

On November 7, the following fall, Lewis and Clark gazed upon the Pacific Ocean. Near the mouth of the Columbia River they constructed Fort Clatsop and spent the winter. They started home on March 18, 1806, later to learn the *Lydia,* a brig out of Boston, stood anchored in an estuary at the mouth of the Columbia in November and had been informed of the expedition's presence. But after an initial sighting, the crew never found the expedition. The ship remained off the coast until August 1806, by which time the expedition was nearly back to St. Louis. The *Lydia* docked in Boston a year later.

On September 23, 1806, the expedition rowed once again into St. Louis. Gunfire celebrated the momentous occasion. During the nearly eight thousand–mile, twenty-eight-month journey, the expedition lost only one member, Sergeant Charles Floyd, to appendicitis. It returned with the great Mandan chief, Sheheke, whom Lewis accompanied to Washington, where they were received by the "Great White Father" (Jefferson) in the White House on January 10, 1807 (Sheheke finally returned to his home village again in autumn 1809). While one of the great adventures in American history was concluded, the door had been opened for others.

Yet what had they seen of the West? Although they passed through the tallgrass and short-grass prairies and the semiarid high plains, they experienced little of those landscapes. Instead they remained within the narrow band, the slender thread, of the river corridor for virtually the entire transcontinental journey. They traveled through the most unusual landscape within the Great Plains—the riparian landscape, a more lush landscape than the surrounding seas of prairie. And even though they ventured out of the river corridors to hunt and explore the adjacent prairie, they always returned to the river. The river was their base, their home. Their perception of the prairie was from afar, while their perception of the river corridor was from within. And, too, they saw the prairies during abnormally wet years.

Lewis and Clark's voluminous reports documented their trials and tribulations in great detail. The journals today are among the classics of American history. Like the trappers, traders, and explorers who first blazed the wilderness, as well as the pioneers who followed, the expedition suffered hardships we can scarcely imagine. The journals' stirring account of life in the wilderness tells an amazing tale of courage, determination, resiliency, and faith—faith in themselves, faith in the future. Their collection of botanical specimens, as well as the meteorological and geographical measurements they made, were also remarkable achievements. Of the more than 150 plants they collected, only a handful were known at the time. Far more were lost as a result of various accidents, so their collection scarcely scratched the surface of the landscape's true diversity.

Their reports told of a strange land of opportunity. They reported the Indians to be cooperative, some even eager for trade, with the exception of the more skeptical and hostile Sioux and Blackfeet peoples. They felt they found a region ripe for commercial exploitation, especially for fur trading, even though the 50-mile portage they had hoped to find turned out to be 340 miles, 140 miles of it over the heart of the Rocky Mountains. Lewis believed the route the expedition followed offered American trading companies a competitive advantage versus the routes across Canada used by the giant English trading companies. Most furs he felt should be shipped to Canton, China, and exchanged for other commodities, and that the latter should be shipped to eastern U.S. markets via clipper ship around Cape Horn to yield double profits. He also felt a portion of the furs, as well as commodities that were not bulky, brittle, or perishable, could be transported overland from a Columbia River port to the Atlantic Coast faster and cheaper than the Hudson Bay or North West companies could bring furs to European and American markets. So the long-sought Northwest Passage, and the long-awaited way to capitalize on America's advantageous central location between Asian and European markets—what Walt Whitman called the passage to India—had, in part, been found.

In effect, if more psychologically than practically, the expedition linked the isolated, islandlike sliver of Pacific America with Atlantic America. The continent finally had been crossed, three centuries after Columbus found it, and a way shown for national expansion to follow. Dreams of a continental "nation," whether of one political entity, or of several populated with people of American descent, became palpable. But it would be other routes—the Oregon, Santa Fe, and California Trails, and the transcontinental railroads—that would make the linkage real decades later.

Even before Lewis and Clark returned to St. Louis, several other expeditions were sent to survey the unknown, the largest launched amid much intrigue by General James Wilkinson, Jefferson's newly appointed governor of the Louisiana Territory. Acting on Jefferson's desire, but without his specific instruction, Wilkinson dispatched Lieutenant Zebulon M. Pike and a party of twenty-three soldiers and scientists, accompanied by fifty-one Indian porters, to cross the heart of the Great Plains en route to the Rocky Mountains. Wilkinson was a close confederate of Jefferson's first vice president, the infamous Aaron Burr, so his motives, while responsive to Jefferson's interests, may have been muddied by a hidden agenda.

Pike's party left St. Louis by boat on July 15, 1806, rowing partway up the Kansas River. It then transferred to horseback and proceeded parallel to the Arkansas River across the Kansas Territory. Like his commander and confederate, Pike was also a colorful character, albeit somewhat suspect. That association tainted his reputation, as did his conduct of the expedition. His party was taken prisoner by the Spanish along the Rio Grande drainage of the San Luis Valley after it turned southward into Spanish territory to escape the harsh winter conditions in the Rockies of south-central Colorado. The group was escorted to Santa Fe, the principal Spanish outpost in the Southwest on March 3, 1807. There, Pike made close contact and questionable "arrangements" with his captors. The party was eventually taken farther south to Chihuahua, the military

command of northern Mexico, then escorted through Spanish Texas to
the Louisiana border and released. Despite the expedition's curious outcome, it nonetheless provided one of the first direct descriptions of the
Great Plains since Coronado's more than two and a half centuries earlier. Pike, like Coronado, thought it an alien, uninhabitable landscape, a
hostile desertlike landscape of little use.

In the decade after Pike, government attention was diverted from further exploration of the Great Plains for many reasons, including the perception of the region's general undesirability. Meanwhile, interest flourished in the Pacific Northwest and development of an American fur trade
to challenge that of the British. It was a period of rapid growth in the industry, led by John Jacob Astor, a New York merchant and real estate
tycoon. The role of the beaver, otter, seal pelt in the settlement of the
American continent was profound. Pity those defenseless creatures for
whose fur European and American fashion developed an insatiable appetite that helped shape a continent by triggering westward expansion,
especially into the Oregon Territory.

A decade after Pike's survey, U.S. interest in the Great Plains returned,
barely. The third major U.S. expedition was led by Major Stephen H.
Long. President James Monroe and his secretary of war, John C. Calhoun, originally dispatched Colonel Henry Atkinson in 1819 to establish military posts in the Yellowstone and upper Missouri basins to protect the American fur trade and counter British presence in the region.
The massive expedition included eleven hundred men and five steamboats. Many saw it as the first step in a concerted American effort to establish control of the region, based on Calhoun's plan for a line of forts
along the Plains border of the Northwest Territory.

The massive force bogged down the first year with delays and the on-
set of winter, still a thousand miles from its destination. During its winter encampment near Council Bluff, Iowa, three hundred soldiers died
of scurvy and other diseases. When Congress, already pressed financially
by a depressed economy, learned of the expedition's graft and ineptness,
it withdrew support, arguing that the expedition was too slow, too expansive, and too ineffectual. A trimmed-down party with Major Long
in command was then proposed.

Long was to proceed with twenty men up the Platte River to its headwaters at the base of the Rockies, then march south along the front range
of the mountains to the Arkansas River, and return down the river. The
expedition reached the base of the mountains near present-day Denver

on July 5, 1820. Within two weeks it reached the Arkansas River and split into two groups. One group continued down the Arkansas and arrived at Fort Smith (on the Arkansas-Oklahoma border) on September 9. The other, led by Long, set out over land for the Red River to the south. Unknowingly it found the Canadian River first, which it descended, not discovering the mistake until the group arrived at Fort Smith only four days after the first. From there the combined party made the short trip downriver to the Mississippi. The first group became notorious because two members had deserted, taking with them the records and notes kept by the expedition's zoologist and assistant topographer.

Despite the loss, the expedition produced a wealth of information about the Plains. And like Pike and Coronado before him, Long perceived the landscape as a wasteland. Such language filled Long's massive report, prepared in 1823 with the assistance of Dr. Edwin James, the expedition's botanist. One map labeled the region extending eastward nearly five hundred miles from the base of the Rockies as "The Great American Desert." The report read:

> These vast plains of the western hemisphere may become in time as celebrated as the sandy deserts of Africa; for I saw in my route, in various places, tracts of many leagues where the wind had thrown up the sand in all the fanciful form of the ocean's rolling wave, and on which not a speck of vegetable matter existed. . . . But from these immense prairies may arise one great advantage to the United States, viz.: The restriction of our population to some certain limits, and thereby a continuation of the Union. Our citizens being so prone to rambling and extending themselves on the frontiers will, through necessity, be constrained to limit their extent on the west to the borders of the Missouri and Mississippi, while they leave the prairies incapable of cultivation to the wandering and uncivilized aborigines of the country.[11]

The Great American Desert designation reinforced the prevailing western myths. Despite the vast herds of buffalo that swept across it, in the American mind the Plains became a real desert, a vast treeless wasteland inhabitable only by nomadic bands of hostile Indians mounted on horseback. That was what Coronado meant when he wrote, "It was the Lord's pleasure that, after having journeyed across these deserts seventy-seven days, I arrived at the province they call Quivira [Kansas]."[12] That was what Pike meant too. One easterner traveling from Illinois to Oregon in 1839 described the landscape stretching hundreds of miles east of the Rocky Mountains as a "burnt and arid desert, whose solemn silence is seldom broken by the tread of any other animal than the wolf or the starved and thirsty horse which bears the traveller across its wastes."[13]

In 1849, twenty-six-year-old Francis Parkman echoed the same message in *The California and Oregon Trail,* republished about twenty years later as *The Oregon Trail.* The book was adapted from the young Brahmin's journal kept during his western travels several years earlier. Part history, part literature and personal philosophy, the book became perhaps the period's most influential work about the West. As an easterner familiar with the wet, humid, fertile, and forested landscapes along the Atlantic Coast, his reaction to the relative desolation and dryness of the Plains and prairies was, like that of Pike and Long, unsympathetic. After following the Platte River valley across the Plains on his 1846 trip, he described the vast landscape east of the Rockies as an unbroken, barren wilderness:

> For league after league a plain as level as a frozen lake was outspread beneath us; here and there the Platte, divided into a dozen threadlike sluices, was traversing it, and an occasional clump of wood, rising in the midst like a shadowy island, relieved the monotony of the waste. No living thing was moving throughout the vast landscape, except the lizards that darted over the sand and through the rank grass and prickly pear just at our feet. . . . [This was a] barren, trackless waste—The Great American Desert—extending for hundreds of miles to the Arkansas on the one side, and the Missouri on the other. Before us and behind us, the level monotony of the plain was unbroken as far as the eye could reach. Sometimes it glared in the sun, an expanse of hot, bare sand; sometimes it was veiled by long coarse grass. Huge skulls and whitening bones of buffalo were scattered every where.[14]

No need to settle in the Great American Desert; it was not possible, as the first waves of pioneers discovered when they crossed the Mississippi before midcentury and reached the edge of the prairie sea. There they found the turf too tough to plow. There was too little water for livestock and too little timber for building and fuel. And the weather was violent and unpredictable. Horace Greeley, editor of the *New York Tribune,* crossed the wasteland in 1859 and reported,

> The plains are nearly destitute of human inhabitants. Aside from the buffalo-range—which has been steadily narrowing . . . and is now hardly two hundred miles wide—it [the plains] affords little sustenance and less shelter to man. . . .
> Wood and water—the prime necessities of the traveler as of the settler—are in adequate though not abundant supply for a hundred miles or more on this [the west side] as they are throughout on the other side of the buffalo-range; at length they gradually fail, and we are in a desert indeed. No spring, no brook, for a distance of thirty to sixty miles (which would be stretched to more than a hundred, if the few tracks called roads were not all

run so as to secure water so far as possible)—rivers which have each had fifty to a hundred miles of its course gradually parched up by force of sun and wind.[15]

It simply made more sense to leap over the hostile ocean, jumping the two thousand miles westward to settle the more fertile and familiar Willamette Valley of Oregon and, a bit later, the coastal lowlands of California. There they could recreate the basic settlement pattern of the eastern United States. Stuart originally found the way over South Pass into Oregon in 1813. Ashley and Smith rediscovered it in 1822. And John Charles Frémont helped blaze the Oregon and California Trails for the first fleets of Conestoga wagons in the early 1840s. For nearly two generations, the center of the continent was passed over by the wave of settlement. The wave broke instead along the Pacific Coast.

CHAPTER 5

Landscape

Myth and Reality

All the pulses of the world,
Falling in they beat for us, with the Western movement beat,
Holding single or together, steady moving to the front, all for us,
Pioneers! O pioneers!

Walt Whitman

Was the Plains hostile, unfertile, and uninhabitable? Was it really the wasteland that Coronado, Pike, and Long perceived? How did it, the continent's core, differ from the eastern landscape? Could the settlement pattern used east of the Mississippi work there? Would Kilbourne's values transfer there? Would Jefferson's and Hamilton's designs fit there? If not, what modifications were needed to those values and that settlement pattern? And what type of landscape, both physical and cultural, would result from the modifications? The eventual answers to those questions mixed myth and reality.

The fact was, the Plains was fundamentally different from the East as Walter Prescott Webb described in his classic history of the region, *The Great Plains*. Beginning several hundred miles west of the Mississippi River, or roughly along a line north to south from the Dakota-Minnesota border to east Texas, the Great Plains extended across virtually the entire western half of the continent, interrupted only by islandlike mountain ranges and coastal valleys such as the Willamette. Compared to the eastern forest, the Plains was in fact the more typical American landscape.

Whereas the East was mostly forested, the Plains was mostly treeless. Whereas the East was gently rolling and of low elevation, the Plains descended gradually in elevation from over seven thousand feet above sea level along the eastern face of the Rockies to below five hundred feet along the eastern boundary. Whereas the East was drained by a dense network of navigable rivers meandering within narrow, well-defined floodplains, the Great Plains was bisected by relatively few rivers of intermittent flow within wide, ambiguous corridors (although some rivers, such as the Colorado and Columbia, cut deep canyons). Whereas the East suffered few violent storms, the serenity of the Plains was often shattered by blizzards, hail, and thunderous rain. Whereas the East received precipitation more evenly distributed throughout the year, and more consistently from year to year, the Great Plains suffered more frequently from the vagaries of nature. Whereas the East received ample precipitation for forest growth and row crop agriculture, the Plains received enough to support forest growth mostly along drainage ways, with prairie and scrub elsewhere. But the absence of precipitation told only part of the story.

Wind, unobstructed by trivialities such as trees, blew incessantly across the defenseless prairies with its hot breath in the summer and crisp bite in the winter. Higher wind speeds and the absence of shade combined to make evaporation rates greater on the Plains as well. Many parts required nearly twice the precipitation of the forested Midwest to yield roughly equivalent amounts of usable moisture during the growing season.

Wind, rain, sun, and soil yielded a Plains landscape clothed in broad swaths of vegetation. Tallgrass prairie formed the eastern swath. There the moisture was sufficient to support a community of taller prairie plants. Forest filled bottomland and other "wet" areas. Where rainfall was insufficient for even tallgrass prairie, a short-grass prairie nearly devoid of trees blanketed the central Plains from the mid-Dakotas, western Nebraska, and Kansas, and the Texas-Oklahoma panhandles west to the Rockies. Semiarid rangeland or arid desert formed the fabric through which forested mountains such as the Sierra Nevada and Rockies protruded.

Despite the scarcity of precipitation and plant growth, some prairie soils were remarkably fertile, with higher levels of organic matter and deeper profiles than some of the eastern glacial till soils. Many prairie plants root deeper in search of water than plants in wetter climates; when the plants die and the roots decompose underground, subsurface organic matter amasses deeper into the soil profile than in eastern soils. With less standing biomass aboveground than in the dense deciduous forests, more of the nutrients remain in the soil.

Even so, the Plains guarded that treasure more closely than did the eastern landscape. The East welcomed row crops and pasture. With few natural constraints, it accommodated agriculture and readily forgave the disturbance. But to tap the western wealth, major obstacles had to be overcome. Water was one. Its scarcity limited the potential usefulness of the soils for intensive row cropping. Yet where sufficient moisture was available, many areas yielded vigorous growth. And the tangle of thick roots made the prairie turf impenetrable by the wood bullnose or iron blade plows common in the East. To mine the soil, new tools were also needed.

Animal life of the Great Plains was as alien as its plant life. The abundance of that life reflected the reality of the landscape's true fertility, in stark contrast to the myth of its sterility. Single buffalo herds containing over a million head were known to cover more than one hundred square miles. Coronado wrote, "I found so many . . . [buffalo] that it would be impossible to estimate their number. For in traveling over the plains, there was not a single day, until my return, that I lost sight of them."[1] Three hundred years later, 40 million head might still have roamed the continent. Then, too, perhaps as many as 5 billion prairie dogs inhabited the Plains. One prairie dog "town" in west Texas covered twenty-five thousand square miles and was estimated to contain 400 million individuals. As late as 1900 Texas was believed to harbor 800 million of the large rodents, a population that required as much grass to feed as nearly 3.2 million cattle. While estimates vary widely of wildlife populations before European contact, one well-known estimate placed the number of elk, pronghorn sheep, bighorn sheep, and mule deer at, respectively, 10 million, 30 to 40 million, 1.5 to 2.0 million, and 10 million. On seeing such abundance near Fort Union on the upper Missouri, John James Audubon enthusiastically noted, "An immensity of Game of all description." The plenitude in the Platte River valley moved John Charles Frémont to comment, "In the sight of such a mass of life, the traveler feels a strange emotion of grandeur."[2]

As in every landscape, though, broad environmental patterns obscure great diversity. That diversity was perhaps less recognized but more critical in the Great Plains than the more familiar midwestern and eastern till plains. Liebig's law of the minimum dictated life on the Plains more obviously, as Elliott West described in *The Way to the West*. Resources were not so uniformly available there. This concentrated species at specific places and times for obtaining the needed resources, making the limited supply especially critical for survival and vulnerable to disturbance.

Liebig's law dictated the life of the Great Plains people too. Like the eastern forests, the Plains was populated with native peoples whose cultures developed and differentiated for thousands of years: the Assiniboin, Blackfeet, and Gros Ventre in the northern Plains; the Crow, Cheyenne, Sioux, and Teton-Dakota in the central Plains; and the Arapaho, Cheyenne, Kiowa, Apache, and Comanche in the southern Plains. The Southwest was home to the Navajo, Hopi, Zuni, and other pueblo peoples; the western Plains to the Ute, Shoshone, and Paiute; and the Pacific Northwest to the Yakima and Nez Perce.

The lifestyles and cultures of the native Plains people varied with the natural variation of the landscape and differed from those of their eastern relatives. The pueblo people in the Southwest relied on agriculture (maize) while the native people in the Pacific Northwest relied on salmon fishing. Those in the central Plains, unlike their distant cousins, were more nomadic and relied on hunting the buffalo and some indigenous cattle. Their lifestyle and culture were inextricably linked to the beast from which they derived more than meat.

The buffalo clothed and sheltered the Plains Indians against the harsh environment and provided them with some of the luxuries of prairie life. In return, the Plains people respected, even worshipped, the buffalo. Its movements across the prairie sea drew them along in tow. As they roamed, they were the only American Indians to use beasts of burden: first dogs, then the horse.

The effect of the horse on the life of Plains people was rapid and profound. Few "technological" advances have triggered the same speed or magnitude of cultural change. Prior to the widespread acquisition of the horse, the Plains Indians struggled to survive in the harsh landscape. Without the technology to farm in the dry conditions, agriculture was impractical. Hunting was little easier. Plains wildlife, such as the elk, antelope, and jack rabbit, possessed keen senses and great speed. Hunting the wary creatures on foot in the open landscape was extremely difficult. Necessities, such as shelter and water, were often scarce too. Some were thinly dispersed over immense areas, others were available only seasonally in specific areas, so the Indians suffered frequent shortages as they wandered on foot to find them.

The horse, however, gave the Indians the ability to hunt with far greater speed and effectiveness, and the ability to reach these scarce resources. It gave them mobility in a vast landscape where mobility was the key to survival. Acquisition of the horse also magnified cultural characteristics. Mobility made them less agriculture-based, and it promoted a warrior culture as tribes more frequently came into contact and conflict over resources. American settlers frequently referred to the

Plains Indians as "wild" Indians while their eastern relatives were "civ-ilized" Indians.

Around 1680, only 140 years after a mere handful of horses from Spanish sources, including Coronado's expedition, found freedom in the Plains, the Kiowa and Missouri Indians were mounted, the Pawnee by 1700, the Comanche by 1714, the Cree and Arikara by 1738, and the Assiniboin, Crow, Mandan, and others several years later. By the 1780s the horse had dispersed throughout the Plains, and the adapta-tion of the native peoples to them manifest. Most Plains Indians had been mounted for barely 100 years when Euro-Americans returned to the prairies. The horse became more than a means of transportation: it also became an integral part of the Indian culture and self-identity. The horse, as in many other mounted cultures, including the Spanish and the American cowboy, was an extension of the individual, a source of pride and personal identity.

Not only did the horse make survival easier for the Plains Indians, it also made them more difficult to displace. The Spanish learned that les-son long before, a lesson that further dampened their interest in colo-nizing northward from Mexico. A nomadic people of skilled horsemen and mounted hunters posed different obstacles to subjugation than those who were more stationary and agrarian based.

So the Great Plains was vastly different from the eastern forests fa-miliar to the settlers. It suggested a fundamentally different form of set-tlement and a new set of landscape values. Eastern values reflected the eastern landscape's abundant rainfall, dense network of rivers, fertile soils, and pervasive forest. It was a landscape where individual-based set-tlement and resource exploitation, and individual-oriented institutions, were practical. It was a landscape conducive to small-scale parceling that made a patchwork landscape, a mosaic of small pieces. The East was a closed landscape, one contained and enveloped by the forest, which had to be cleared to enable settlement. The forest was the heart of the eastern landscape. It formed the essence of place: from it the central character of the landscape was derived. The forest dominated all else.

The Great Plains was a vast, open landscape, one where the sky dome arched uninterrupted overhead from horizon to horizon. It was an ocean-like landscape over which powerful currents and forces flowed unim-peded. Water, weather, wind, and sun lay at the heart of the American West. Their effects formed a landscape clothed only by a thin veil of veg-etation. But the story of the Plains and the West was mostly one of wa-ter, its scarcity, its variety of forms, its flow, and its erosive force. Water dominated the western landscape far more than the much wetter East. To know the western landscape, to truly see it, one had to first understand the role of water there. From it, the character and essence of the western landscape derived. The result was a large-scale landscape of great expanses, widely dispersed resources, and severe limitations, a landscape that dictated collective effort and cooperation to enable the small parcel-based form of development and exploitation seen in the East. More intensive development required giant parcels and powerful corporate and government institutions. But the West was not uninhabitable wasteland. The Great American Desert was a myth founded on ignorance and bias, not understanding.

As America acquired the Great Plains from France and began its initial exploration with Lewis and Clark, Pike, and Long, the Plains posed a stumbling block to the nation's westward expansion. Why did settlers continue to move westward? Why not in-fill the more familiar and more hospitable landscape east of the Plains? Why was the nation so intent on expansion across the Plains to the Pacific?

The reasons were little changed from the motivations that drove men and women westward over the Appalachians into the Northwest Territory. For most, like James Kilbourne, the many motivations included idealism, freedom, and opportunity. Jefferson, Madison, Monroe, and Adams remained intent on westward expansion to promote economic growth and development, to transfer the public domain to freehold farmers, and to secure the borders. Presidents Andrew Jackson, Martin Van Buren, William Henry Harrison, and James K. Polk continued the westward push. Entrepreneurs such as Astor and railroad magnates saw expansion as a nearly unlimited business opportunity. Others saw it as an opportunity to affect the slavery issue by the addition of slave or free states.

Beneath those rationales was a fundamental belief in the future of the United States, a fundamental belief in its destiny. Fulfillment of Franklin and Jefferson's vision of American people settled from the Atlantic to the Pacific loomed more clearly ahead after the Louisiana Purchase. It was the nation's destiny, its unavoidable, inescapable destiny—its manifest destiny. John L. O'Sullivan inscribed the phrase into the American consciousness in 1845. A proponent of the avid, expansion-oriented Young America movement, and editor of the *United State Magazine and Democratic Review,* O'Sullivan wrote that the annexation of Texas was "the fulfillment of our manifest destiny to overspread the continent allotted by providence for the free development of our yearly multiplying millions."[3] His phrase crystallized and embraced the loose assortment of beliefs about the nation's future, and their supporting rationales, which evolved into a national credo by the midpoint of the century.

It was America's manifest destiny to be a great continental nation. It was inescapable to many; after all, America was strategically located in the center between the two great markets of Europe and Asia. Even though the fabled Northwest Passage had not been found, America could still capitalize on its advantageous geographic position in the same manner that Muslims had profited earlier from their position between the same two continents. Except now, rather than caravans of camels, clipper ships integrally linked to a network of railroads served as the means of trade. Goods from Asia would be shipped to Pacific ports, transported overland across the continent, then shipped across the Atlantic to Europe. European goods would follow the same sequence in reverse. In 1829, Caleb Atwater expressed the dreams of Jefferson, Astor, Missouri senator Thomas Hart Benton, and many others when he wrote of the South Pass route to Oregon, "That this will be the route to China within fifty years from this time, scarcely admits of a doubt." Perhaps Atwater's excitement about the discovery of the wagon trail was a bit optimistic for international commerce, but wagon trains could be replaced by railroads. Had Columbus found the passage to India after all? "There is the East, there lies the road to India," proclaimed Benton in a famous 1825 speech favoring the military occupation of Oregon.[4]

Benton, Henry Nash Smith asserted in *Virgin Land,* was the father of the Manifest Destiny philosophy. Raised in North Carolina and Tennessee, he served with Jackson's Tennessee Militia in the War of 1812. The war drove Benton to repudiate the East Coast due to its stifling English influence and heritage. In contrast, the West, he came to believe, reflected the nation's future. Central to his view was the ongoing struggle between America and its principal cultural and economic nemesis, England.

The key to unlocking America's future, Benton believed, was the development of trade with China and Asia. He felt its control had for centuries been at the core of Western history and was the foundation for recent English dominance. America's geographic location, he thought, would inevitably enable it to undercut English commercial interests in the Far East and thus control Eastern trade. That would inflict a devastating blow to the heart of the world's great mercantile power and

extract a vengeance. With American control of Eastern trade, Benton envisioned that history would repeat itself as the United States would replace England as the dominant Western power. America would finally
be free of the stifling yoke of its English heritage.

Benton argued that the network of western rivers, the Yellowstone,
Platte, and Arkansas, or later, a network of railroads, would do for
America what the key rivers of earlier great European civilizations had
done: "what the Euphrates, the Oxus, the Phasis, and the Cyrus were
to the ancient Romans, lines of communication with eastern Asia, and
channels for the rich commerce which, for forty centuries, has created
so much wealth and power wherever it has flowed."5 To do that, the
United States had to first colonize and annex the Oregon Territory, then
under English dominance, and California, then under Spanish control.

The Anglo-American Joint Occupation Agreement for the Oregon
Territory (1818), and the Adams-Onís Treaty with Spain (1819), outraged Benton. Surely the government, he railed, had sold out the nation's future by negotiating agreements that blocked unobstructed U.S.
expansion to the Pacific Coast. Not only that, it had done so to two
of the nation's historical rivals for continental supremacy. His call for
the forceful annexation of Oregon was not answered. Yet by 1846 his
dream for a continental nation was realized, peacefully, with the acquisition of the Oregon Territory. Only two years later the Mexican cession
brought California and the remainder of the Pacific Coast under American control.

Ironically, just months after California was ceded to the United States,
James W. Marshall's curiosity was aroused by gleaming flecks he noticed in a stream as he tested the undershot wheel of the new sawmill he
had built on the American River for his employer, John Sutter. Sutter, a
Swiss adventurer, had fast-talked the Mexican governor into granting
him Mexican citizenship and a fifty thousand–acre tract to settle and develop in the sparse Sacramento Valley. What Spain so desperately sought
for centuries was found, and the great California gold rush began.

By year's end, the quiet valley teemed with prospectors and promoters in a wild, speculative frenzy. Compelled by Manifest Destiny and carried by opportunism and greed, the fifty thousand hopefuls who arrived
in the first wave drowned the lifestyle of the region's ten thousand inhabitants. Wave after wave followed to reap the treasures of the Mother
Lode. Ten years later, as these veins were drained of wealth, the search
spread to the Colorado Rockies. The boom-and-bust cycle of mining ravaged the Sierra and Rockies, along with tens of thousands of lives, leav-

ing a devastated physical and social landscape in its wake. Still, the boom had brought torrents of people to those regions, many of whom remained and found other, more permanent occupations.

Franklin and Jefferson smiled from above. By 1848, only sixty-five years after the Revolution, all the major pieces of the American continental puzzle were in place. The young nation had doubled in size after the Revolution, doubled again with the Louisiana Purchase, and then grown by nearly another two-thirds at the giddy, rarefied height of Manifest Destiny.

Those acquisitions provided further proof of the geopolitical philosophy. They reinforced it and engrained it deeper into popular American thought. In a single lifetime, America spanned the continent. It swept aside each of the great European powers, the French, the Spanish, and the English. The nation's destiny was unfolding. From the 1820s until his death in 1858, Benton championed the cause of westward expansion with unparalleled passion and rallied support for the transcontinental railroad that would carry that westward movement.

Benton's brand of Manifest Destiny was based on more than America's certain victory over the English in the commercial competition for Asia. The ascension of America was also assured by another historical force. Benton subscribed to the popular theory that the center of Western civilization inexorably marched westward—Greece, Italy, France, Spain, England, and America. Destiny drove the course of empire westward along a "hereditary line of progress." Westward, he stated, was "the course of the heavenly bodies, of the human race, and of science, civilization, and national power following in their train."[6] The events of the first half of the nineteenth century did nothing to dispel that belief.

Key roots of Benton's philosophy were found in the philosophy of his mentor, Thomas Jefferson. Benton's interests in Pacific trade surely grew from Jefferson's, as did his belief in the desirability of an agrarian nation. He too foresaw the farmer as an integral cog in the nation's westward expansion and economic development. Yet, like Hamilton, Benton also pictured an integrated network of (transcontinental) highways. Farmers would move westward in the great migration to the nation's interior carried by a network of roads and railroads. Farms would blossom along transportation corridors that would provide the conveyance for farm necessities as well as the fruits of the farmers' labor.

Propelled by Manifest Destiny, the nation followed Horace Greeley's 1837 charge to the urban poor and unemployed to "Go West, young man, go forth into the Country" as a means of lessening chronic urban problems. The West was seen as a release valve for pressures brought by immigration and surging urban populations. By the thousands, Americans sought their destiny in western gold, furs, and farmland. The Plains was the first landscape they encountered, and as they reached that chasm, Manifest Destiny propelled them over the great gulf to the Pacific Coast. Another philosophy, another attitude toward the Plains would be necessary to entice them into its interior to settle and stay—an attitude that countered the Great American Desert myth, and one that overlooked the disastrous attempts to farm the prairies. The Plains was, after all, not a desert but the Land of Gilpin.

William Gilpin was Benton's disciple, just as Benton was Jefferson's. And as Benton had further shaped and molded Jefferson's philosophy, blending to Jefferson's visions his own to yield Manifest Destiny, Gilpin added his vision to that of Benton's. Gilpin transformed Manifest Destiny from a motivation that pushed settlement westward, that drove it across the Plains, to one that instead pulled it to the Plains. Benton's brand had been that of the politician; Gilpin's that of the salesman or evangelical preacher. Gilpin wrote:

> The *untransacted* destiny of the American people is to subdue the continent—to rush over this vast field to the Pacific Ocean—to animate the many hundred millions of its people, and to cheer them upward . . . —to agitate these herculean masses—to establish a new order in human affairs . . . —to regenerate superannuated nations— . . . to stir up the sleep of a hundred centuries—to teach old nations a new civilization—to confirm the destiny of the human race—to carry the career of mankind to its culminating point—to cause a stagnant people to be reborn—to perfect science—to emblazon history with the conquest of peace—to shed a new and resplendent glory upon mankind—to unite the world in one social family—to dissolve the spell of tyranny and exalt charity—to absolve the curse that weighs down humanity, and to shed blessings round the world![7]

From the 1840s through the 1890s, Gilpin was one of the nation's most influential promoters of the West. Part salesman, preacher, and politician, he was also a soldier, explorer, and entrepreneur—a renaissance man of unbounded energy and optimism. A prolific writer and mesmerizing orator, he joined Walt Whitman and others in transforming Manifest Destiny from an abstract economic and geopolitical philosophy into a social cause célèbre. His version was emotional, patriotic, and personal. It blended pseudoscience with good old-fashioned snake oil hucksterism. He metamorphosed the Great American Desert into the Land of Gilpin—a lush land of unlimited opportunity.

The quest for opportunity, he knew, lay at the center of the American psyche. A half century earlier Alexis de Tocqueville had seen this: "To clear, to till, and to transform the vast uninhabited continent which is his domain, the American requires the daily support of an energetic passion; that passion is the love of wealth. . . . "[8] That quest had brought the Spanish to the Caribbean and sent Coronado in search of Cíbola. It initially had propelled the French to Canada in search of fish and fur, and English companies to the mid-Atlantic Coast in search of profit. Later it had driven armies of pioneers, like James Kilbourne, westward across the continent. Certainly other motivations were at work as well, but land and economic opportunity were intertwined at the core of the American experience; so, Gilpin preached an old, resonant theme.

William Gilpin was born in 1813 to a prominent Philadelphia family. Andrew Jackson was a close family friend. Politics and influence permeated the household. One brother became a Democratic appointee to the consular service, another Van Buren's attorney general. William's career began on a different course, though, when President Jackson appointed him to West Point. He never finished. His was an adventurous, footloose personality. Having served briefly as a volunteer in the Seminole War, Gilpin went west in 1838 as editor of the *Missouri Argus* to promote Thomas Hart Benton's Senate reelection. Several years later his association with Benton broadened.

While exploring the West in search of source material to write about, Gilpin accidentally met John Charles Frémont on the Oregon Trail in 1843. Frémont was Benton's son-in-law and a disciple of sorts as well. Gilpin accompanied Frémont to Walla Walla, Washington, then separated and went on to Fort Vancouver as the guest of the Hudson Bay Company. Even though he stayed only two short months, the British colossus soon regretted its hospitality. Gilpin actively participated in the Americans' burgeoning campaign for independence when in March 1844 they held a convention for the establishment of a government. Armed with a petition from the settlers requesting American occupation of the territory, Gilpin returned to Washington, where echoes of Benton's speech to the Senate twenty years earlier still reverberated. There Gilpin advised Benton, Buchanan, Polk, and other politicians during the mid-1840s, the height of national expansion driven by Manifest Destiny.

Within a year he emerged as a major in Alexander Doniphan's First Missouri Volunteers' famous victory in the war with Mexico. He later commanded troops against the Pawnee and Comanche as he participated in the realization of the philosophy he preached. Over the next decade he continued to write, speak, and lobby for his vision. Having cast the only vote for Lincoln in Jackson County, Missouri, he was one of thirteen men who accompanied the president-elect from Springfield to Washington. There he was one of the president's one hundred volunteer guards in the White House during the first tense weeks of the new administration. Lincoln rewarded Gilpin's loyalty by appointing him the first governor of the Colorado Territory. By the 1860s he had a platform from which to preach his new brand of Manifest Destiny.

Gilpin embellished Benton's beliefs in the inevitable westward progression of economic and political power from Europe to America. He envisioned that progression would culminate in the "Republican Empire of North America," the center of Western civilization. This, he insisted, was supported by Alexander von Humboldt's recent theory of isothermal zodiacs. The theory proposed that the Northern Hemisphere was encircled by "a serpentine zone" that alternated about the fortieth parallel. The zone reflected continental and oceanic climatic patterns, and the area within shared common geophysical characteristics. China, India, Persia, Greece, Spain, England, and America all fell within. From Gilpin's perspective the theory merely reinforced what was obvious—America was destined to be the world's next great empire, perhaps even its ultimate empire.

As the course of Western civilization marched continually westward, Gilpin felt the moral and cultural level of each dominant civilization reached a progressively higher plane. America was preordained to be not just economically and militarily superior to the other nations of the world, as Benton believed, but to be morally and culturally superior as well. To some advocates, the latest revelations in science supported this. By twisting Darwin's new theory about survival of the fittest into a pseudoscience called Social Darwinism, they rationalized imperialistic national behavior based on the apparent superiority of the victors over the inferior, vanquished people.

Most of all, Gilpin was a salesman of the West. To him, as it was for many other promoters of western expansion, the West was the land of opportunity. The Land of Gilpin was a place where "rain follows the plow," as one popular claim pronounced in an effort to attract settlers. The Plains did not suffer from a shortage of rainfall or from a shortage of trees. "The PLAINS are not *deserts*," Gilpin wrote in one of his popular books, "but the OPPOSITE, and the cardinal basis for the future empire now erecting itself upon the North American continent."[9] All one had to do was break the prairie soil and plant a crop; the rain would follow, insisted local boosters. As evidence, they pointed to recent weather data. The Kansas Board of Agriculture reported in 1888 a statewide average of 44.17 inches of precipitation, and 43.99 inches the following year. Never mind that it probably had not rained that much before (or since); facts do not lie, they said. The noted climatologist Professor Cyrus Thomas concluded in 1869, "Since the territory [of Colorado] has begun to be settled, towns and cities built up, farms cultivated, mines opened, and roads made and travelled, there has been a gradual increase in moisture. . . . I therefore give it as my firm conviction that this increase is of a permanent nature, and not periodical, and that has commenced within eight years past, and that it is in some way connected to the settlement of the country, and that as population increases the moisture will increase."[10]

The famous explorer and scientist Ferdinand V. Hayden concurred. The Kansas Bureau of Immigration proclaimed the state's climate to be the finest in America, where summer lingered into November and "at the close of February we are reminded by a soft, gentle breeze from the South, that winter is gone."[11] The Plains states and territories waged a promotional war of words to entice settlers to the paradise found only within its borders. Each was more wondrous than the rest. Land speculators, individual entrepreneurs, business organizations, local and state governments all beckoned the naive easterners. Come to the Plains; come to paradise.

The railroads were undoubtedly the champion boosters; their motivation, profit. J. J. Hill, founder of the Great Northern Railroad, typified their attitude when he remarked to an acquaintance, "You can lay track through the Garden of Eden. But why bother if the only inhabitants are Adam and Eve?" The Rio Grande and Western Railroad compared the Utah landscape to the Promised Land. A Northern Pacific circular claimed Montana to be so healthful that not one case of illness, other than indigestion caused by overeating, had been recorded the preceding year. "Why emigrate to Kansas?" asked a satisfied resident in *Western Trail,* the Rock Island Railroad's gazette. "Because it is the garden spot of the world. Because it will grow anything that any other country will grow, and with less work. Because it rains here more than in any other place, and at just the right time."[12] With money to be made and money to spend on promotion, the Land of Gilpin was indeed paradise.

The Great Plains, Gilpin argued, was part of the basin of the Mississippi, a "vast amphitheater, opening toward heaven." Centered in Humboldt's zodiac, the basin extended from the Rocky Mountains to the Appalachian Mountains—the watershed of the Mississippi River. Gilpin

claimed the basin would one day be inhabited by more than 1.3 billion
people, because it would "receive and fuse harmoniously whatever en-
ters within its rim."[13] That vast, fertile landscape uninterrupted by natural barriers or borders, yet linked by a network of rivers and railroads,
would be the heart of the North American empire. Nowhere else on earth
did such a landscape exist, one so large, so fertile, so rich in resources,
so ripe for human endeavor. "What an immense geography has been revealed," he shouted to the frenzied Fenian Brotherhood in Denver, "what
infinite hives of population and laboratories of industry have been elec-
trified and set in motion! The great sea has rolled away its somber veil.
Asia is found and has become our neighbor. . . . North America is known
to our own people. Its concave form and homogeneous structure are
revealed. Our continental mission is set to its perennial frame. . . . "[14]
That was the nation's untransacted destiny; that was its Manifest Destiny.

CHAPTER 6

Pilgrims' Progress

Poet and romancer, as well as hunter and tourist, have
lamented that in so short a time the wild West would be a
thing of the past. . . . Let them be reassured. The wild West
will continue wild for centuries. . . . The mountain Territories
will long remain the abode of romance; and "Western Wilds"
will be celebrated in song and story, while generation suc-
ceeds generation.

> J. H. Beadle, *Western Wilds and the Men
> Who Redeem Them*, 1878

The landscape puzzle of continental America was completed in the second quarter of the 1800s as the last European claims were banished
and the 350-year struggle for supremacy of the continent concluded. The
United States stretched from coast to coast. The most important milestone in our landscape history had passed. One era ended and another
began. From that point forward the story of the American landscape
was no longer one of exploration, expansion, and international competition. It no longer focused on the transfer of little-changed European
landscape values and land management practices, nor their leap from
the East Coast across the nation's interior to the West Coast. From the
mid-nineteenth century forward, the story instead focused on in-fill and
adaptation in the heartland.

The U.S. population continued its phenomenal growth in the mid-
1800s, surging by nearly a third from 1840 to 1850. Manifest Destiny
infected popular American thought as wave upon wave of pioneers set
out across the Great Plains in search of opportunity in Oregon and
California. California gained statehood in 1850, Oregon in 1859. Still,
permanent settlement bypassed the Plains as the cultural and technological adaptations necessary to successfully implant an agrarian lifestyle on the prairies had yet to be made. Over the next twenty-five years,
those requisite changes began and settlement slowly started, although
statehood would not cover the region for another half-century: the

Dakotas and Montana in 1889; Idaho and Wyoming in 1890; Utah in 1896; and Oklahoma, nearly fifty years after Oregon, in 1907. The (semi)arid southwestern states of Arizona and New Mexico would not join the Union until 1912. The Great American Desert was gradually buried beneath the Land of Gilpin as pioneers, driven westward by Manifest Destiny and enticed westward by speculative hype, stopped and stayed on the Plains. The story of the American landscape in the third quarter of the nineteenth century is the one with which we most identify. It is the singular defining element of our national self-image. It was the story of cowboys and Indians, buffalo and horses, pioneers and homesteaders—Euro-American settlement of the Great Plains.

Settlement was possible, Webb detailed, only after an assortment of adaptations occurred, disconnected adjustments that made settlement possible only in combination, not individually. The indigenous plants, animals, and people were changed, and eastern agricultural practices were modified to suit the distinct character of the Plains. Those practices were not scrapped in favor of new ones specifically tailored to the Plains. Instead, only the minimum adjustments necessary to superimpose "the plow and the cow" on the western landscape were made. Nor were the dominant landscape values that underlaid those land management practices modified: the belief that humans exist apart from nature, not as a part of nature; the belief that the threatening wilderness—the uncivilized landscape occupied by Native Americans—was merely a commodity to be reshaped to better suit Euro-American wants and needs; and the belief in private property and economic prosperity. Today, after a century of western settlement, as Donald Worster explained in *Rivers of Empire,* significant differences in the form and forces governing eastern and western development have emerged. Yet underlying the differences, our general approach to landscape organization and exploitation, and our traditional landscape values, remain remarkably unaltered.

The first requisite modification was the elimination of competition, both human and wildlife. If eastern pioneers were to settle the Great Plains to farm or raise livestock, the Indians had to be concentrated into small enclaves rather than permitted to remain widely dispersed across the land. Close intermixing and peaceful coexistence of whites and Indians were less practicable on the prairies than in the East, for the Plains Indians were less settled and agrarian than their "civilized" eastern relatives. Their concentration, though, posed a major obstacle. The challenges it presented to the U.S. military differed significantly from those confronted in the East. The initial use of eastern tactics proved disastrous. Troops had to become mounted and to acquire horsemanship skills. The U.S. cavalry had to emerge as an effective fighting force adapted to the size and conditions of its new theater. Even logistical support posed serious problems as the army tried to cope with the vastness of the unfamiliar landscape. An archipelago of islandlike forts had to be constructed across the prairie sea.

Weaponry also had to improve in response to the swiftness of engagement and absence of cover. Fighting on the Plains was often in the open, sweeping across broad landscapes; at other times, it was channelized in steep-walled canyons, or at close distances across rocky outcrops. Winchester, Remington, and Colt revolutionized cavalry effectiveness with accurate, repeating rifles and the six-shooter pistol. What advantages the Indians had in motivation, horsemanship, tactics, and knowledge of the terrain were overcome by technology and numbers.

Other forces affecting Indian occupancy of the Plains were at work as well. As in the East, European diseases decimated the Plains people well in advance of white settlement, seriously disrupting traditional lifestyles. The decline in population was offset by the influx of Indians displaced by whites in other regions. In fact, the Indian population increased on the central Plains in the early 1800s, affecting the lives of resident and refugee alike. Competition for scarce resources intensified as demand grew. The delicate balance between Plains people and the environment was broken before whites arrived.

That arrival was at first linear and fluid rather than fixed. As Elliott West detailed, the first significant white presence on the Plains was the wagon trains on the Oregon and Santa Fe Trails. Their passing brought profound change to the landscape and native peoples, extending far beyond the corridors' relatively narrow width. They imported plants, animals, and opportunities for trade and interaction, which altered the region's landscape and its inhabitants. They also devastated the land by cutting timber, grazing livestock, and trampling grasses. The effects were catastrophic for the native people and wildlife, since those corridors were critical seasonal habitat for both. Liebig's law told the tale.

The other major competitor to the advance of Euro-Americans could not resist. Wildlife was helpless to protect its claim to the landscape, already weakened by the invasion of people, plants, and animals. The vast herds of elk, sheep, deer, and other Plains species were defenseless to changes to habitat and to the onslaught of hunters who sought profit and pleasure in advance of settlement. By the early 1900s, that carnage and those changes had radically reduced their populations. The mule deer population was cut by as much as 95 percent, and the elk population by more than 99 percent. Others fared even worse.

For millennia the North American bison had dominated the Great Plains. Perhaps nowhere on earth was so large a landscape so dominated by a single species. Herds that stretched from horizon to horizon thundered across the prairie sea. Unfortunately the fixed, enclosed form of eastern settlement was not compatible with free-flowing currents of buffalo herds that trampled the landscape in their passing. Nor did whites value the buffalo as a source of beef, although its hide and fur were widely used. Beef, easterners insisted, came from cattle. So the buffalo had to be eliminated. Millions were slaughtered in the ensuing war for no other purpose than their elimination. Harvesting by Indians for food, by-products, and trade also increased. Human encroachment on critical habitat, changes in vegetation, and climate fluctuations hastened the decline. By 1875, the buffalo was at the brink of extinction. And with it the cultures of the indigenous Plains people were cast asunder.

In place of the native inhabitants of the Plains, a new people, a new animal, and a new culture arose. The people were white ranchers; the animal, the longhorn cow; and the culture, that of the cowboy. Together they formed what Webb called the "cattle kingdom." Overlapping the demise of the buffalo and the confinement of the Indians, the cattle kingdom spread across the Great Plains from 1850 to 1875 like the wildfires that swept the prairies, then just as quickly vanished. By 1885 it too had neared extinction, replaced in the succession of prairie settlements by another form. Yet its short presence forever changed American culture and the Plains landscape.

Cattle ranching developed in Texas south of San Antonio along the Mexican border and the Nueces River valley. There it evolved from ranching practices derived from the Spanish. The process began slowly at first as Texans gradually acquired horses and longhorn cattle from their southern neighbors, along with a form of herding that relied on the open range and the cowboy mounted on horseback. The Texas Territory's independence from Mexico in 1836 hastened the process. As an outcome, Mexican-controlled ranch lands and large unclaimed herds of Spanish cattle were transferred to Texan ownership.

The Lone Star Republic's tumultuous ten-year life span witnessed the slow, steady spread of the fledgling ranching system. The number of cattle exploded. With no place to sell the cattle, most were simply left wild to roam and reproduce on the range. In 1830 the number of cattle in Texas was around one hundred thousand head. By 1860, a conservative estimate counted 3.5 million head. The longhorns threatened to overrun the region. Yet sales of the wild and ornery animal remained inconsequential. The thirty years loose on the Plains had not bred a gentle creature, and there was little market for them in the East as a result of excessive shipping costs. By 1870, their number approached five million head. The economics of the situation, however, suggested opportunity.

In 1865 a longhorn could be purchased in Texas for three to four dollars. The same cow garnered thirty to fifty dollars in other markets as demand for meat began to grow after the California and Colorado gold rushes and the Civil War. The math was simple and the potential profits incredible if the supply in Texas could reach the demand. But how to get it there? Lengthy cattle drives to Denver, Kansas City, or St. Louis proved too costly to man and beast. Most shipping points were also located in settled, parceled landscapes partitioned with fences and fields. To move a herd across the landscape one needed open range where it could graze along the way and a clear route that avoided other land uses. In short, drives were limited to the Plains, then becoming vacant as the Indians were confined to reservations and the buffalo removed. But that emptiness was also the problem—no markets or transshipment points. The solution was obvious to Joseph G. McCoy, who operated a large livestock shipping business in Illinois with his two brothers.

McCoy was a dreamer and visionary entrepreneur. He proposed to build destinations in the heart of the Plains where the cowboy could drive his herd and sell it to an eastern buyer, who could then load the herd on a train for shipment to an eastern slaughterhouse. Simple. And refrigerated boxcars had just been developed, enabling slaughtered meat to be shipped without rapid spoilage. The potential profits were enormous. McCoy proposed the creation of the "cow town."

Critical details remained. Where should the cow town be located? Could a town on a rail line be transformed, or would a new town need to be built from scratch; would the railroad have to extend a spur to an existing town? The rail system, even after the surge in construction associated with the war, extended little past St. Louis, several hundred miles east of the Plains. With so little settlement, the choice of a town, particularly along a rail line, was limited. And the entire proposition would likely require a large speculative investment in hopes of profit.

The president of the Kansas Pacific Railroad was lukewarm to McCoy's proposal, promising assistance should the other pieces come together. The president of the Missouri Railroad threw McCoy out of his office. Soon, though, McCoy signed a contract with the Hannibal and St. Joe Railroad granting him favorable rates on its lines from the Missouri River to Chicago. One piece of the puzzle had been found, one with profound implications since it directed the cattle business to Chicago instead of St. Louis, dramatically affecting the future of both cities. With the Kansas Pacific's lines running from the edge of the prairie in Kansas and across Missouri to St. Louis, he had the second piece. The last remaining piece was the cow town itself, preferably one on or near a Kansas Pacific line. Salina and Solomon City said no. But, like McCoy, the few residents of Abilene saw the future and bought into the proposition. The quiet, slow-paced life in the tiny seat of Dickinson County was about to change. McCoy wrote of the first cow town, "Abilene in 1867 was a very small, dead place, consisting of about one dozen log huts, low, small, rude affairs, four fifths of which were covered with dirt for roofing; indeed, but one shingle roof could be seen in the whole city. The business of the burg was conducted in two small rooms, mere log huts, and of course the inevitable saloon also in a log hut."

Other than because of its willingness to join the scheme, why Abilene? McCoy explained, "Abilene was selected because the country was entirely unsettled, well watered, excellent grass, and nearly the entire area of country was adapted to holding cattle. And it was the [farthest] point east at which a good depot for cattle business could have been made."[1] Within sixty days, stockyards, pens, and loading chutes sufficient to hold three thousand head were built with lumber shipped in from Hannibal, Missouri, and Lenape, Kansas. All was ready; only the cattle were missing.

To round up business, McCoy sent a rider into the Plains in search of herds. The rider rode two hundred miles west into the Indian territory, finding one herd after another wandering aimlessly since the drovers had no regular marketplace as a destination. Word of a safe, convenient place to sell the herds spread like wildfire, and the cattle began to arrive. On September 5, 1867, the first herd was shipped by rail from Abilene to Chicago; and with it, the era of the cattle kingdom began. By the end of that first short season, thirty-five thousand head had arrived in Abilene; one thousand cars of cattle had been shipped, all but seventeen going over the Kansas Pacific to Chicago. The next year, seventy-five thousand head would be shipped; and by 1871, seven hundred thousand head. Abilene boomed.

What only months before had been a desolate backwater stop was

transformed into a boomtown bursting at the seams. Abilene epitomized the much-romanticized Wild West town. It was the archetype—
dance halls, brothels, and saloons filled with cowboys fresh from the
long drive and flush with fat paychecks. Drinking. Gambling. Gunfights.
And cattle, everywhere cattle. But it worked. Abilene and the cow towns
made the cattle business on the Great Plains. And that business spread
across the Plains, covering parts of twelve states in a heartbeat.

Cattle ranching on the Great Plains rested on managing a hardy stock
(the longhorn) accustomed to the Plains, let loose to feed and breed
freely. The cattle were rounded up periodically by cowboys mounted on
horseback, and the unclaimed head branded (claimed). The herd was
then driven to the cow town, where it was sold and shipped east by rail.
If not sold, the herd was grazed on the extensive prairies surrounding
the cow town until the market matured. Underlying the system was the
free use of the open range, the endless prairies of the Great Plains. They
provided the vast, uninterrupted ocean of grass that fed the cattle; and
they served as the space through which herds moved without barrier. In
effect, the cattle rancher had virtually no investment. The cattle were
free and the land was free. His only costs were the scant wages paid the
cowboy, the cost of a few good horses, and the construction of a ranch
house as a base of operations. The potential profits were staggering. No
wonder cattle ranching spread. No permits were necessary and no one's
permission was needed. No business skills were required and little up-
front capital. Almost all you needed was gumption and an adventurous,
indefatigable spirit. From 1865 to 1880 more than 5 million head were
driven north out of Texas; still, the number in Texas continued to rise.
In the early 1880s the cattle kingdom peaked.

Yet even at its height, cattle ranching on the Plains accounted for less
than one-third of the total U.S. beef production. By 1885, the cattle
kingdom began to implode, rapidly mutating into a revised form of
ranching called stock farming and merging with a new form of dryland
farming practiced by the scores of pioneers and homesteaders settling
the prairies. The cattle kingdom was only a precursor to Plains settlement by people of European descent.

The forces that triggered the demise of the cattle kingdom, like those
that had triggered its rise, were not necessarily connected to or specific
to cattle ranching. In combination, they extinguished its short, spectacular life. Some were political programs, others were environmental. Technical innovations contributed too. Many of these factors began before
the cattle kingdom developed.

Perhaps the death warrant of cattle ranching was signed by President Lincoln in 1862, five years before it even began. On May 20, Lincoln signed into law the Homestead Act and, only weeks later, the Pacific Railway Act on July 1. Lincoln considered them among his greatest achievements. The Homestead Act granted title to 160 acres of the public domain for $1.25 per acre after six months' occupancy to men and women over twenty-one years of age, or for only a registration fee of about $30 following five years' occupancy. Occupancy, rooted in the English Common Law doctrine of ownership, meant the construction of a house and cultivation of a minimum percentage of the acreage. The act was partly the product of larger national political forces wherein the Republicans, desperately in need of congressional and popular support to maintain the Civil War effort, sought to garner support from western interests who stood to profit from settlement. Long-standing opposition from southern states had become moot due to their secession.

The Homestead Act modified the seventy-five-year legacy of fee simple purchase as the principal means of distributing the public domain. Although the minimum size of each Land Ordinance parcel gradually decreased from 640 acres to 80 acres (and even 40 acres) in response to the size needed for a viable farm in the East, and although the minimum sale price and terms of sale fluctuated, the basis of distribution remained actual purchase. Meanwhile, the means of parceling remained division according to the national grid. The Homestead Act remained loyal to Jefferson's grid, but it abandoned Hamilton's fee simple purchase and doubled the minimum parcel size. At its heart, the act continued the Jeffersonian infatuation with yeoman farmers transforming wilderness into a pastoral paradise. America was still to be agrarian. And the act perpetuated the perception of land as property, where one piece was not different from another.

The Lincoln administration had several reasons for the change in pricing policy and minimum parcel size. Because expansion into the Great Plains was so slow, additional incentives were necessary. Proponents believed spin-off revenues generated by the more rapid settlement promoted by the free land would more than replace revenues lost from discontinuing fee simple purchase. The minimum size of parcel was increased to 160 acres to reflect the drier conditions and lower productivity of the prairies and the consequent need for larger tracts to constitute an economically viable farm.

The basic premise of the Homestead Act was carried further by subsequent acts. The Timber Culture Act (1873) granted the homesteader title to 160 additional acres after eight years if 40 acres (later dropped to 10) were planted and maintained in trees. The act was revised several times, then repealed in 1891, in part as a result of widespread abuse.

The Desert Land Act (1877) granted settlers title to 640-acre parcels throughout the dry regions of the West for $1.25 per acre (25¢ down, and the remaining dollar within three years), provided they irrigate a specified portion within the three years. Although reasonable on the surface, the program was suspect for it sold "desert" land at the same time more fertile and productive land was available through the Homestead Act. The level of irrigation it required was also virtually impossible to achieve and certainly uneconomical at the time. Still, land was land. Abuse of the Desert Land Act was widespread as people simply ignored its requirements and placed claims anyway. An 1890 modification in the law reduced the parcel size to 320 acres, and another in 1891 refined the requirements for desert improvements, yet these new requirements were easily circumvented too.

The Enlarged Homestead Act (1910) raised the minimum parcel size of the original Homestead Act from 160 acres to 360 acres for dryland farming in semiarid regions; and the Stockraising Homestead Act (1916) raised it again to 640 acres for desert grazing. Both retained the other provisions of the original.

Consistent with the history of land distribution, rampant abuse and speculation swirled around all those acts, and most assisted big business more than the individual yeoman. The disposal of the public domain was further complicated by generous government grants to states and companies, and by other, often overlapping or contradictory means of legal and illegal land acquisition. In short, the process was a tangled mess. Government remained unable or unwilling to effectively reform it. Little had changed in the seventy-five years since the Land Ordinance. By 1900, the various homestead acts and land programs had granted claims for approximately 80 million acres of the public domain. Settlers poured onto the Plains and competed with cattle ranchers for land. But the two forms of settlement were incompatible, and the numbers favored the settlers.

While up to 90 percent of the land claims made under the various programs were likely fraudulent, the government's general purpose was achieved as millions of people moved west of the Mississippi River. Unfortunately, few homestead claims were patented (finalized) as nearly two-thirds of the homesteaders failed and returned East. No more than four hundred thousand families, about 2 million people, successfully retained their homestead claims. Life on the Plains still posed serious challenges to the settlers' preferred lifestyle, one founded on eastern landscape values and conceits. And much of the land available for homesteading was isolated and inferior to the plentiful lands available from railroads or other programs. Government grants to railroads were several times greater than the amount set aside for homesteading, and grants to states were nearly double. Overall, though, the number of permanent settlers on the Plains sprouted from virtually none to several million between 1860 and 1900. Yet the total U.S. population grew from 31 million to 76 million persons during the same period, so the overwhelming majority remained east of the Mississippi, despite the surge in the number living on the Plains.

The Pacific Railway Act was the other component of the Lincoln administration's death warrant for the cattle kingdom. Following a well-established federal precedent to give generous land grants to private companies to promote turnpike, canal, and early railroad construction, the act granted public land to private companies to build railroads across the Great Plains and Rockies to the Pacific Coast. The tracks would link the nation and make a continental empire. The five transcontinental railroad companies received just over 100 million acres to lay less than twenty thousand miles of track. The Northern Pacific Railroad Company, for example, alone received 45 million acres for a twenty-one-hundred-mile-long line from Duluth to Portland and Tacoma, creating a swath of private land that it was free to use as it pleased eighty miles wide in western states and one hundred twenty miles wide in territories. As John Opie noted in *The Law of the Land*, net congressional grants to private railroads, under various programs, in total exceeded 127 million acres; with state grants, the amount reached 213 million acres. The railroads then leased or sold much of the land to settlers and ranchers as a means of generating immediate income and greater future business as the settlers' prosperity fueled the demand for rail services.

The railroads were essential for western expansion. They were the armature around which settlement wrapped, just as Benton and Gilpin had proposed. While the Homestead Act gave land away, unless the land was near a railroad, it usually had little value since it was too isolated; consequently the railroads sold more land at premium prices along their rights-of-way than the Homestead Act gave away.

The railroads were also the most vocal promoters of that expansion, yet their self-interest and evangelist-like zeal were in many ways good for the nation as a whole. The western railroads were powerful engines for national growth and development. Before their presence on the Plains, settlement progressed only along the few riparian corridors or in the scant wooded areas. Arrival of the railroads meant settlement could follow wherever the tracks went. Towns often vied fiercely to have the life-giving right-of-ways pass nearby; and new towns, often located in response to the fuel and water necessities of the engines, sprang from the prairie sod virtually overnight along the lines. Like arteries, the railroad right-of-ways became the principal conduits for western expansion and growth.

When California governor Leland Stanford drove the last spike into the track at Promontory Point, Utah, on May 10, 1869, it represented a far more fundamental event in American history than merely the linkage of the Union and Central Pacific Railroads. It fulfilled many dreams—those of Franklin, Jefferson, and Hamilton; of Ledyard and Lewis and Clark; of Astor, Benton, Gilpin, and so many others. The linkage completed by that golden spike fulfilled old dreams of a Northwest Passage and new dreams of a continental nation.

The railroad, the most obvious manifestation of industrialization, also added a new perception in the nation's landscape psyche.[2] As the powerful machines penetrated the nation's recesses, they despoiled deep-rooted perceptions of wilderness as a virgin utopia. Technologies like the railroads also shattered the long-standing perception of agrarian America as an idyllic, peaceful garden, a plentiful Eden-like paradise where people tilled the land with hand and beast. The railroad transfigured the Jeffersonian pastoral landscape populated with noble yeoman farmers, what Henry Nash Smith termed the "myth of the garden," into a new landscape myth of the Industrial Age. Timothy Flint's description of settlement in the (mid)western wilderness in *The History and Geography of the Mississippi Valley* (1833) typified that traditional garden myth. He wrote that with the redbud in flower, flocks of paroquets glittering in the trees, gray squirrels skipping from branch to branch, and the call of the chanticleer echoing in the woods and welcoming the pioneer to the wilderness, "pleasing reflections and happy associations are naturally connected with the contemplation of these beginnings of social toil in the wilderness." He continued:

> In the midst of these solitary and primeval scenes the patient and laborious father fixes his family. In a few days a comfortable cabin and other out buildings are erected. The first year gives a plentiful crop of corn, and common and sweet potatoes, melons, squashes, turnips, and other garden vegetables. The next year a field of wheat is added, and lines of thrifty apple

trees show among the deadened trees. If the immigrant possess any touch of
horticultural taste, the finer kinds of pear, plum, cherry, peach, nectarine and
apricot trees are found in the garden. In ten years the log buildings will all
have disappeared. The shrub and forest trees will be gone. The Arcadian aspect of humble and retired abundance and comfort will have given place to a
brick house, or a painted frame house, with fences and out buildings very
like those, that surround abodes in the olden countries.

It is a wise arrangement of providence, that different minds are endowed
with different tastes and predilections, that lead some to choose the town,
others manufactures, and the village callings. It seems to us that no condition, in itself considered, promises more comfort, and tends more to virtue
and independence, than that of these western yeomen, with their numerous,
healthy and happy children about them; with the ample abundance of their
granaries; their habitation surrounded by orchards, the branches of which
must be propped to sustain their fruit, beside their beautiful streams and
cool beech woods, and the prospect of settling each of their children on similar farms directly around them. Their manners may have something of the
roughness imparted by living in solitude among the trees; but it is kindly,
hospitable, frank, and associated with the traits, that constitute the stability
of our republic. We apprehend, such farmers would hardly be willing to exchange this plenty, and this range of their simple domains, their well filled
granaries, and their droves of domestic animals for any mode of life, that a
town can.[3]

Mechanization and technology transformed that mythical landscape
(and life) into a new one in which the land became a production machine run by companies and land speculators, especially in the West,
where development became more contingent on big business, big government, and big machines to overcome the landscape's inherent limitations. Agrarian and industrial images contradicted and collided with one
another in the mind's eye. Increased yields fed by mechanization and
innovation inevitably led to further development, contradicting the Jeffersonian vision of a static agrarian nation. As the garden became more
productive, attracting more people, more towns, more roads and more
tracks, it lost its romantic, pristine character that, ironically, so often
contributed to its initial attraction. The Homestead Act suffered in that
change.

Conceived as the embodiment of the former myth, it failed to meet expectations that it would help the poor immigrant and laborer in crowded
eastern cities find freedom and virtue in western land ownership, and
failed to help the individual western farmer grow and prosper. The
Homestead Act was an idealistic anachronism. It was founded on several illusions: of the farmer and farming practices; of the physical char-

acteristics of the West; and of the western settlement process. It was the manifestation of a Jeffersonian philosophy long since out of step with reality. Yet that philosophy resonates in the American mind today.

In the near term, however, the Homestead Act meant free land, and the railroads promised to link that land with the rest of the country, to integrate it into the nation's economic and social fabric. As those two Lincoln administration programs took effect across the Plains, and as a segment of the nation's surging population was driven west by Manifest Destiny or enticed west by fast-talking promoters, the era of open range was doomed. Still, a form of agriculture-based homesteading suitable for the prairies was yet to be found.

One gradually evolved as the Industrial Revolution flooded the nation with new technologies, some of which resolved old problems associated with settlement in the prairies. Since the failure of initial settlement attempts years earlier, two problems had limited western settlement—the lack of wood and water. Wood was needed to construct homes, barns, and fences, and as fuel for heating and cooking. While no one found a complete substitute for wood, more efficient ways of using that scarce Plains' resource were discovered. Ironically, the struggle in the East had been to clear forest—there was too much wood—while in the West, it was the opposite. Development in the 1830s of balloon-framed construction techniques in the Chicago area, and inexpensive wire nails, enabled houses and barns to be built with less wood than required by traditional eastern construction techniques such as the log cabin or mortise-and-tenon framing. The balloon frame relied on thin, two-inch-by-four-inch studs and joists milled from whole logs and attached by metal nails. The frame's strength came from the whole, not the individual members. Today it remains the basis of framing techniques for most American homes.

Water was required for irrigation and general human and livestock needs. Development of effective, low-cost windmills and deep water-well drilling techniques after the Civil War enabled the homesteader to sink a well to tap into the plentiful groundwater and obtain a steady stream of water for household use, livestock, and limited irrigation. Even still, most homesteaders lacked sufficient water to irrigate enough land to farm in the eastern manner.

The lack of water was not the only factor that limited the transfer of traditional eastern row crop agriculture. Other fundamental adaptations were necessary if eastern agriculture was to become viable on the prairies. One vexing problem was the inability of the wooden or iron plows used in the East to break the tough prairie turf. Even a team of eight oxen was not strong enough to pull a bullnose plow through it. John Deere's slippery steel moldboard plow, invented in 1837, solved the problem, and became widely available as steel production exploded after the war.

With the turf turned and water available for limited irrigation, the problem remained that common eastern crops were too fragile for conditions on the Plains. The soil dried and cracked during the long, parched summers, and the corn withered and died, desiccated by the constant wind and baked by the relentless sun. Traditional tillage practices used in the East exacerbated those problems. In response, new dryland farming practices gradually emerged by trial and error. By the mid-1800s they centered on two adaptations: a change in crops to wheat and sorghum, which were hardier and more compatible with conditions on the Plains; and, more important, a change in the timing of plowing and pattern of cultivation.

To make the most efficient use of every precious drop of water, tillage on the prairies was carefully timed to anticipate spring rains. It was critical for fields to be turned just before the rains so they could soak up and retain as much moisture as possible. Once planted, the fields could not be allowed to dry and crack, creating fissures in the soil column that exposed the profile to additional evaporation. Instead, the surface was permitted to dry and crust over, forming a hard, shell-like barrier that shielded the moisture below from the sun and wind. Then, just before cracking, the top layer was cultivated to maintain an even surface. That technique was supplemented with the practice of sowing alternate, parallel strips of crops separated by a strip of fallow land. The idle strips, cultivated to control weeds, stored moisture for the next cropping season, when they would be sown. While dryland farming practices varied across the Plains as a result of the natural variation in soil and precipitation, its permutations combined with the other technological advances to make the prairies inhabitable.

Yet one last innovation, one key piece, was required to connect the others. Joseph F. Glidden, a farmer and part-time inventor from De Kalb, Illinois, found that key in 1873. After three failed attempts to patent his invention, on November 24, 1874, U.S. Patent No. 157,124 was granted for "a twisted fence-wire having the transverse spur wire D bent at its middle portion about one of its strands A of said fence-wire, and clamped in position and place by the other wire Z, twisted upon its fellow, substantially as specified."⁴ The long-sought method to fence the Plains had been found in the invention of double-strand barbed wire.

As settlement began in the Plains, the need for inexpensive fencing that required little wood was apparent. The lack of water made it necessary to fence cattle out of irrigated fields or streams, rather than to keep them in a pasture as in the East. By the 1860s hundreds of patents for improved fencing were granted each year. Slowly, gradually, it was refined. Versions without barbs were too easily trampled by the cattle. Versions with barbs mounted on wood cross-members still used too much wood. Versions with metal posts or cross-members were too heavy and expensive; and versions with single-wire strands were too flimsy. Glidden's patent marked the culmination of an intensely competitive process of experimentation and innovation.

Barbed wire revolutionized ranching and farming on the Plains. In the six weeks after its patent Glidden's business produced ten thousand pounds. By 1880, total sales reached 80.5 million pounds. As sales and production increased, the price plummeted from $20.00 per hundred pounds in 1874, to $10.00 per hundred in 1880, and $3.45 by 1890. The price finally hit bottom in 1897 at $1.80.

The effect of barbed wire on settlement and cattle ranching on the Plains was immediate. An old trail captain explained,

> In those days [before barbed wire] there was no fencing along the trails to the North, and we had lots of range to graze on. Now there is so much land taken up and fenced in that the trail for most of the way is little better than a crooked lane, and we have hard lines to find enough range to feed on. These fellows [homesteaders] from Ohio, Indiana, and other northern and western states—the "bone and sinew of the country," as politicians call them—have made farms, enclosed pastures, and fenced in water holes until you can't rest; and I say, D—n such bone and sinew! They are the ruin of the country, and have everlastingly, eternally, now and forever, destroyed the best grazing-land in the world.⁵

Barbed wire was the last of the many pieces that, combined, brought a catastrophic end to the cattle kingdom's wild reign in the West. Fencing enabled homesteaders to protect their small, irrigated fields from roaming cattle. Fencing partitioned the Plains into a vast checkerboard of quarter sections, imprisoning the open range. Ranchers resisted the conversion as long as possible and waged a brief range war over fencing, but the ranchers were few. The homesteaders, backed by the government, swept over the prairie like locusts. Dryland farming of wheat and sorghum made agriculture possible for pioneers attracted to the Plains by free land and promotional hype. Plows transformed the Great American Desert myth into the Land of Gilpin. And the new track laid by the railroads linked the homesteaders who filled the land made vacant by the removal of the buffalo and confinement of the Indian.

Cattle profits and the supply of cattle boomed at the height of the kingdom, triggered by a nationwide economic boom and several years of favorable weather on the Plains. This further fueled the speculative binge. By 1885 the cycles reversed. Drought devastated the prairie grasses, already overgrazed and badly stressed by the surge in cattle. Hard winters killed cattle by the thousands. Prices and profits plummeted. The economics of free range and free cattle, of high demand and low costs, which had made the place so irresistibly profitable, collapsed like a house of cards.

The Great Plains, like the East before, was transformed into an agrarian landscape that blended row cropping with livestock management. Stock farming replaced cattle ranching as the economics became much different. That made the new system more stable. Now farmers had greater personal investment in their operations even though the land was often free. Stock farms raised crops and cattle. Grazing land was fenced. Wells and windmills were constructed. Small acreages were irrigated. And herds were bred and more intensively managed to yield a higher return on the farmer's investment. The backbreaking reality of life on the Plains replaced the romantic, glamorous Wild West days of the cow town.

In hindsight, the cattle kingdom was doomed from the beginning because it was based on values incompatible with prevailing ones. Open land, common land without legal or physical boundaries, contradicted the deep-rooted preference for a partitioned landscape of private property. Competition, not cooperation, remained dominant. The Homestead Act and the railroads simply reflected those traditional values and facilitated their spread across the continent.

Homesteading based on the many legal and illegal forms of land acquisition dominated the last quarter of the 1800s as the Plains filled with small farms. People made small concessions, small adaptations to the inherent conditions of the prairies, but only to the minimum extent necessary. Technology filled the gap between landscape perception and landscape reality. It intervened between human values and the physical world. In the 1900s other more fundamental changes shifted the settlement story of the West from the individual to big business and big government. But

first, the Plains yielded to the plow and cow of the homesteader. The
prairie wilderness was, from an eastern viewpoint, at long last civilized.

Nearly a half century after wagon trains set off across the Oregon
Trail, bypassing the uninhabitable Plains of the Great American Desert,
an endless line of white canvas-topped prairie schooners lined up for
miles at the border outside Arkansas City and Caldwell, Kansas, waiting the opportunity to settle on the prairie. At high noon on April 22,
1889, the Oklahoma Land Rush began: outrageous, optimistic, wild, and
wonderful. Fifty thousand homesteaders joined in "Harrison's Horse
Race" across the prairie in a mad dash to claim one of the twelve thousand previously surveyed quarter sections, or the smaller but potentially
more valuable sites in the soon-to-be towns of Guthrie and Oklahoma
City (the population of the latter grew from nearly zero to 10,000 that
day). When the cannon boomed at noon, a human wave backed up for
miles at the border was let loose on the 2 million acres of vacant land
in the heart of Oklahoma. No event was more distinctly American. No
event better encapsulated the many parts to the settlement story of the
Great Plains and the American West.

The unassigned block of prairie, eighty miles long by fifty miles wide,
sat in the middle of the territory surrounded by lands granted to the
Indians by sacred treaty with the United States. The "white man's island" was connected by a rail line that bisected the block, running north
to south, through the planned towns of Guthrie and Oklahoma City.
Never mind that the unassigned lands were originally part of the Indian
land but were taken back ostensibly as punishment for the Indians' support of the South in the Civil War. Never mind that the entire Oklahoma
Territory, the "Indian Territory," had been set aside for the Kiowa, Comanche, Caddo, and Wichita. Never mind that it too had been given to
the five "civilized" Eastern tribes (Cherokee, Choctaw, Chickasaw, Creek,
and Seminole) for "as long as the waters run," as they were displaced
from their forested homelands and marched westward over the Trail of
Tears to the prairie. Soon much of the remaining Indian land would be
taken. That shameful episode was not the whole story. The land rush reflected much more about America and the West.

It was also a story of optimism and faith in the future for oneself
and the nation. By the time President Benjamin Harrison announced the
strange event on March 23, 1889, speculation in the preceding months
had already made it a national and international phenomenon. For this
harsh landscape, many pioneers left the comfort and familiarity of their

eastern and European homes in search of freedom and opportunity—
free land and a fair chance to make a better life (the land was not really
free: to receive title the homesteader had to meet the same requirements as the Homestead Act and pay $1.25 per acre, but the fee was
often ignored). They came by wagon, on horseback, on foot, even on
bicycle to join the tidal wave of humanity surging across the prairies to
stake a claim on the future. In the days preceding the rush, sleepy border towns like Arkansas City were transformed with the influx of hopefuls: hucksters stood at every street corner; brothels and saloons sprang
up in tents on vacant lots; tent hotels filled others; and at night, new ar-
rivals simply unrolled blankets to sleep on the ground wherever space
permitted.

In " 'Sooners' or 'Goners,' They Were Hellbent on Grabbing Free
Land," author Robert Day described the scene:

> Many wagons were stacked high with survival gear: water barrels, pots and
> pans, and furniture. Since the conditions for holding onto the land included
> living on it and improving it by planting crops or putting up a house, there
> were tools for building and heavy metal-edged plows, too. . . . Riders, buck-
> boards, oxen and mules choked the streets. Some Boomers rode bicycles à la
> Paul Newman in *Butch Cassidy and the Sundance Kid*. There was a rumor
> that some enterprising souls had a balloon in which they planned to ride the
> local zephyrs to a free farm all their own. . . . Naturally, real estate men
> popped up ready to sell you shares in land companies at sites that were, or
> were likely to be, designated as towns. That way, at a price, you could get a
> piece of the action without making that messy dash across the prairie.[6]

The land rush was the Homestead Act and the Land Ordinance. It
was Jefferson, Benton, and Gilpin. It was the rampant speculation of
the Ohio Company and the railroads. It was "Boomers," those who un-
scrupulously promoted and speculated in land sales in the territory, usually prematurely and often illegally. It was "Sooners," who jumped the
gun and snuck in to stake claims before the official opening. And it was
"Goners," people who pulled up stakes that laid claim to their quarter
section and moved on to await the opening of other lands, often Indian
lands.

Colorful stories of settlers' exploits during the first frenzied days of
the rush paint the picture of the time. Some staked claims on the poorer
land near the border. One woman staked her claim on the spot where
she landed after her son kicked her off the back of their racing buck-
board so as not to slow his pace forward for another parcel over the
next hill. Another man who fell from his horse in the rush jumped up

yelling, "My claim, my claim!" Others raced to the town sites. On the first day of the rush, Hamilton Wicks leaped from the train as it entered Guthrie to find this scene:

> I joined the wild scramble for a town lot up the sloping hillside. . . . There were several thousand people converging on the same plot of ground, each eager for a town lot which was to be acquired without cost or without price. . . .
>
> The race was not over when you reached the particular lot you were content to select for your possession. The contest still was who should drive their stakes first, who would erect their little tents soonest, and then, who would quickest build a little wooden shanty.
>
> The situation was so peculiar that it is difficult to convey correct impressions of the situation. It reminded me of playing blind-man's-bluff. One did not know how far to go before stopping . . . and it was a puzzle whether to turn to the right hand or left. Every one appeared dazed and all for the most part acted like a flock of stray sheep.[7]

Wicks found a corner lot and, having set his stakes, he "threw a couple of my blankets over the cot, and staked them securely into the ground on either side. Thus I had a claim that was unjumpable because of substantial improvements. . . . " That night he heard gunfire and "haloos and shoutings" and general mayhem; fortunately sunrise revealed no corpses, and the guns had been put away in favor of hammers and saws. After three months Wicks described the new prairie community: "Guthrie presents the appearance of a model Western city, with broad and regular streets and alleys; with handsome stores and office buildings: with a system of parks and boulevards[,] . . . with a system of water-works that furnishes hydrants at the corners of all the principal streets . . . ; with an electric-light plant on the Westinghouse system of alternating currents, capable not only of thoroughly lighting the whole city, but of furnishing the power for running an electric railway, for which the charter has already been granted by the city council."[8]

In the end the essence of the Oklahoma Land Rush was common people scratching the prairie soil to start small farms. It was the story of the Plains itself—the interplay of wind, water, soil, and sun. And it was the story of human struggle, perseverance, failure, and success. For many settlers, reality met fiction in Oklahoma, and that reality was often much different from the frequently told fiction of the West as a utopian garden, or the Jeffersonian fiction of the yeoman farmer as an American hero(ine). Hamlin Garland and Willa Cather told the more realistic story.

CHAPTER 7

Looking Ahead,
Looking Back

Practically all values inhere in the water.
>   John Wesley Powell, *Report on the Lands
>   of the Arid Region of the United States*

From the eastern standpoint, substantial adaptations were made in the eastern settlement practices and landscape values in response to the inherent differences in the Great Plains. Land law was changed with the Homestead Act and its permutations, while dryland farming, irrigation, and an array of technological innovations tailored land management practices to suit the conditions of the Plains. The wave of American settlement that initially balked at the prairies gradually made the adaptations necessary to survive in the new landscape. The nation rushed triumphantly to fill the void in its heartland during the last quarter of the 1800s.

From another perspective, the true essence of the western landscape had yet to be seen. To Major John Wesley Powell, explorer, scientist, and civil servant, the American perception of the West remained a mix of fact and fantasy. Misperceptions, he recognized, would continue to frustrate and doom many homesteaders to failure until adaptations for successful settlement of the Plains had been made.

Rarely can one person successfully stand alone to challenge public perception and divert it to a contrary view, especially when the public has deeply vested interests in the prevailing view and is not yet ready to see an alternative. John Wesley Powell, as Wallace Stegner described in his extraordinary biography, *Beyond the Hundredth Meridian,* stood alone at the end of the 1800s, prescribing a new vision of the West, a more realistic, pragmatic vision, one sharply in focus, free from the tints added by history and tradition, ignorance, and biased self-interest. But it was too radical a vision for the nation, requiring an alteration in traditional landscape values and perceptions that was too significant. Powell's vision was not widely shared then, nor is it today.

Powell was born in Mount Morris, New York, on March 24, 1834, the eldest son of an immigrant Wesleyan circuit rider. The family moved four years later to Jackson, Ohio, a small town nestled in the foothills of the Appalachian Mountains in the southeastern part of the state. Like so many American families, the Powells later moved farther westward, to a frontier farm in Walworth County, Wisconsin, in 1846, and yet again to Bonus Prairie, Illinois, in 1851. In Jackson, though, the course of John Wesley Powell's life was charted.

There Powell learned what it was like to think counter to prevailing public opinion when he suffered a stoning for his father's abolitionist views. There too he saw settlement transform the landscape as the community was hewn from the surrounding hills. He would watch that same settlement process take place in the forested till plains of Wisconsin and the tallgrass prairie of Illinois. And in Jackson his curiosity drove him—like Lincoln and so many others settling the frontier, people denied convenient access to and the benefits of formal education—to seek his own explanations of the surrounding world, a search all the more passionate as a result of its isolation. Reading was a luxury for Powell, one he took full advantage of. The subject did not matter. Just to get a book, any book on any subject, was a gift. Yet most people eventually need another person, a sounding board, to focus and crystallize their thinking. For Powell that person was an abolitionist active in the Underground Railroad.

George Crookham, a successful farmer, was also a self-taught naturalist and scientist. His close friends included William Mather, the state geologist, Salmon P. Chase, one of the state's and nation's most influential politicians, and Charles Finney, the president of Oberlin College. Over the years Crookham amassed an immense collection of Indian relics and geological, botanical, and biological specimens that he kept in a private museum. He also acquired an extensive library of natural philosophy books (natural philosophy was the traditional term for the "sciences" before they emerged in modern form). But more important, he shared his collections and knowledge with anyone interested in learning from them. Powell was keenly interested.

Crookham's joyful wonder of nature infected young Powell as they explored the wooded hillsides, rocky ravines, and small creeks around

Jackson. The excursions were enriched by his friends who periodically joined the fascinating outings, particularly Mather, who published an extensive geological survey of the state in 1838. Crookham's collections and the surrounding landscape provided a learning laboratory unsurpassed by the best books or facilities available in the finest, most modern schools. Although he tutored young Powell for only several years, his influence on him lasted a lifetime. Tragically, a gang of proslavery thugs torched the library and museum.

After the move to Walworth County, Powell spent his adolescence engaged not in school but in the backbreaking physical labor and daily drudgery of cutting the forest, pulling the stumps, clearing the glacial erratics, and setting the fields for the family farm. Although Powell's father was an educated man, he objected to his son's interest and readings in natural history and science. Such subjects conflicted with Methodist doctrine. So the boy's mental excursions were limited to the late night hours. After working from sunup to sundown in the fields, Powell stole sleep from his exhausted body in order to refresh his mind.

His education changed direction as he approached manhood. Solitary reading and periodic rambles into the landscape to collect specimens had filled the void of formal education well, but not completely. Powell next sought to further his education at college. In 1853 he registered at Illinois Institute (later to become Wheaton College), which his father helped organize. He soon left after discovering it offered few science courses. Next he tried Illinois College in 1855, but soon left it too to go on a four-month collecting trip through Wisconsin. Classrooms proved too confined in content and too separated physically from the subject matter he cared about. In 1856 he descended the Mississippi River in a small skiff from the Falls of St. Anthony in Minneapolis to New Orleans. While drifting silently along on his Huck Finn–like adventure, Powell passed many ordinary steamboats, each no different than the rest. One he passed, Stegner speculated, may have been piloted by Horace Bixby, who was instructing an eager young boy soon to be known as Mark Twain on the powerful ways of the river.

The next spring Powell took a train to Pittsburgh, then floated down the Ohio River to St. Louis, following the route Lewis and Clark had traveled before their departure up the Missouri River. That fall he rambled into Missouri, collecting fossils, then he tried Illinois Institute again. With spring, he was off once more, rowing down the Illinois River, then up the Des Moines River into central Iowa. Finally he tried one last term of classes, in the fall of 1858 at Oberlin College. That ended his formal education. In total he left with a smattering of things like Greek and Latin for which he had little interest.

Illinois Institute, Illinois College, and Oberlin were typical of hundreds of small colleges scattered across America in the mid-1800s. Like the grade school and the church, they were often among the first institutions created as a community formed at the frontier, reflecting the pioneers' fundamental belief in the value of education. Kilbourne's group, recall, set aside land in Worthington to support a school even before it left Connecticut. Schools were a source of community pride and, as in Worthington, closely reflected local religious beliefs. Most colleges were church affiliated and stressed a liberal education in the "classics." The recent theories of Lyell, Humboldt, Darwin, and Marsh, and the modern sciences they represented, were only slowly incorporated. To most people, an educated person was one who could read and write Greek and Latin, a person who had studied classic literature and philosophy. Powell's interests, though, lay in a different curriculum.

Powell began to teach part-time in rural schools at age eighteen to support himself during periodic rambles and sporadic college attendance. By 1858, he had settled down a bit and begun to teach full-time in Hennepin, Illinois. But he couldn't sit still. In 1860, he interrupted his tranquil classroom duties for a brief stint lecturing on geology and geography throughout the lyceum circuit in Kentucky, Tennessee, and Mississippi. On return he was named the principal of the Hennepin public schools. And by then he had been elected the first secretary of the state's new Natural History Society. Only twenty-six, Powell was well-known and respected within the state's natural science community. The Civil War did not care.

In the spring of 1861, Powell enlisted as a private in the Twentieth Illinois Volunteer Infantry. By June he rose to second lieutenant, and by November to captain with expertise on fortifications on Ulysses S. Grant's staff. The rapid-fire pace of events in his life continued. On April 5, 1862, at Shiloh, a minié ball shattered his right arm, which was then amputated above the elbow three days later. After recuperation he returned to active duty, serving as an artillery officer with Grant, Sherman, and Thomas, and eventually commanded the artillery for the Seventeenth Army Corps. Powell finally resigned on January 2, 1865, as a brevet lieutenant colonel. A battle-hardened veteran, he weighed only 110 pounds and was in poor health from his old wound.

By fall he returned to academia as a professor of geology at Illinois Wesleyan University in Bloomington. He taught as Crookham had taught him—wandering with his students outdoors, using the landscape as his laboratory. The next year he moved next door to Illinois State University in Normal. He also resumed his activities with the Natural History Society, successfully lobbying the legislature for operating funds and becoming curator of the society's museum. Those modest lobbying efforts on behalf of the small bureaucracy foreshadowed his later lobbying of Congress for radical revisions in national policy.

The first major step toward that government career began in 1867, when Powell led his first expedition west. Running on a shoestring budget with aid obtained from the museum and a consortium of several universities, the railroads, and the federal government (thanks to his friendship with Grant), Powell's small party set off for the Badlands in the Dakotas to collect specimens. Once it reached Council Bluffs, Iowa, where it was to acquire a military escort through the hostile Indian territory, General Sherman, the territorial commander, directed it south to the Colorado Rockies to avoid potential trouble with the Sioux. The party eventually returned home with sufficient specimens from the Front Range of the Colorado Rockies to significantly expand the museum's modest collection (ironically, the group climbed Pikes Peak, named after a mythmaker whose legacy Powell would soon struggle to reverse). But beyond Powell's small circle of colleagues in Illinois, the expedition garnered little attention.

His next, however, the Rocky Mountain Scientific Exploring Expedition, would capture the nation's attention in a manner equaled only by Lindbergh's solo flight across the Atlantic and the first Apollo landing on the Moon. Neither Powell, Lindbergh, nor Armstrong was well-known before his feat, yet each was the first to cross a physical and psychological gulf of almost mythical proportions.

The expedition spent the summer of 1868 in the field surveying the Rockies around the Middle Park region in western Colorado and hardening itself to life in the wild. As the relatively easy life of summer yielded to the autumn winds and the pending harshness of winter, camp was relocated to what is now named Powell Bottoms in the White River valley near the present-day town of Meeker. There the tempering of eastern and midwestern civilities continued.

The expedition's real adventure began the next spring in a small town built where the Union Pacific railroad crossed the Green River in Wyoming, 175 miles to the northwest. With the river successfully bridged by the railroad only months earlier, the town had quickly shrunk from about two thousand workers to several hundred people and a handful of shacks. There, at noon on May 24, 1869, the dwindling population of Green River gathered on the riverbank to bid farewell as four small wooden boats carrying ten men, including a bearded, one-armed leader, set adrift, destined for the Grand Canyon. There too, less than two weeks before, a train had crossed the new wooden trestle spanning the river. It was the first train on its way east from the ceremony at Promontory Point, Utah, a few hundred miles to the west. The golden spike driven to secure the final rail the day before the train left symbolized the psychological fusion of the East and West into a single nation spanning the continent. The great gulf formed by the Plains, the great prairie chasm that divided the nation into east and west, had been overcome.

Yet as that train crossed the trestle on its way east, its comfortable passengers might have looked down with some puzzlement at a small group of men camped along the bank of the river below. There, Powell's expedition prepared to explore the last remaining unknown region of the nation's interior. For although the East had been stitched, albeit tenuously, to the West, a huge hole in the nation's fabric still remained. The new Green River trestle metaphorically bridged the gap epitomized by the wild, unknown reaches of the Colorado River that flowed beneath. Between the Americas of the Atlantic and Pacific Coasts, a vast landscape remained shrouded in myth, mystery, and misperception. That was the landscape into which Powell's expedition set adrift.

For thirty-seven days, there was no news of the men and their boats, the *Emma Dean,* the *Maid of the Canyon,* the *Kitty Clyde's Sister,* and the *No-Name.* On June 30, a tabloid in Utah reported that all but one brave soul had drowned in fearsome rapids. By July 2, the story, then greatly elaborated, appeared in Omaha, and on July 4, in Chicago. By week's end, authoritative accounts from the actual survivor accompanied the story as it flashed across the country. Why was the nation so captivated by news that a small party of unknown adventurers had drowned in the fabled rapids of the Colorado River? Was it because the Colorado remained the last great challenge of the West? Or was it because the western explorer, like the trapper, the cowboy, the homesteader, and the pioneer, had been transformed into a national hero?

That transformation began in the late 1700s with books recounting the stirring adventures of Daniel Boone. The process continued in the 1820s with James Fenimore Cooper's Leatherstocking stories of Natty Bumppo. Colorful descriptions of the incredible exploits of mountain men such as Kit Carson and Jedediah Smith honed it further. By midcentury the genre bridged fine literature and popular reading. Beadle's

Dime Library novels sold millions in the East, sensationalizing real and imaginary heroes and heroines such as Calamity Jane, Seth Jones, and Deadwood Dick.

America was infatuated with the Wild West, not the real West. Certainly there were voices of reason, and pioneers prepared as best they could for the challenges confronting western migration. Fact and fiction, though, were hardly distinguishable. Even the facts were cast with an eastern bias. That was the market William Gilpin and the western boosters fed upon. The California gold rush had yielded a new wealth of publicity, and the cattle kingdom provided the "Mother Lode"—cowboys, "wild" Indians, cattle drives, and cow towns. Later, events like the land rushes and stories of valiant homesteaders had provided the press with another rich vein. "Authentic" traveling Wild West shows such as Buffalo Bill's and Annie Oakley's, and carnival sideshows, had enthralled huge crowds. Speakers toured the lyceum lecture circuits; no matter whether they actually were western "experts" as they claimed, the eastern public was mesmerized. Giant, wall-sized paintings that romanticized the West toured the nation and attracted throngs who paid two bits just to gaze at one for a few moments.

The work of artists such as Albert Bierstadt, George Caleb Bingham, George Catlin, Thomas Moran, and Frederic Remington not only reflected popular values and attitudes toward the West but also shaped them. These artists expressed the nation's perception of the West, gave the West color, shape, and form. They gave it character and made it tangible to those who had never seen it firsthand. They depicted what America wanted to see, and what they wanted Americans to see. Writers of popular western novels had a hand in this phenomenon, too, which continued into the early twentieth century. The works of writers like Louis L'Amour and Zane Grey, including *Riders of the Purple Sage*, sold millions of copies in the early 1900s. Together the many authors and artists, even poets and musicians, helped define our national psyche, our self-imagine of who and what we are as a people. Much of that image was, and is, based on our collective interpretation of the nation's westward expansion and settlement and on our collective image of the Wild West. But to those struggling in the mountains or on the prairies, reality was often much different.

With Manifest Destiny and the Land of Gilpin, with the Homestead Act and the railroads, settlement of the western frontier became a national crusade. It was how the nation perceived itself in the latter half of the 1800s. Powell reflected those hopes and dreams and the collective destiny. His struggle to survive the boiling rapids embodied the common pioneer's struggle to survive on the prairie, and the nation's quest to conquer and tame a hostile landscape. His journey into the unknown reaches of the lower Colorado River represented the inexorable march of America ever westward across the continent. It was a compelling story. It made great press. However, little of it was true.

Twice more the expedition's demise was widely reported, but on August 30, six intrepid men reappeared into the light of public adulation (the others had left the expedition as it proceeded). Powell had descended the Colorado River through the nation's last remaining wilderness. He had brought light to a landscape clothed in darkness since Coronado's men gazed across the Grand Canyon nearly 330 years earlier. He had opened to inquiry an unexplored region as large as the nation of France. And he had refuted the vacuous claims and pompous boasts of people like William Gilpin. In so doing, he finally dispelled the last vestiges of the Northwest Passage myth. Some, including Gilpin, still believed or hoped the Colorado River passed through California and emptied into the Pacific Ocean and, therefore, would prove to be the long-fabled missing link in a navigable route to China. In some ways, Powell's expedition finally ended the search begun by Columbus. The outrageous stories of the boaters' deaths now fed the public's fascination with their survival. Powell was a national hero of the first order. His story became legend almost before it was told.

In the aftermath of his success, Congress funded the creation of the United States Geographical and Geological Survey of the Rocky Mountain Region, J. W. Powell in charge, on July 12, 1870. Unsuccessful in gaining congressional funding for his two previous expeditions as an unknown amateur, Powell's newfound fame garnered him ten thousand dollars to continue his work in the Southwest while remaining on the faculty at Illinois Normal. He formally accepted government employment two years later. His new survey closely followed three others organized after the Civil War to produce a wealth of resource, geologic, topographic, and ethnographic information about the West.

The first of the three other surveys was led by Clarence King, a brilliant young geologist from California. His United States Geological Survey of the Fortieth Parallel began in 1867 under the supervision of the War Department. It studied the geology and mineralogy along a hundred-mile-wide swath centered on the fortieth parallel, roughly the route of the Union and Central Pacific Railroads across the intermountain region of the Rockies, north of Powell's later focus.

Dr. Ferdinand V. Hayden's United States Geological and Geographical Survey of the Territories was the second survey. Also established in 1867, but by the Department of the Interior, it set out to inventory the geology, geography, archeology, and paleobiology of the northern Rockies. Hayden was ambitious, energetic, and well-connected in both the political and scientific communities, as well as an old hand in the wilderness of the northern Rockies and Plains. Hayden's surveys would later prove instrumental in the establishment of the first national park, Yellowstone.

The third survey, called the Geographical Surveys West of the 100th Meridian, began in 1869. Like King's survey, it was also supported by the War Department. However, unlike the others, Lieutenant George M. Wheeler's survey sought mostly to make detailed topographic maps of the central Rockies. All four surveys, including Powell's, continued off and on for several years, making repeated trips into the field. And each made notable contributions to the nation's understanding of the West. The King and Powell surveys in particular, and Hayden's to a lesser degree, made significant scientific contributions in geology, geomorphology, physiography, and ethnology.

So, unlike those of Lewis and Clark, Pike, and Long, the four later surveys were not expeditions; rather, they were full-fledged, properly equipped scientific ventures staffed with qualified scientists (Powell, however, once again relied on amateurs as a result of limited funding). Their goals also differed from those of the previous expeditions. Lewis and Clark, Pike, and Long set out to initially explore unknown lands for economic and geopolitical reasons. The new surveys, as follow-ups, set out to inventory and map the regions in detail. The motivations were many: for the sake of pure science; to satisfy the practical need for basic topographic mapping; to identify potential resources for exploitation; and to lay the foundation for future settlement.

Powell's survey of the southern Rocky Mountain region around what is today the junction of Arizona, Utah, New Mexico and Colorado ran off and on until 1879, and it included a second descent of the Colorado River. Over its life, the survey was criticized for its slow, diligent pace and the lack of product, reflecting its leader's meticulous nature. Nonetheless it eventually produced several exceptional reports that detailed the region's geography, geology, flora, fauna, and native peoples.

While the King, Hayden, Wheeler, and Powell surveys ran, business interests and politics permeated every aspect of western settlement, as did the jumbled mix of values and perceptions of the West. The 1860s and 1870s were a tumultuous time in Washington. The political and economic stakes of the western land grab were high, just as the stakes for control of the nation's public domain had been since Jefferson and Hamilton waged philosophical war over the future of the nation's landscape. Ambitious personalities jostled for power and influence over the settlement process. Land laws and Indian policy were changed again and again in response. Consequently the four field surveys, the focus of much attention in Washington as well as nationwide, were constantly buffeted. Powerful people wrestled for their control as attempts were made to administer them from a single new agency.

Powell's interest was science not politics, although his fame and personality did not allow him to remain very far outside the political struggle. Since the 1876 publication of his personal account of his first Colorado River descent, he too had an agenda he promoted with all his skill and influence. But unlike most others, his agenda was not about personal power. Powell had no interest in the infighting over direction of the proposed new agency. He openly supported Clarence King for director. Powell's agenda instead was to redirect national policy regarding the West. It was to enlighten government about the true West and to reform national land policy to reflect that truth.

Powell was not a politician skilled at compromise. He was not one who could manipulate others with anything but straight talk and truth. He could not tolerate incompetence or dogmatic narrow-mindedness. Nor could he shy away from fighting for his beliefs. His *Report on the Lands of the Arid Region of the United States* reflected those characteristics. Only two hundred pages long, the report was one of the pivotal documents in American landscape history. In his introduction to a recent edition, Stegner wrote, "It would ultimately be recognized as one of the most important books ever written about the West, and it was understood at once to be loaded with dynamite."[1] Unfortunately its explosive message fell largely on deaf ears.

Powell spent seven years writing this report. He delivered it to Carl Schurz, the secretary of the interior, on April 1, 1878. Like Powell, the report was straightforward, factual, and blunt. Its content went well beyond the initial charge given his survey and beyond the scientific data the survey gathered. In short, it was a manifesto more than a report, as it called for a revolutionary change in American landscape action and perception. Powell proposed to reform the West from the grassroots and, by changing land and water laws in advance of further settlement, change its whole institutional base. He saw what most Americans overlooked and nearly all Westerners denied—that much of the West was arid land where recurrent drought would doom homesteaders then

invading the dry country beyond the ninety-eighth meridian. He was
convinced that settlement had already reached as far west as it could go
without grave risks to both the settlers and the land.[2]

The report called for drastic revisions in the nation's nearly sacred
homestead and rectangular land survey laws when applied to arid lands.
It also called for the reform of the General Land Office, the realignment
of responsibility for cadastral surveying and mapping, and the central-
ization of geological and geographical surveys into one agency within
the Department of the Interior. The report exploded in Washington.
There, in one shattering blow, Powell had, according to Stegner,

> alienated the War Department, the General Land Office, Hayden, Wheeler,
> and all their several congressional supporters. Add to those the powerful
> land, cattle, mining, and timber interests of the West, the speculators and
> peculators busy under the old land laws gobbling the Public Domain by
> fraud, grab, appropriation, barbed wire, guns, and the election of political
> yes-men. And add to those the myth-bound citizens for whom the 160-acre
> homestead, the sturdy pioneer farmer, the freehold yeoman, the "Garden of
> the World," and the other shibboleths had all the force of revealed truth. It
> is possible that Powell did not yet appreciate how strong the coalition could
> be, and in particular how potent was the force of myth. But even if he had
> estimated the political opposition at something like its true strength, he would
> have had to challenge it, because he could see the lands beyond the 98th me-
> ridian going in venal, mistaken, or monopolistic ways to great land and wa-
> ter barons, or being chopped to ruinous bits by the advancing front of the
> rectangular surveys and the tradition-bound, hopeful, ignorant, and doomed
> homesteaders.[3]

The report was a frontal attack on powerful political factions and
government agencies, in particular the "notoriously corrupt" General
Land Office that had managed the distribution of the public domain
since the agency's creation in 1812. It was the Land Office that had implemented the Land Ordinance and the Homestead Act, and it was the
Land Office that supervised the settlement of the Great Plains. From its
inception the Land Office had been one of the government's most powerful agencies. And from its inception it had both reflected the best American ideals and suffered from the nation's worst faults. It reflected both
sides of the landscape settlement process: freedom, optimism, and opportunity; incompetence, ignorance, corruption, greed, and bigotry.

How could Powell reform seventy-five years of entrenched government
thinking and fundamental public values? He proposed nothing less than
a new blueprint for the settlement and governance of nearly half the na-

tion. His plan was based on the inherent character of the landscape, not on abstract social philosophy or geometry. He proposed to discard the eastern values and settlement concepts being applied with little modification to the West in favor of new ones designed for the West.

The starting point was the recognition that the West differed fundamentally from the East. It argued that 40 percent of the nation was too dry for eastern forms of land management, particularly row crop agriculture. He called for the recognition of water as the central truth of the western landscape. Its conservation and wise use should dictate the settlement pattern. The western landscape, he proposed, should be classified in three categories: grazing lands, bottomlands suitable for irrigation along drainages, and timberlands in the mountains. He argued that the 160-acre parcel size of the Homestead Act was nonsense in the West: too much land to irrigate and not enough land for grazing. Instead, he proposed the minimum parcel size be enlarged sixteen times to 2,560 acres, four full sections for standard farms, and reduced by half to 80 acres for irrigated farms along the streams or upland reservoirs. All grazing land would be open (unfenced) range. In a similar fashion, laws and regulations would be developed for special-use areas such as timberlands and mining areas usually found in elevations above the grazing land.

He argued too that the precise geometry of Jefferson's grid made no sense in the West (or really any landscape). Land parcels should reflect the physical character of the land—its topography, soils, vegetation, and hydrology. The grid imposed an artificial settlement pattern on the landscape, forcing people to modify the landscape to make it fit their wants and needs. Powell argued it was far wiser for human settlement to adapt to the land. Let the homesteaders stake their own property lines. Property and political boundaries, he said, should reflect watershed boundaries, the dominant physical feature in the landscape.

He proposed that political structure be organized around cooperative irrigation districts, since the principal role of local government in the West would be the management of water resources. Water laws, he also argued, should be revised so that each 2,560-acre parcel could include 20 acres of irrigable land, with the water rights attached to the land, not assigned according to the "prior appropriation" doctrine. Those fields would provide a base commercial crop for the homesteader and additional feed for the livestock. Last, he proposed the establishment of a new agency to build reservoirs for irrigation in the West.

As copies of the report circulated the corridors of power in Washington, it was entered into the swirling debate over the future of the four surveys, as well as into sensitive debates over other key land issues. The political atmosphere was already highly charged when Powell's bombshell exploded. A friendly National Academy of Science, then developing a recommendation to Congress regarding the surveys and the various land issues, was generally supportive. But the vested interests and old behaviors in Congress proved too deeply entrenched to permit adoption of the bold changes Powell proposed. The Land of Gilpin held sway. In 1879 the four surveys were finally consolidated in the Department of the Interior to form the U.S. Geological Survey, with Clarence King as director. Otherwise, Congress only accepted the report's recommendation that a Bureau of Ethnology be created. Powell founded the new bureau within the Smithsonian Institution in 1879 and, as the first head, began the study of Native American culture.

In 1881, after only two years as director, Clarence King resigned from the Geological Survey. Although in some circles King was considered the "best and brightest of his generation," his meteoric rise to social and political prominence faded just as quickly, extinguished by personal greed and scandal.[4] John Wesley Powell was named his successor. During Powell's subsequent tenure, he reorganized the Geological Survey's structure and refocused its efforts on the comprehensive, unified mapping of the nation (this led to the Geological Survey's topographic quadrangle maps, a landmark series of maps as important today as ever).

Over the next decade Powell remained one of the most influential scientists in government, although his power and responsibilities changed with each shift in the political tides. He continued to serve as a voice of reason and a man of vision, struggling in particular to expand government-sponsored irrigation programs. In 1894, he retired amid a congressional witch-hunt led by the Gilpinists who still stubbornly clung to the past. Little had changed out West. His lifelong quest finally yielded perhaps its only real success just before his death when the Newlands Act created the U.S. Bureau of Reclamation in 1902 to construct reservoirs for irrigation, just as Powell had proposed a generation earlier. In the nearly twenty-five years since he delivered his plan, little ground had been gained in his passionate fight to realize the remainder.

As the congressional hearings that drove Powell from government service proceeded—hearings driven by the myths of the past and the wounded ambitions of powerful people—another landmark event took place in Chicago that glorified the western attitudes that Powell sought to exorcise. The Columbian Exposition, the 1893 Chicago World's Fair, was a celebration of America's ascension to the pinnacle of the Western world, and an unbridled public affirmation of Manifest Destiny. Late on July 12, a blonde, handsome, athletic thirty-two-year-old geography professor from the University of Wisconsin presented the last of five papers at an evening session in the new Art Institute on the lakefront. The speaker's modest reputation as an eloquent, engaging speaker and interesting scholar was sufficient to keep the audience in place to hear him even after the long session.

Frederick Jackson Turner's hour-long presentation that night culminated a century that witnessed the creation of a continental nation spanning from the Atlantic to the Pacific, a nation many perceived as the American empire. Turner's talk, titled "The Significance of the Frontier in American History," redefined the nation's history and self-image as he distilled the essence of the American experience into the Frontier Thesis. Turner told America who it was and how it came to be. Not since Lincoln's Gettysburg Address had a single speech done so. The thesis was simple and elegant. It tapped common dreams and noble aspirations:

> From the conditions of frontier life came intellectual traits of profound importance. The works of travelers along each frontier from colonial days onward describe certain common traits, and these traits have, while softening down, still persisted as survivals in the place of their origin, even when a higher social organization succeeded. The result is that to the frontier the American intellect owes its striking characteristics. That coarseness and strength combined with acuteness and inquisitiveness; that practical, inventive turn of mind, quick to find expedients; that masterful grasp of material things, lacking in the artistic but powerful to effect great ends; that restless, nervous energy; that dominant individualism, working for the good and for evil, and withal that buoyancy and exuberance which comes with freedom— these are the traits of the frontier, or traits called out elsewhere because of the existence of the frontier.[5]

Turner proposed that true American culture and institutions, those things truly American, arose at the frontier. There, the struggle to forge civilization from the western wilderness forced the pioneers to shed, to strip away and discard, the values and institutions that tied them to their European heritage. The essence of America was the process of pushing the frontier across the continent, especially across the western plains and prairies. That process, that struggle, gave birth to American culture and served as the wellspring for the most noble American values, he said: "Up to our own day American history has been in a large degree the history of the colonization of the Great West. The existence of an area of free land, its continuous recession, and the advance of American settlement westward, explain American development."[6]

Unfortunately, according to the 1890 Census, the frontier was now closed. The unbroken line drawn each decade by the census, marking the frontier's westward march across the continent, no longer separated civilization and wilderness into two discrete zones. The frontier was reduced to a plain spotted with rapidly expanding nodes of civilization that would soon consume and occupy the entire land area. To Turner that meant not only the closing of the frontier but also the drying up of the wellspring of American values and culture. What would fill the void, he wondered. He continued:

> Since the days when the fleet of Columbus sailed into the waters of the New World, America has been another name for opportunity, and the people of the United States have taken their tone from the incessant expansion which has not only been open but has ever been forced upon them. He would be a rash prophet who would assert that the expansive character of American life has now entirely ceased. Movement has been its dominant fact, and, unless this training has no effect upon a people, the American energy will continually demand a wide field for its exercise. But never again will such gifts of free land offer themselves. For a moment, at the frontier, the bonds of custom are broken and unrestraint is triumphant. There is not [sic] tabula rasa. The stubborn American environment is there with its imperious summons to accept its conditions; the inherited ways of doing things are also there; and yet, in spite of environment, and in spite of custom, each frontier did indeed furnish a new field of opportunity, a gate of escape from the bondage of the past; and freshness, and confidence, and scorn of older society, impatience of its restraints and its ideas, and indifference to its lessons, have accompanied the frontier. What the Mediterranean Sea was to the Greeks, breaking the bond of custom, offering new experiences, calling out new institutions and activities, that, and more, the ever retreating frontier has been to the United States directly, and to the nations of Europe more remotely. And now, four centuries from the discovery of America, at the end of a hundred years of life under the Constitution, the frontier has gone, and with its going has closed the first period of American history.[7]

Calling on Jeffersonian-like philosophies, Turner glorified the effects of free land and agrarian settlement, and the resulting rise of democracy, while discounting the effects of immigration and urbanization. Overtones of Benton and Gilpin were heard as well. His revisionist view of American history challenged the two popular historical explanations during his lifetime: slavery and European "germs" implanted in the New World during colonization. But his view struck an accord with American thinking. It was what Americans wanted to believe about themselves. It was colorful mountain men and cowboys, and plain pioneers and homesteaders. It encapsulated the many myths and illusions about the

West. The general picture it painted fit the prevailing public opinion, as did popular painting and literature of the West. Yet the thesis collapsed under close academic scrutiny. Scholars soon attacked it. While the frontier was important, and his thesis had significant insight, it oversimplified history. Distinctly American institutions and values arose not because of western settlement, nor did frontier conditions alone make them arise. Instead, they arose in response to many forces, some frontier- and western-oriented, others eastern and urban. The ennobling values Turner assigned to the American character, values forged at the frontier, were perhaps self-serving too. Still the thesis rapidly infected popular thinking. It gained acceptance not because it was true, but because Americans believed it to be.[8]

The nineteenth century concluded with American landscape values and perceptions confused between myth and reality, fact and fiction. The third quarter of the century, the era of the cattle kingdom and the beginning of permanent settlement on the Plains, still dominates our national landscape consciousness. Why does the "Marlboro Man" mentality so permeate our culture? Why are we so infatuated with the Turner image of that transient period in our history? Why have we not seen the western landscape, as Powell did, for what it really was? And why do we not better see the rest of the American landscape? The answers may lie in the continued effect of the landscape values held by Kilbourne and Jefferson, and the way they have evolved since early settlement.

· Out of Sight, Out of Mind ·

Rush Run is a small creek barely maintaining a quiet trickle for its tadpole pools during dry summer months. Snags of fallen trees form small falls; glacial erratics clog its path. Large sycamores and cottonwoods and a gravel bar mark its outflow. Blue herons, Canada geese, mallards, and wood ducks like the spot. Tracks record the passage of other wildlife: deer, mink, rabbit, and raccoon. From its headwaters on a broad upland plain, the creek gently winds through a ravine about two miles to its mouth. Along the way, the creek eroded bluffs fifty feet high into the till and outwash material sitting atop the 350-million-year-old Devonian shale bedrock. Its small floodplain and side slopes remain a tangle of vines and young trees.

Erosion formed Rush Run since the Wisconsinan glacier left the area about fifteen thousand years ago. The glacier had scoured the region for several thousand years, obliterating the previous landscape and leaving in its wake the genesis of the current one. Advancing and retreating, pulsating in response to climatic fluctuations far to the north, the glacier again and again deposited materials across the land, and then flushed them away. Like all landscapes, this one was born of catastrophic forces we can scarcely imagine.

The mammoth ice sheet at one time may have been about one thousand feet thick where Rush Run now flows (and nearly three *miles* thick in northern Canada). The ice sheet blanketed much of the Midwest and virtually all of Alaska and Canada, while other ice sheets extended into Scandinavia, Siberia, and northern Europe. For millennia, the Rush Run landscape lay locked in suspended animation, buried beneath the frozen mass.

As the glacier gradually receded, the landscape left in its immediate wake was cold, wet, and barren, a desolate, devastated place. A thick layer of till and outwash covered the moonlike world. Shallow, braided rivers raced across the newly leveled surface, discharging the torrent of meltwater. The process that would form river corridor and floodplain began again, often over segments of previous river valleys buried beneath the latest deposit of glacial material. The new rivers began again to wash away bit by bit the till and outwash that had filled them, their flow sorting and depositing the sediments downstream. Eventually the many braids would coalesce, forming the rivers seen today. Erosion also began to recontour the topography, cutting ravines like Rush Run.

Walking a half mile upstream from its mouth, one has little sense that Rush Run is a remnant piece of "wildness" in the midst of the city crowding in several hundred yards on either side. That reality sets in where a brutal erosion control structure shores up the bluff to compensate for the surge of runoff that scours the creek bed and side slopes after storms, runoff collected from the rooftops, parking lots, and streets of the recent development upstream. Rush Run, like most urban drainage ways, functions as an engineered storm water discharge channel the remainder of its short length.

Several years ago, Jim Davis drew a wonderful Garfield cartoon that illustrated this point. One rainy day, a curious Garfield was accidentally washed down a curbside drain. He tumbled over and over as the surge of water carried him through the city's subterranean storm water system until he eventually popped out at its discharge point into the river. As he recovered his senses, he found himself surrounded by several other felines. "So this is where dishwater goes," said one; "so, this is where bathwater goes," said another; and a third pledged, "I'll never drink from the toilet again" (most household drains now empty into sanitary sewers, not storm water sewers). Their curiosity, fortunately, didn't lead to their demise after all. Runoff captured by curbs and gutters eventually finds its way into the network of rivers and streams, as does much of the loose material we leave on our streets and parking lots, and most of the chemicals we spread on our lawns and farms.

This type of storm water control is a complicated and expensive problem. Where are the streams that once drained the landscape on which our cities sit? Virtually every one has been placed in a pipe underground. The removal of vegetative cover in city and farm has dramatically increased the volume of runoff and greatly concentrated its flow. Where before rivers and streams rose more gradually after storms, and rose not as high, urbanization and agriculture exaggerate the hydrographic reading. To compensate, we build expensive and ecologically disruptive dams and dikes, channels and pipes. Despite these, repeated floods remind us of the dangers and drawbacks inherent in our attempts to control nature.

I remember a TV commercial for an insurance company that aired years ago that always struck me as a bit strange. It was filmed following a flood in a small city nestled in a steep valley of the Appalachian Mountains in western Pennsylvania. Policyholders devastated by the tragedy extolled the caring responsiveness of the insurance company in promptly settling claims, and proudly, defiantly, declared they would overcome the latest natural disaster and rebuild in the same place, just as they had done after each of the previous floods. Flood insurance and floodplain regulations limit this now. Still, we too often miss the message carried by floodwater.

Ravines like Rush Run, small remnants of the region's character before suburbanization, have a positive effect on adjacent residential neighborhoods. They lend the neighborhoods a sense of nature that often enhances property values. Today, neighborhoods throughout the country work to establish or reestablish a tangible connection to the environment similar to that enjoyed along lower Rush Run, while

also avoiding costly urban storm water drainage problems frequently
incurred where the natural surface drainage system has been ignored
and replaced by an engineered system.

About halfway from mouth to headwater, a Levittown-like subdivision was built in the 1950s along Rush Run. Little else stood along
its way. A generation of baby boomers grew up playing in "Devil's
Creek," the neighborhood name for Rush Run, and crawling within a
mysterious labyrinth formed by the knurled roots of an old sycamore
tree clinging precariously to the bluff. They learned many simple
things about the landscape at Devil's Creek and used that tiny piece
of nature to gauge their growth and social assimilation. Years later the
city cleared that small world, straightening and channelizing the creek
to discharge greater peak volumes of storm water runoff generated by
suburban development upstream. They also graded out the bluff. In
the process, the thousand-root tree was lost and the creek converted
to a drainage ditch.

Lost in the process was far more than a small stream that once
drained the landscape. Lost, as streams such as this are channelized or
buried in landscapes where soil and vegetation blanket bedrock, is the
most visible evidence of the city's geophysical heritage. Lost is a sense
of the way physical forces shape the land. Rush Run reminds us of
their powerful stirring ever so slowly beneath our feet. Intellectually
we know the surface we see is a direct reflection of its subterranean
layers and the titanic forces that shaped them. Cities tend to mask
that fact. Imagine, people might have once stalked mammoths and
mastodons through a boreal-like landscape along a small creek where
a little boy would later seek adventure beneath the tangled roots of
an old sycamore.

CHAPTER 8

The View from Afar

I wish to speak a word for Nature, for absolute freedom and
wildness, as contrasted with a freedom and culture merely
civil—to regard man as an inhabitant, or a part and parcel
of Nature, rather than a member of society.

Henry David Thoreau, "Walking"

"Walking," an essay by Henry David Thoreau that appeared in the
June 1862 issue of the *Atlantic Monthly*, was a quintessential expression of American transcendentalism. The essay stitched together two of
Thoreau's favorite topics from the numerous public presentations he had
given over the previous dozen years—one on the joy of walking in the
fields and forests around his Concord, Massachusetts, home, and the
other on the value of wildness. He had traveled the New England lyceum
circuit speaking in hope of supplementing his meager income and spreading his personal philosophy regarding life and nature. Neither goal met
great success. The essay stated, "In Wilderness is the preservation of the
world." To Thoreau, "hope and the future for me are not in lawns and
cultivated fields, not in towns and cities, but in the impervious and quak-
ing swamps" to which he frequently walked. There, in the raw, untouched
beauty of the swamp and surrounding forest he reconnected with nature and refreshed his soul.[1]

Ironically, the essay appeared after the transcendental movement
had nearly expired, and just a month after Thoreau's tragic death due
to tuberculosis on May 6. He was forty-four; the transcendental movement he championed had an even shorter life, lasting scarcely thirty
years. Yet the movement marked a major milestone in American environmental thought and landscape perception. Transcendentalism reflected a softening in the antipathy toward wilderness associated with
traditional Euro-American landscape values. Wilderness was not simply

a source of resources, nor just a raw paradise awaiting transformation into an Eden-like garden. Transcendentalism saw new values in untouched nature.

Up to the nineteenth century the forest, the wilderness, was, for most Americans, a dark, fearful, unholy place. Nathaniel Hawthorne depicted that disdain in *The Scarlet Letter* (1850). To his seventeenth-century Puritans of Salem, Massachusetts, the forest wilderness was a place of refuge and respite only to Satan, savages (Indians), and outcasts such as Hester Prynne and her bastard child, Pearl. It was a place to be cleared and cultivated, to be civilized, as quickly as possible. Only after the clearing of the forest and, in its place, the creation of settlement would enlightenment shine down upon the land and its people. In the following passage Hester and Pearl walk into the "dismal" forest to escape their public torment. Here Hawthorne depicted the primeval New England wilderness as a cold, dark, ominous place broken only intermittently by the hopeful, impish quality of light:

> [The footpath] straggled onward into the mystery of the primeval forest. This hemmed it in so narrowly, and stood so black and dense on either side, and disclosed such imperfect glimpses of the sky above, that, to Hester's mind, it imaged not amiss the moral wilderness in which she had so long been wandering. The day was chill and sombre. Overhead was a gray expanse of cloud, slightly stirred, however, by a breeze; so that a gleam of flickering sunshine might now and then be seen at its solitary play along the path. This flittering cheerfulness was always at the farther extremity of some long vista through the forest. The sportive sunlight—feebly sportive, at best, in the predominant pensiveness of the day and scene—withdrew itself as they came nigh, and left the spots where it had danced the drearier, because they had hoped to find them bright.[2]

Hawthorne continued to depict the forest as a melancholy, mysterious place as Hester and Pearl then rest by a small, meandering brook:

> Here they sat down on a luxuriant heap of moss; which, at some epoch of the preceding century, had been a gigantic pine, with its roots and trunk in the darksome shade, and its head aloft in the upper atmosphere. It was a little dell where they had seated themselves, with leaf-strewn bank rising gently on either side, and a brook flowing through the midst, over a bed of fallen and drowned leaves. The trees impending over it had flung down great branches, from time to time, which choked up the current, and compelled it to form eddies and black depths at some points. . . . Letting the eyes follow the course of the stream, they could catch the reflected light from its waters . . . but soon lost all traces of it amid the bewilderment of tree-trunks and underbrush. . . . All these giant trees and boulders of granite seemed intent on making a mystery of the course of this small brook; fearing, perhaps,

that, with its never-ceasing loquacity, it should whisper tales out of the heart of the old forest whence it flowed, or mirror its revelations on the smooth surface of a pool. Continually, indeed, as it stole onward, the streamlet kept up a babble, kind, quiet, soothing, but melancholy, like the voice of a young child that was spending its infancy without playfulness, and knew not how to be merry among sad acquaintance and events of sombre hue.

"O brook! O foolish and tiresome little brook!" cried Pearl, after listening awhile to its talk. "Why art thou so sad? Pluck up a spirit, and do not be all the time sighing and murmuring!"

But the brook, in the course of its little lifetime among the forest-trees, had gone through so solemn an experience that it could not help talking about it, and seemed to have nothing else to say.[3]

Such imagery reflected the prevailing landscape values in Hawthorne's day, as well as Hester Prynne's. Popular nineteenth-century landscape painting often conveyed the same message in scenes where a beam of hopeful, golden sunlight shone down through a narrow parting in a thick blanket of gray clouds looming threateningly above, to illuminate a rough-hewn cabin set in a pastoral clearing carved from the surrounding forest wilderness. The cabin and clearing represented the advance of enlightened civilization. The righteous light represented God's blessing. Those were the common views of nature and wilderness from the colonization of the New World through the early 1800s. Those were the values held by pioneers like James Kilbourne.

Similar depictions are seen in another literary classic of the time, Herman Melville's *Moby-Dick* (1851). In one pivotal passage, Melville presented two of the book's central themes as Ishmael described the complex mixture of feelings he harbored toward the great white whale. At one point, Ishmael compared the whale to the "White Steed of the Prairies":

[The steed was] a magnificent milk-white charger, large-eyed, small-headed, bluff-chested, and with the dignity of a thousand monarchs in his lofty, over-scorning carriage. He was the elected Xerxes of vast herds of wild horses, whose pastures in those days were only fenced by the Rocky Mountains and the Alleghenies. At their flaming head he westward trooped it like that chosen star which every evening leads on the hosts of light. The flashing cascade of his mane, the curving comet of his tail, invested him with housings more resplendent than gold and silver-beaters could have furnished him. A most imperial and archangelical apparition of that unfallen, western world, which to the eyes of the old trappers and hunters revived the glories of those primeval times when Adam walked majestic as a god, bluff-bowed and fearless as this mighty steed. Whether marching amid his aides and marshals in the van of countless cohorts that endlessly streamed it over the plains, like an Ohio; or whether with his circumambient subjects browsing all around at the horizon, the White Steed gallopingly reviewed them with warm nostrils reddening through his cool milkiness; in whatever aspect he presented himself, always to the bravest Indians he was the object of trembling reverence and awe. Nor can it be questioned from what stands on legendary record of this noble horse, that it was his spiritual whiteness chiefly, which so clothed him with divineness; and that this divineness had that in it which, though commanding worship, at the same time enforced a certain nameless terror.[4]

To Ishmael the leviathan represented wildness, all of nature beyond the dominion of humankind; thus Captain Ahab and the *Pequod*'s hunt to conquer it, to kill it, represented a righteous conquest in the advance of civilization. Melville, echoing Benton's Manifest Destiny and foreshadowing Turner's Frontier Thesis, saw virtue in that conquest. Yet Ishmael's feelings of fear and revulsion for the demonic creature were also mixed with a respect, a sympathy, an attraction to it as well. The whale and the White Steed (nature) were not just menacing, they were at the same time free and wild in the positive sense.

Ralph Waldo Emerson, a leader of the transcendental movement, presented a much different view of nature in his poem "The Apology":

Think me not unkind and rude
That I walk alone in grove and glen;
I go to the god of the wood
To fetch his word to men.

Tax not my sloth that I
Fold my arms beside the brook;
Each cloud that floated in the sky
Writes a letter in my book.

Chide me not, laborious band,
For the idle flowers I brought;
Every aster in my hand
goes home loaded with a thought.

There was never mystery
But 'tis figured in the flowers;
Was never secret history
But birds tell it in the bowers.

One harvest from thy field
Homeward brought the oxen strong;
A second crop thine acres yield,
Which I gather in a song.[5]

Emerson replaced the foreboding imagery of Hawthorne and Melville with a sense of wonderment and fanciful joy. Here, nature was a place of refreshment and contemplation, a place of beauty, a place to know

God and see his handiwork, a place to transcend the stifling confines of the day-to-day world in order to find the truth about one's self and the cosmos. Where before colonists like James Kilbourne, driven by a puritanical work ethic and set of landscape values, felt compelled to clear the wilderness in order to create a civilized landscape, Emerson, Thoreau, and the other transcendentalists unapologetically extolled the virtues of nature as a place for idle relaxation, for pensive reflection and meditation, for "walking." Undisturbed nature, they believed, served as the best medium for one's personal quest for enlightenment. To transcendentalists, the quest for truth was the centerpiece of their philosophy and the ultimate goal in life.

They believed, like the European romantics, that to find truth one had to use intuition and emotion rather than the classicism and empiricism popularized by the Enlightenment. As Donald Worster described in *Nature's Economy,* transcendentalists such as Emerson and Thoreau believed in the existence of an "Oversoul," a benign, godlike spiritual force that permeated and animated all things, whether material or immaterial. They believed that, using intuition rather than reason and science, one could transcend physical appearances and ordinary understanding of the imperfect material world to obtain understanding of that ultimate truth—one could achieve a oneness, a holism, with the universe and the fundamental force that animates it. Yet even though nature, as part of the imperfect physical world, was to be transcended to enable discovery of the Oversoul, it was to be celebrated as the Oversoul's most direct earthly expression.

In their quest the transcendentalists and romantics saw the trappings of society only as shields hiding the truth; so they sought a simpler life and saw in more primitive landscapes, lifestyles, and societies, a beauty, a purity, a correctness long lost in the hustle and bustle of the Industrial Revolution and Enlightenment. Some argued that people living simply in nature were inherently good, and that civilization tended to corrupt that natural goodness, a goodness that stemmed from emotion. Reason, they felt, often resulted in the very evils that plagued society. Therefore, intuition and emotion were better guides to morality than reason was. And a return to a simpler life in as close contact with nature as possible enabled one to better discover the truth. The transcendentalists focused on the individual, not society, stressing personal freedom from most societal constraints, including many laws and regulations they felt were unnecessary intrusions or unwarranted restrictions on personal liberty. For that reason Thoreau chose jail instead of payment of a tax levied to finance the war with Mexico, a tax and a war with which he disagreed. For some, civil disobedience was a natural extension of the transcendental philosophy.

And whereas wilderness was ugly to enlightened classicism, the romantics and transcendentalists drew upon emerging concepts of scenery and the picturesque that perceived a raw, primitive beauty in nature. Wilderness scenery to some Americans, long suffering from a sense of intellectual and aesthetic inferiority vis-à-vis their European cousins, also offered an opportunity to best them in a long-running game of cultural "one-upmanship." While America might not possess the refined gardens, parks, and historical landscapes of enlightened Europe, it could claim an advantage in one type of scenery—wilderness scenery—if wilderness was perceived positively rather than negatively. America was still rich with wondrous wilderness; Europe had little.

For all those reasons, then, Thoreau went to the woods around Walden Pond on the Fourth of July 1845 "to live deliberately, to front only the essential facts of life[,] . . . to live deep and suck out all the marrow of life."[6] He went to simplify his life, to free himself from the shackles of society, to return to nature and, in so doing, to use intimate contact with nature, the clearest reflection of God's handiwork, as the vehicle to transcend the limits imposed by the physical world, in order to understand the Oversoul and find the ultimate truth.

Walden Pond, however, was not truly wild. He chose that "wilderness" refuge, really a young second-growth forest only a mile or so from the village, because it was near the safety, comforts, and conveniences of his civilized world. He later experienced real wildness on three trips deep into the virgin Maine woods around Mount Katahdin, but found the privations distasteful and too threatening. The wilderness Thoreau, like the other transcendentalists, exalted most often meant a place like Walden Pond, that is, merely a place out-of-doors, a quiet, peaceful place in the woods, but a place still within reach of civilization. It meant a place where people still retained some semblance of control. The transcendentalists did not advocate total rejection of civilization in favor of real wildness, for they still retained aspects of the traditional antipathies. Instead they sought balance, a life between the extremes that mixed selected "wilderness" contact with a steady diet of the fruits of human society and culture. Theirs was an intellectualized, romanticized view of nature as opposed to one immersed in firsthand, detailed study. It was a view where the *idea* of nature was far more important than nature *in fact*. And theirs was an anthropocentric view that valued nature, like a commodity or thing, as a medium for thought, as opposed to a view that valued nature intrinsically. Despite the positive values transcendentalism assigned to nature, the transcendentalist's progressive view remained a view from afar.

Consequently, most of their writing, including that of Emerson, was full of ecological inaccuracies. Most transcendentalists were satisfied to derive their understanding from supposition and intuition rather than careful observation and study. After all, they believed nature was only an imperfect reflection or manifestation of the Oversoul. Understanding was not necessary or perhaps even desirable. Worster stated, "The transcendentalist movement placed little value on nature in and of herself; indeed the transcendentalist was as often repulsed by this slimy, beastly world as any good Christian. The lower order was not coequal with the higher realm of spirit; it was inferior, blemished, incomplete. Rather than looking deeper into nature to find the divine spark, the transcendentalist raised his eyes above this unsatisfying life toward a vision of serene and immortal harmony."[7]

Thoreau, on the other hand, drew inspiration and insight not only from intuition but also from extensive field study. That he supplemented with the leading scientific work of the period from people like Charles Lyell, Alexander von Humboldt, Charles Darwin, Gilbert White, and, especially, Carolus Linnaeus. Unlike other transcendentalists, Thoreau was an excellent naturalist. His keen understanding of the natural world permeated his writing. Foreshadowing the later landmark work of George Perkins Marsh, Thoreau studied Euro-American modification of the New England landscape, especially the extensive changes settlement had wrought on the forests. His study had scientific and practical value. One lecture he presented to the Middlesex Agricultural Society's annual cattle show in Concord, titled "The Succession of Forest Trees," was later published in the Society's annual *Transactions,* as well as in the *New York Weekly Tribune, Century, New-England Farmer,* and the *Annual Report of the Massachusetts Board of Agriculture.*

Hence transcendentalism was an amalgam of ideas, a synthesis of many influences in the early 1800s. It was a reaction against the stifling Puritan and Calvinist doctrine of the New England church. It was an outgrowth of European romanticism as a reaction against the classicism, the cold logic and empiricism of the Enlightenment. And it drew upon Neoplatonism and ideas regarding the holistic relationship between people and nature found in Eastern religions and philosophies. Coupled with those were other, less esoteric, factors too.

By the beginning of the 1800s the physical threat posed by the wilderness to those east of the Appalachians had greatly diminished as the frontier was pushed to the edge of the Great Plains well to the west, and as the vast eastern forest had been engulfed for the most part by a rapidly expanding sea of settlement. To be sure, life was still a struggle for pioneers at the forefront of settlement as they spread westward across the Northwest Territory toward the Mississippi. Yet life in the Atlantic states was relatively secure from the environmental threats that challenged those who had colonized the coast more than a century before. Such security made it easier for some to question, and even cast aside, their traditional, deep-rooted environmental values in favor of values more sympathetic to and appreciative of wilderness.

For some in the East, the growing scarcity of wilderness contributed to its newly perceived value and virtue. Perhaps it wasn't so bad after all. Some considered wilderness nostalgically, even romantically in the sense that one considers only the positive aspects while ignoring the negative aspects. And its rugged beauty, its very challenge and threat, became a source of potential pride for the young nation—pride in its landscape, its people, and the struggle to carve a civilized world out of the wilderness. Only in America could one experience the raw splendor of wilderness, and only here could one, individually and collectively, be truly tested by its unique challenges. Transcendentalism fed on such attitudes, as did Manifest Destiny and the Frontier Thesis.

Transcendentalism blossomed from within those secure eastern circles. Its budding was marked by the publication of Emerson's *Nature* (1836), its full flowering by the publication of Thoreau's *Walden* (1854) and Walt Whitman's *Leaves of Grass* (1855), and its passing as an organized movement by Thoreau's death in 1862. Beyond those three leaders, most of the movement's members were literary dilettantes publishing in their short-lived, little read magazine, *The Dial* (1840–44), or giving public lectures at the local lyceums. Mostly they met for evening discussions in the parlors around Boston, Massachusetts. Transcendentalism grew in the safety and comfort of New England drawing rooms where the wealthy and well-educated met to discuss art, nature, politics, philosophy, religion, and science, much as the Enlightenment had grown in the fashionable salons and cafés of Europe. Emerson, Thoreau, and Whitman then spread it beyond the closed circles of the New England intelligentsia.

Transcendentalism was never a mainstream belief in America. During its brief period many other ideas and landscape perceptions flashed across the American mind, competing for attention and cross-fertilizing one another. In "Walking," Thoreau asserted that "life consists of wildness. The most alive is the wildest. Not yet subdued to man, its presence refreshes him. . . . In short, all good things are wild and free," he proclaimed.[8] America, though, was not quite ready to go "walking" with him into the wilderness. The wild landscape, particularly in the West, still posed too much of an obstacle to the wave of settlement driven toward the Pacific by Manifest Destiny and the mixture of landscape myths and misunderstanding that clouded the nation's vision. The landscape also offered too much economic opportunity. As transcendentalism blossomed in the mid-1800s, it flowered at the height of those compulsions, so its passionate pleas fell mostly on ears listening to the call westward, the call to growth and development, the call to expansion and extraction, the call to conquer, clear, and cultivate.

CHAPTER 9

The View from Within

Even open-eyed Thoreau would perhaps have done well had
he extended his walks westward to what God had to show in
the lofty sunset mountains.

John Muir

The fact that people disturb the landscape, that our actions have environmental consequences, is common knowledge and common sense. We
may dispute the extent of the relationship and the ethical implications
of our actions and outcomes, but most people today accept the relationship as fact. That recognition is relatively recent in American thinking.
James Kilbourne, remember, left his Connecticut home in 1803 for the
wilderness of central Ohio in part for the economic opportunity it offered: it was a paradise in the rough that still possessed clean water, fertile soils, thick forests, and abundant wildlife—the natural resources
with which he could prosper. Those basic resources had been (ab)used
in his Connecticut landscape, as they had throughout much of the original colonies, to the extent that parts of the region were nearly as inhospitable as the mountains where Elzéard Bouffier worked his miracle
planting trees to heal the impoverished landscape of Provence. While
common sense suggested environmental actions had direct consequences,
and while the evidence of that was all around pioneers like Kilbourne,
they were often blinded to that link by their other landscape values: values based on the belief that people existed separate from and were superior to an environment perceived as a threat, so that human actions
were disconnected from and inconsequential to its operation; values
based on the belief that land was property and an inexhaustible commodity for human use; and values that held that the land, even if depleted by human (ab)use, quickly healed itself and soon restored its original abundance.

Our landscape thinking has progressed from that personified by James Kilbourne. Understanding of our intimate relationship with nature has grown slowly, incrementally, spreading and deepening. The mid-1800s was pivotal in that progression as a number of philosophies, people, and programs buffeted the nation's physical and psychological landscape. The period was full of milestones that marked movement in our landscape thinking: Pike and Long, Benton and Gilpin; the Homestead Act and the Pacific Railways Act, each signed by Lincoln the same year Thoreau died; and several years later, the first shipment of cattle from Abilene.

All of these milestones marked change, but none marked change in the core landscape values typified by Kilbourne. Americans still adhered to the traditional values, those articulated by Hawthorne and Melville. Transcendentalism's less hostile view of wilderness first changed that core. The emergence of science and its effect on environmental thought, denoted by two events, one in 1859 and the other in 1864, marked the second change in the core.

In 1859 Alexander von Humboldt died and Charles Darwin finally published *On the Origin of Species*. The overlap of these events marked the end of one epoch in Western environmental thought and the beginning of another: the break between amateur and professional, generalist and specialist; but most important, the breakpoint between the traditional Platonic, parochial natural philosophies and the new secular science.[1] The revolutionary message of *Origin*—the new relationship it suggested between humans and nature—propelled the Western world into a fundamental reconsideration of the traditional landscape values that had emerged from Grecian and Judeo-Christian legacies. That reconsideration continues today. Humboldt and Darwin set American landscape thought farther along Thoreau's path as an alternate from the traditional direction of illusion, dominance, and disconnection followed by Kilbourne, Benton, and Gilpin. They propelled American thought farther toward a view of nature from within.

In 1864 George Perkins Marsh moved it even farther when his *Man and Nature* made the connections between human action and environmental outcome clear to an American audience ready, at least in part, to see the obvious. In the preface, he stated the book's purpose:

> to indicate the character and, approximately, the extent of the changes produced by human action in the physical conditions of the globe we inhabit; to point out the dangers of imprudence and the necessity of caution in all operations which, on a large scale, interfere with the spontaneous arrangements

of the organic or the inorganic world; to suggest the possibility and the importance of the restoration of disturbed harmonies and the material improvement of waste and exhausted regions; and, incidentally, to illustrate the doctrine, that man is, in both kind and degree, a power of a higher order than any of the other forms of animated life, which, like him, are nourished at the table of bounteous nature.[2]

Marsh believed nature, when undisturbed by human interference, always moved toward a balance, a harmony, that promoted diversity and stability. Humans, he believed, interfered with that harmony, often resulting in waste and destruction. Yet he believed humans existed separate from all other life and were given the earth as home by divine right. Consequently, he believed people had a responsibility not to squander that patrimony: "Man has too long forgotten that the earth was given to him for usufruct alone, not for consumption, still less for profligate waste. Nature has provided against the absolute destruction of any of her elementary matter. . . . But she has left it within the power of man irreparably to derange the combinations of inorganic matter and of organic life, which through the night of æons she had been proportioning and balancing, to prepare the earth for his habitation, when, in the fulness [sic] of time, his Creator should call him forth to enter into its possession."[3]

Man and Nature argued that human societies disturb the landscape and disrupt its natural harmony. That was simply human nature. Yet the resulting disruption often contributed to the eventual collapse of the society. Where great civilizations once flourished in fertile lands around the Mediterranean, Marsh detected the legacy of devastated landscapes and weakened peoples during his diplomatic travels (including those in the Provence region in southern France, Elzéard Bouffier's homeland). Of such regions he wrote, "Vast forests have disappeared from mountain spurs and ridges; the vegetable earth accumulated beneath the trees . . . [and] the soil of alpine pastures . . . are washed away; . . . rivers famous in history and song have shrunk to humble brooklets . . . [, and] harbors . . . are shoaled by the deposits of rivers at whose mouths they lie."[4] He ascribed such abuse as a contributing factor in the decline of many ancient civilizations, including those of classical Greece and Rome. Deforestation, overgrazing, and wasteful agricultural practices all led to excessive erosion and other devastating environmental effects that helped undermine those civilizations.[5]

Although Marsh believed that people remained fundamentally separate from and superior to nature, he felt they could still act as agents of positive change to at least partially remedy the imbalance:

The extension of agricultural and pastoral industry involves an enlargement of the sphere of man's domain, by encroachment upon the forests which once covered the greater part of the earth's surface otherwise adapted to his occupation. The felling of the woods has been attended with momentous consequences to the drainage of the soil, to the external configuration of its surface, and probably, also, to local climate; and the importance of human life as a transforming power is, perhaps, more clearly demonstrable in the influence man has thus exerted upon superficial geography than in any other result of his material effort. . . .

But man is everywhere a disturbing agent. Wherever he plants his foot, the harmonies of nature are turned to discords. The proportions and accommodations which insured the stability of existing arrangements are overthrown. Indigenous vegetable and animal species are extirpated, and supplanted by others of foreign origin, spontaneous production is forbidden or restricted, and the face of the earth is either laid bare or covered with a new and reluctant growth of vegetable forms, and with alien tribes of life. . . .

The fact that, of all organic beings, man alone is to be regarded as essentially a destructive power, and that he wields energies to resist which, nature—that nature whom all material life and all inorganic substance obey—is wholly impotent, tends to prove that, though living in physical nature, he is not of her, that he is of more exalted parentage, and belongs to a higher order of existences than those born of her womb and submissive to her dictates.[6]

In *Man and Nature,* Marsh traced the effects of widespread deforestation of European and New England watersheds on soil fertility, water quality, even climate, at a time when Americans believed the landscape was an endless, indestructible resource, one they had an obligation to civilize. Marsh proposed forest management practices that would lessen the destructive consequences, practices that better maintained the natural harmonies vital to human well-being. His findings foreshadowed Jean Giono's story in nearly every detail. Marsh's book was based on firsthand observations of the profound change occurring in European and New England landscapes, observations made during years of travel for his political and professional duties. Those observations he then supported with the latest scientific thinking of Lyell, Humboldt, Darwin, Carl Ritter, and other leading naturalists-scientists of the period. Biographer David Lowenthal noted, "Until the twentieth century, no one except Marsh perceived the problem of conservation as one of interdependent social and environmental relationships. *Man and Nature* . . . made men aware of the ways their own behavior, unconscious as well as conscious, transformed the lands in which they lived."[7]

Before *Man and Nature,* most Americans, Lowenthal observed, "believed in the plenitude of nature and the inexhaustibility of natural resources, and had serene confidence in their ability to master the environment; indeed they considered this conquest the American destiny."[8] But the age-old notion of a separation, a disconnection, between human action and environmental outcome was not rapidly cast aside in America, despite Marsh's monumental work. Ours was a continent still ripe with resources. Ours was a landscape of seemingly unlimited opportunity. Americans then, and now, struggle to fully comprehend the profound implications of our relationship with nature. Nearly one hundred years after *Man and Nature,* the publication of Rachel Carson's *Silent Spring* would make the connection more compelling.[9]

Man and Nature was a clarion call for environmental stewardship. Considered the "fountain-head of the conservation movement" by the prominent twentieth-century historian and urbanist Lewis Mumford, the book marked the birth of environmental planning and conservation in the United States, for it dispelled the notion that human action and environmental outcome were separate.[10] Stewart Udall, the former U.S. secretary of the interior, believed the book's publication marked "the beginning of land wisdom in this country."[11] In effect, *Man and Nature* called into question the prevailing stewardship attitudes and proposed a new set based on alternate responses to three fundamental components: to whom is the stewardship obligation or responsibility held; what is the thing held in trust; and how, or for what purpose, is that trust to be managed? Whereas the prevailing attitudes identified no real obligation, hence little basis for stewardship existed, Marsh believed God had entrusted humankind with the responsibility to husband the environment to foster human growth and development. Consideration of those three components continues today.

Man and Nature was an immediate international bestseller. It made the obvious visible—people affect the environment and it affects us. The message was most clearly heard in Europe, where the effects of human disturbance on the landscape had had a longer legacy as a result of the centuries of settlement. In particular, the book's influence was most immediate on forestry practices there. By the end of the century, it had redirected American forestry as well. Gifford Pinchot, the father of the U.S. Forest Service and a leader of the conservation movement, called *Man and Nature* "epoch-making"; in 1874, *The Nation* said it was "one of the most useful and suggestive works ever published," a work whose message had "come with the force of a revelation."[12] Its findings remain as timely today as they were over a century ago.

Much of the book's impact stemmed from its timing as it appeared in the midst of a fervent period when the Western world was reconsidering the fundamental relationship between God, humankind, science, and nature. Books like Lyell's *Principles of Geology,* Humboldt's *Personal Narratives* and *Kosmos,* and Darwin's *On the Origin of the Species* contributed to that milieu, as did the profound social changes triggered by the Industrial Revolution and the Civil War. Marsh was on the cusp of the shift. His revelations reflected the new scientific order, but were still shaped by the old theology.

The great naturalists and scientists of the mid-1800s, people like Humboldt and Marsh, even John Wesley Powell, were the last of the self-taught generalists. They were the last generation of amateurs as opposed to the professional scientists of today. Yet they embraced the new science. They saw nature and the landscape from within, not from afar as had those before. One who joined Thoreau in issuing the call to go "walking" in the wilderness was also such a person. John Muir began his journey in 1864, and it proved to have a profound effect on the nation's attitudes toward the land.

The politics of war in 1864 forced twenty-six-year-old Muir to leave the familiar confines of his family's farm in south-central Wisconsin. Raised in a strict fundamentalist Christian home, his pacifism was incompatible with the violence of the Civil War. And the remote war was too distant, too disconnected from those struggling to carve a farm from the forests and prairies of Marquette County. It had no meaning, no significance to immigrants like Muir and his neighbors from Britain, people yet to be assimilated into their new homeland. Muir considered himself more of a Scotsman than an American. So he followed in his youngest brother's path and fled northward for the wilds of Canada to avoid the draft, as would other conscientious objectors a century later during another divisive war.

Muir's family had immigrated to America fifteen years earlier from Dunbar, Scotland, a small town facing the North Sea near the mouth of the Firth of Forth, thirty miles outside Edinburgh. As Stephen Fox recounted in his fascinating account of Muir's life, *The American Conservation Movement,* John was the third of eight children. His father, Daniel, was a hardworking, Christian zealot who ran a grocery business successful enough for the family to afford servants. Ann was as affectionate, quiet, and gentle as her husband was harsh and austere. As the eldest son, John was doted on by his adoring older sisters as well as the grandparents who lived across the street. His grandfather often took the youngster on walks along the coastline or into the countryside around Dunbar, walks that fed the boy's growing fascination with nature.

The sweet flavor of John's childhood changed abruptly as he entered school. The Scottish schools, as Fox portrayed, were strict, sour places of rote learning enforced by stern, cane-wielding martinets who believed the road to enlightenment, the path to salvation, passed through the skin—when applied in copious amounts, corporal punishment, they believed, would lead the most obstinate and the least inclined to discover the truth. Between school's steady diet of religion and classics, young Muir found brief, tantalizing tastes of natural history in descriptions of America's wildlife by Audubon and Alexander Wilson.

Life outside the classroom paralleled the physical persuasion applied within. After school, small gangs of hooligans caterwauled about, fighting and making trouble well into the night. The next morning usually brought the unruly boys, including John, a thorough thrashing by their fathers. The disciplinary beatings, though, never seemed to dissuade the mischievousness; instead, their behavior reflected some fundamental aspect of the Scottish psyche that tolerated, or even reveled in, the never-ending cycle of rowdiness, retribution, and repentance. Home life became equally severe with constant religious lessons that meant memorizing the entire Bible, verse by verse, word for word. And once again, the flesh paid the price for poor performance.

Eventually Daniel Muir's religious fervor led the family away from that life; away from the prosperity, away from the schools, away from the young ruffians and their cycle of juvenile crime and punishment, and away from Scotland. The elder Muir broke from the local Presbyterian Church, rejecting its institutional structure and portions of its Calvinist doctrine, and joined the more fundamentalist Campbellite Disciples of Christ.[13] When the Campbellites moved their mission to the wilds of North America in February 1849, the Muirs moved too. John was not quite eleven. Like the Kilbournes and so many American families, the Muirs left behind one life for the promise of another. The movement of people, their settlement practices, and their landscape values across the Atlantic, then across the continent, was the genesis of the American landscape; a genesis followed by the evolution of those practices and values as the people adapted to their new world, and as they adapted it to their vision.

The Muirs landed in central Wisconsin as the wave of settlement was transforming the wilderness. There Daniel bought an eighty-acre parcel, a half-quarter section, four miles from the nearest neighbor in a county sprinkled with Scottish settlements. A dense oak-hickory forest blanketed his land. A beautiful glacial lake formed one edge. It was fertile land for farming as well as the gospel. It was a paradise landscape much like James Kilbourne's home in central Ohio.

As had pioneers all across the Northwest Territory, the Muirs set about the backbreaking work of clearing the land to create the fields of their Fountain Lake farm. They quickly built a fine eight-room house. The work felling trees, pulling stumps, and removing erratics continued for years. Fountain Lake became both heaven and hell to young John. All around was unspoiled wilderness. The small lake, highlighted by white pond lilies, wore a collar of rushes and reeds. Willows, mosses, and ferns trimmed a brook that flowed north as it crossed a meadow of wildflowers. His new homeland's phenomenal abundance and diversity, its raw plenitude of life, bedazzled his senses, just as it had the first colonists of the Atlantic Coast two hundred years before. Here he hunted hare, muskrat, gopher, duck, and prairie chicken. Here the wind carried the haunting call of the loon, and the spring and fall sky were blackened for hours as migrating flocks of passenger pigeons passed overhead.

The new land extracted its price too. The endless toil and suffocating isolation wore heavy on John's soul. Old enough and strong enough to help his father, John no longer joined his younger siblings at school; instead he stayed behind to work all day in the fields. And there, with no buffer between father and son, Daniel's zealotry extracted its price from him as well. Every task was accompanied by a sermon, and every act of disobedience or forgetfulness, whether real or imagined, was accompanied by a beating.

Over time Daniel's fanaticism deepened and his obsessive behavior became even more peculiar. The family's only relief came when he rode around the county as it quickly filled with settlers, mostly from the British Isles, distributing religious literature or preaching to any congregation who would hear him. During his absences, the pastimes he prohibited as blasphemous—innocent things such as embroidery, singing and dancing to the traditional fare of the Scottish Highlands, and general lighthearted foolishness—filled the household. No one else, however, found freedom outside the isolation of Fountain Lake on a regular basis. John rarely left the farm. On occasion he went to Portage, sixteen miles away, to fetch supplies or deliver a crop. Beyond that, the outside world was an unknown place. Isolated in a physical and moral wilderness, John turned more and more to nature for refuge and respite.

Unfortunately, skill as a grocer did not make Daniel a good farmer. The abundance of the land gave little cause for husbandry, even to a frugal Scotsman. Like so many city dwellers or tradespeople who came to the New World for opportunity and freedom, people who began their new lives extracting a living from the soil, Daniel did not know how to farm, how to care for the piece of paradise he had purchased so easily. In eight years, after the forest was cut and the fields cleared, his poor farming practices had depleted the land, leaving it, like the Connecticut landscape Kilbourne fled, barren and infertile. Despite the endless work in the fields, the yields of wheat and corn dropped from twenty-five bushels per acre to barely five.

Daniel's response was not to restore the land; instead, he abandoned it in search of inexpensive, fertile new land. This time Daniel purchased a full half section, 320 acres, six miles to the southeast. There on Hickory Hill Farm the process began all over again. That experience left an indelible mark on John and, through his later influence, eventually affected American landscape thinking.

Daniel chose the new land naively. The property occupied a ridge top with no surface source of water. And even worse, the land was underlain by sandstone just ten feet beneath the surface. John, of course, had to dig the well. Day after day he was lowered by bucket to the bottom where he chipped away at the rock with a hammer and chisel. Months later, at ninety feet, he finally struck water. The price of his work in the airless hole was a permanent throat problem. For many Americans all across the country, the price of farming before mechanization was paid with broken bodies.

In his middle teens the war of wills between John and his father moderated a bit. For years the powerful force of Daniel's convictions had always been met with John's equal internal strength. Daniel was loud and relentless, John was stoic and unyielding. His resistance was not disobedience or disbelief, rather disagreement. He could not accept his father's form of foreboding, austere piety. God created a world too full of wonder and joy for that. So when he began sneaking books and magazines from the Grays, a Scottish family of Campbellites who operated a small lending library nearby, Daniel eventually relented and allowed the teen to rise before work to indulge his interests. John read Milton, Shakespeare, even the Bible, now for pleasure instead of punishment, in moments stolen from sleep in the early hours before sunrise. He also earned extra money selling muskrat skins in order to buy books on most any topic—mathematics, poetry, grammar. An intellectual doorway beyond the isolated confines of his physical world had been opened, a threshold over which many others, like John Wesley Powell, also passed.

By that time, too, another of his talents had become apparent. The younger Muir was a mechanical wizard. He began tinkering with rudimentary materials around the farm to invent gadgets for fun and practical, labor-saving uses. The practical included a self-setting sawmill, waterwheels, an automatic horse feeder, various locks and latches, thermometers, hygrometers, pyrometers, barometers, a lamp lighter, and a fire lighter. The whimsical included an "early-rising machine": he placed his bed on a fulcrum connected to a clock so the bed would be tipped at the designated hour (1:00 A.M.), bringing its occupant to his feet to begin his predawn study. The inventions were amazing. His best thermometer was so accurate it could detect the presence of a single person four feet away using the expansion-contraction of a three-foot iron bar, multiplied 32,000 times by an intricate system of levers. Those toys triggered Muir's freedom from the farm and the start of his journey.

In 1860, Muir was twenty-two and showed little interest in striking out on his own. As the eldest son he may have felt duty-bound to the farm since his father now devoted all his time to proselytizing around the region. Regardless, on the encouragement of a neighbor, that summer Muir put three of his gadgets in a grain sack and, with fifteen dollars in his pocket, caught the train at a nearby town to go to the state agricultural fair in Madison, the state capital. Who knew, perhaps one of his two clocks or the thermometer might win a prize. Perhaps he might land a job in a machine shop. Perhaps he might have an opportunity to leave the farm.

To a wide-eyed, long-haired, and scruffy-bearded young man in homespun clothing making his first real trip off the family farm, Madison was an exotic world full of temptations. Muir stayed focused on the fair. He used one clock to rig an improvised version of his early-rising machine and hired two boys as guinea pigs to demonstrate the contraption. His wizardry was the hit of the fair as fascinated crowds gathered to marvel at his inventiveness. Employment offers from local entrepreneurs poured in. The once remote possibility of leaving the farm had overnight become reality. Unfortunately the offer he accepted proved to be a disastrous one, and after three months working on a harebrained scheme to develop a steam-powered passenger iceboat in Prairie du Chien, a town on the Mississippi River in the western part of the state, he returned to Madison. He had disliked the work and disliked the people.

His next decision seemed equally far-fetched at first. He enrolled at the University of Wisconsin, a bold move for someone whose formal education stopped in Scotland when he was not yet eleven. The university, however, was only a dozen years old and in need of students. He begin in a preparatory program. Odd jobs and an occasional gift of ten or twenty dollars from home paid the bills. He completed the program quickly and was admitted as a regular first-year student before his twenty-third birthday.

Amiable, awkward, brilliant, eccentric, and devout (Muir was still greatly influenced by his father's teaching), Muir spent the next three years in and out of school, depending on finances. Two professors were particularly influential to the energetic student, both émigrés from New England, both disciples of Emerson, and both familiar with the periphery of the transcendental parlors. Professor James Butler taught classics, Professor Ezra Carr geology and chemistry. Carr became Muir's mentor, with much the same impact that George Crookham had on John Wesley Powell. Carr's wife, Jeanne, also befriended the rather peculiar, socially inept young man. Her friendship and guidance provided a stabilizing social counterpoint to Muir's backwoods personality. She was not always successful. At one formal college reception, Muir became intrigued by a large piano in the room. Oblivious to the social code for such festivities, he pried open the top and climbed inside to investigate. Having discovered its operation, he climbed out and rejoined the party, unaware of the curiousness of his actions.

Carr's chemistry lessons suggested to Muir that all life followed a single, harmonious plan. Blending science with touches of transcendentalism, Muir began to sense that that harmony was accessible, even understandable, by scientific investigation. Science, not intuition, held the key to the plan that permeated all nature. Botany was a case in point.

Plant identification and plant taxonomy derived from Linnaeus fascinated him. Perhaps its precise methodology, a mechanical process that disassembled the specimen into constituent parts of leaf, bud, and stem, then classified it into species, genus, family, and order with others that shared common characteristics, appealed to his analytical mind—it was ordered, elegant, functional, purposeful, planned. Scientific analysis made that divine order apparent. Muir had begun to form a new linkage between God and nature.

Summers were spent at Hickory Hill again working the fields from sunup to sundown. Evenings were spent collecting and cataloguing specimens and conducting experiments. Unable or unwilling to constrain his enthusiasm for his newly acquired knowledge, he infected the others with the wonders he was shown in Madison, and he even brought his brother David back to the university with him the second year.

Unfortunately his botanizing and endless tinkering, and the virulent knowledge he spread to the rest of the family, ultimately became intolerable for Daniel. The new sciences that fascinated John, his biology and botany, geology and chemistry, raised too many fundamental questions, blasphemous questions, about the creation of the earth and the divine relationship. Violent arguments could not settle the situation. The infection had to be excised. Daniel and John could no longer remain under the same roof. Thereafter, John usually stayed with his sister at her family's farm nearby. Like many of his day, John was ensnared in the shift in Western thought centered on Darwin and the emergence of science. And like so many others of the period, he struggled to resolve the relationship between God, humankind, and nature, between science and the scripture. In the interim, however, another, more worldly cloud gathered over his life.

With the outbreak of the Civil War, the possibility of being drafted loomed constantly. The youngest brother, Dan, fled for Canada early on. John and David remained behind, perhaps because of school. John considered transferring to the University of Michigan to attend medical school. But in February 1864, when President Lincoln signed a draft order for another five hundred thousand men to be called on March 10, the prospect of conscription was too likely. Too few men remained in Marquette County for John and David to escape the call again. So, just ten days before the call, John caught a northbound train to join Dan. David stayed behind. On March 1, then, John Muir's journey finally began.

Like a fugitive, Muir quickly disappeared into the wilderness forests and swamps above Lake Huron. For six months he followed an erratic path along Huron and Georgian Bay, carefully avoiding any sign of human presence, not even writing home. It was a lonely time for him, one he filled collecting plants and searching for an answer to the troubling spiritual question about nature. He found his answer in an epiphany late one afternoon in June as he wandered about a dense swamp looking for a place to camp.

Suddenly, unexpectedly, he came upon a rare orchid, a *Calypso borealis,* along the bank of a stream. The flower's two white blossoms illuminated by a stray beam of sunlight glistened like snow against the backdrop of yellow moss. "They were alone," he noted; they stood apart

from the other plants, isolated like him from the surrounding world. He
later wrote, "I never before saw a plant so full of life; so perfectly spiritual, it seemed pure enough for the throne of the Creator. I felt as if I
were in the presence of superior beings who loved me and beckoned me
to come. I sat down beside them and wept for joy."[14]

There, lost in the most remote corner of the swamp, was that perfect
flower, a flower that served no purpose other than its own, yet a thing
of wondrous beauty whether or not human eyes even beheld it. "Are
not all plants beautiful," he continued, "or in some way useful? Would
not the world suffer by the banishment of a single weed?"[15] That was
it—there *was* a grand order, a sublime beauty in all things, which results
from the hand of the Creator. Science helped reveal it.

From that point on Muir replaced his fundamental religious beliefs
with a new set. Nature, he believed, was not just a commodity, nor was
it separate from people. The hand of the Creator constructed both humankind and nature. Raw, untouched nature was the most direct reflection of his handiwork. Science did not hide that truth nor shield one from
the Creator; instead it served to highlight his wonder, his order, his purpose. Transcendentalism was only partly right. Nature was good and displayed a universal order, but science, rather than intuition and emotion,
offered the key to understanding. Accurate knowledge of nature only
heightened one's appreciation of the Creator's majestic master plan.

Tracing the shore of Lake Ontario, Muir made his way eastward
across southern Canada and joined Dan at Niagara Falls by September.
The two found work in a sawmill run by a pair of fellow Scotsmen in
the town of Meaford on the southern end of Georgian Bay. Muir rejoiced
in the companionship of, and reconnection with, his family (they learned
two brothers-in-law had also fled to Canada to dodge the draft). When
the war ended a few months later, Dan returned to America. John stayed
on, infatuated with the small town and the opportunity to exercise his
mechanical wizardry in the mill. He had not lost his magic touch and
soon doubled the mill's productivity with his inventions.

Dan's departure left his older brother feeling isolated. A sense of intellectual disconnection again set in. The Meaford people were pleasant
and tolerant of Muir's rather peculiar idiosyncrasies, yet he longed for
someone with whom to share his innermost feelings and scientific interests. Daringly, he wrote Jeanne Carr, fifteen years his elder, in Madison.
She welcomed the correspondence. The two began an intimate relationship of words and ideas, of mutual support and encouragement, that continued for ten years. Muir wrote of his hope to save enough money to

return to school, and an even deeper dream to travel through South America botanizing, as had Alexander von Humboldt. "How intensely I desire to be a Humboldt!" he wrote Carr.[16] Just as he was settling into a happy routine of eighteen-hour workdays, botanizing around the countryside during his scarce spare time, and teaching Sunday School, events cast him off on a second "ramble."

A fire destroyed the Meaford mill in March 1866, wiping out his work and his collection of specimens and notes. It also extinguished his college plans. When the owners offered Muir a partnership to help rebuild the business, he chose instead to accept an IOU for the two hundred dollars due him and returned to America for the first time since the war. He visited brother David in Buffalo; then, driven by an unsettled mood, he left to wander the Midwest, to simply walk about the land collecting plants, stopping to support himself with odd jobs whenever necessary. He set out across New York, Ohio, Indiana, and Illinois. In Indianapolis he stopped. The surrounding deciduous forests intrigued him, as did the commotion of a steam-powered wagon-parts factory, where he took a temporary job.

There too his mechanical genius quickly propelled him from ten dollars a week as a sawyer to twenty-five dollars a week as a supervisor. The money was nice, but it tethered him too tightly to the factory. The long hours cut into his search for plants and his search for personal direction. Letters flashed back and forth between him and Jeanne Carr as he hoped to ease his loneliness and find a direction. That direction came in March 1867, when a sharp file flew up and punctured the cornea of his right eye while he was working on a machine. Helplessly he cupped his hand under the eye as the precious aqueous humor drained out, taking with it his sight.

The convalescence that followed lasted months. His sight slowly, miraculously returned, but his sinewy strength was sapped and his loneliness and sense of longing deepened. A visit to friends in Madison helped, as did a brief return to the family farms in Marquette County. The accident, though, had punctured his sense of immortality, and upset his priorities. At twenty-nine he had no real career, no real home, no real family. The lack of accomplishment and stability weighed heavily on him. So he decided to make one final, grand ramble, one last great gesture of independence, then return to settle down. He decided to walk to Florida, then take a boat to South America to follow in the footsteps of his idol, Alexander von Humboldt. That final fling would satiate his wanderlust. It would satisfy his curiosity and leave him ready for domesticity.

Plans, such as they were for Muir, were quickly made. He returned to Indianapolis, arranged his affairs, and took a train on September 8, 1867, to Louisville to begin his journey. A few miles south of town he spread out a map, charted his course, and set off on a thousand-mile walk through Kentucky, Tennessee, Georgia, and Florida to the Gulf of Mexico. As he had in Canada, he avoided civilization as much as possible, although practical reasons periodically brought him back. Life in the wilds refreshed his body, the weeks of quiet contemplation rejuvenated his soul. And there, lost in the landscape, he found his faith:

> The world we are told was made for man. A presumption that is totally unsupported by facts. There is a very numerous class of men who are cast into painful fits of astonishment whenever they find anything, living or dead, in all God's universe, which they cannot eat or render in some way what they call useful to themselves. . . . Not content with taking all of earth, they also claim the celestial country as the only ones who possess the kind of souls for which that imponderable empire was planned. . . . Nature's object in making animals and plants might possibly be first of all the happiness of each one of them, not the creation of all for the happiness of one. Why ought man to value himself as more than an infinitely small composing unit of the one great unit of creation? . . . The universe would be incomplete without man; but it would also be incomplete without the smallest transmicroscopic creature that dwells beyond our conceitful eyes and knowledge.[17]

Muir arrived physically at his initial destination in Cedar Key, Florida. He also arrived philosophically at a long-sought destination. However, while waiting to catch a boat to Galveston, thence on to South America, he caught malaria instead. The fever nearly killed him. Several months of recuperation followed, after which he changed his plans and went to Cuba, possibly because prospects for passage to South America seemed more likely there. Once on the island, however, he was unable to find passage south. Still weak from his illness, he caught an orange boat north to New York to fully recuperate before pressing forward again. The hustle and bustle of the city proved unbearable for one conditioned to life alone in the landscape. Once again his plans took a sudden, unexpected turn: for forty dollars he abruptly booked steerage passage on a steamer to California. There he intended to stay for a few months before continuing his pursuit of Humboldt.

Arriving in San Francisco in March 1868, Muir asked the first passerby the way out of town; when asked his destination in return, Muir responded, "To any place that is wild."[18] The path he followed first led him to the Central Valley, between the coastal mountains and the Sierra Nevada to the east. He spent the summer botanizing and working at odd

jobs—harvesting crops, breaking horses, shearing sheep. The latter led
to the management of eighteen hundred sheep that fall. The sheep, bred
for fleece, were in his eyes examples of humankind's arrogance toward
nature, since the breeding process resulted in weak animals inferior in
most ways to those of the wild. With the spring thaw, he led the hated
flock, what he called his "hoofed locusts," toward the more succulent
grasses higher in the foothills of the Sierra. So, as John Wesley Powell
set off down the Colorado River in 1869, John Muir first entered the
Yosemite Valley.

He knew instantly he had at last found his true home: the one place
on earth where God meant him to be. He rid himself of the flock that
fall and took a job running a small sawmill he built for his employer,
James Hutchins, at the base of Lower Yosemite Fall. The mill processed
only fallen timber and was powered only by the natural flow of the
stream, so it operated only during the portions of the year when both
resources were available. This left Muir free to wander the mountains,
to botanize, to commune with nature in a virtual temple of wonders.
There, into his temple, he disappeared for days, relying on his strength
of character and knowledge of the land to survive.

Those were really pilgrimages, journeys to his Mecca, to his holy land.
Little in the landscape escaped his keen eye or consuming curiosity during his trips to the altar. He soon knew the mountains intimately. To the
other residents of the valley, Muir was harmless, offbeat, and a bit fa-
natical in his zeal to preach his unique religion in which God was best
known through wilderness. But he made a great guide for those who
could tolerate the endless, enthusiastic proselytizing (he was much like
his father after all).

In May 1871, Muir learned from Jeanne Carr that Ralph Waldo Emerson, the patriarch of American letters, was going to visit Yosemite with
a group of Boston literati.[19] The news was like a blessing from above.
Muir could barely contain his excitement about the great man's visit.
Emerson, Muir thought, was truly a kindred spirit. Surely he was one
who shared his religion. Perhaps, then, Emerson might be convinced to
stay and commune with him at Yosemite's high altar before he returned
to the comfortable parlors of Boston.

Emerson's party stayed at a hotel at the base of the valley, a hotel too
spartan for their refined tastes, but the magnificent scenery made excellent compensation. Muir hovered about the fringes of the group, too shy
to approach the elder Emerson, although Jeanne had extolled one to the
other. However, when Muir learned the group was ready to leave in a

day or two, he left Emerson a note at the hotel making his request: "Do not thus drift away with the mob while the spirits of these rocks and waters hail you after long waiting as their kinsman and persuade you to closer communion. . . . I invite you to join me in a month's worship with Nature in the high temples of the great Sierra Crown beyond our holy Yosemite. It will cost you nothing save the time and very little of that[,] for you will be mostly in eternity. . . . In the name of a hundred cascades that barbarous visitors never see[,] . . . in the name of all the spirit creatures of these rocks and of this whole spiritual atmosphere Do not leave us *now*. With most cordial regards I am yours in Nature, John Muir."[20]

Muir's plea, as Fox noted, struck a chord with the sixty-eight-year-old transcendentalist. The next morning the venerable man and a companion rode out to the mill to meet their brother-in-nature. Muir, by that time, lived in a small room he had constructed that jutted out from one side of the mill, precariously overhanging the stream below. The spectacular views compensated for the room's tiny size. South Dome loomed stately above, and to the west the valley floor lay open like a flower. There, in the "hang-nest," the two met. Muir fussed about the piles of specimens that cluttered every corner of the room, eagerly, gleefully showing Emerson his treasures and answering the questions they raised with a flood of arcane information. Muir reiterated his invitation. Emerson, although tempted, deferred to his companion and the plans of the entire party and declined. Muir later wrote, "His party, full of indoor philosophy, failed to see the natural beauty and fullness of promise of my wild plan and laughed at it in good-natured ignorance, as if it were necessarily amusing to imagine that Boston people might be led to accept Sierra manifestations of God at the price of rough camping."[21]

Muir did accompany Emerson's group through the Mariposa sequoia grove outside the valley on the last day of its visit, bemusing the easterners with his wide-ranging literary tastes while he also identified every tree species they passed. At the end of the day Emerson bid farewell to Muir with a tip of his hat from a distant ridge and, with that polite nod, the leader of the transcendental movement unknowingly passed on the mantle of literary leadership in American landscape philosophy.

Over the next months, still in the afterglow of the encounter, Muir flooded Emerson with letters and packages of specimens and persistently made his plea for the sage to join him in nature's sanctum sanctorum. Emerson responded by inviting Muir, as he had another young protégé named Thoreau some years earlier, to join his group in Concord when the loneliness and solitude of nature became too much. He also sent along two volumes of his collected essays. But surely, Muir reasoned, the wildness of the Sierra had as much or more instruction to offer than the proper parlors of New England. Each remained in his place.

Muir pored over Emerson's work, dissecting it in meticulous detail, and found it filled with technical errors and inaccuracies in its depictions and descriptions of nature. To him, Emerson seemed only able to accept nature as a commodity for human use. Nature was only an imperfect reflection of the Oversoul. From then on, Muir disassociated with formal transcendentalism, although he continued to share many of its philosophical tenets. He continued to admire Emerson too, not so much for his knowledge of nature but, rather, for his quiet, granitelike character and the philosophy expressed in the poetry of his powerful writing. Thoreau's work was more to Muir's liking, especially *Walden* because of its more accurate depiction of nature, which he read for the first time in 1872 after Emerson's visit. Still, he felt Thoreau needed a strong dose of Yosemite. Shortly before his death in 1882, Ralph Waldo Emerson added Muir's name as the last on a list he kept called "My Men."

In the early 1870s, Muir left his wilderness temple in Yosemite and moved from the hang-nest to San Francisco as a result of personal issues that stemmed from his stormy relationship with James Hutchins and family. To make a conventional living, Muir pursued his fledgling writing career, which had begun informally when several short articles on Yosemite were published over the preceding two years. Friends who noted his unique philosophy and eloquent writing skills had encouraged that path. Those first articles were modest pieces for the local newspapers and short articles for the *Overland Monthly* that paid little, but they soon led to more significant articles for New York–based national magazines such as *Harper's* and *Scribner's* who, unlike the *Overland,* promptly paid $150 to $200 per article. His interest in writing and preaching his gospel to millions of potential converts nationwide blossomed, as did his success.

Once Muir had settled in the city, his writing quickly attracted a national audience. His vivid, powerful style appealed to the steadily growing number of Americans, mostly easterners, who read descriptions of nature by authors like Audubon and Thoreau, or those who read wilderness adventure stories like James Fenimore Cooper's Leatherstocking stories, or those who enjoyed travelogues to America's places of great wilderness scenery, places like the Sierra Nevada, called the "California Alps" by some still competing with their European cousins. In an excerpt from *Harper's*, Muir described a sudden thunderstorm on Mt. Shasta: "Presently a vigorous thunder-bolt crashes through the crisp sunny air, ringing like steel on steel, its startling detonation breaking into a spray of echoes among the rocky canyon below. Then down comes a cataract of rain to the wild gardens and groves. The big crystal drops tingle the pine needles, plash and spatter on granite pavements, and pour adown the sides of ridges and domes in a net-work of gray bubbling rills. In a few minutes the firm storm cloud withers to a mesh of dim filaments and disappears, leaving the sky more sunful than before."[22]

Unlike the work of other popular authors, who approached nature as an adversary with suspicion and fear, or those who judged it only according to human use, or those who romanticized and glossed over its infinite detail, Muir's writing embodied a fresh, new perspective—a view, like Humboldt's and Darwin's and Marsh's, truly from within. From the early 1870s until his death in 1914, John Muir's writing advanced American environmental thinking toward a softer, more appreciative, more benevolent and respectful attitude. Following the path marked by the transcendentalists and George Perkins Marsh, Muir and his writing set the next major milestone in the evolution of American landscape thought. In his work and his philosophy, we see the genesis of the American preservation movement.

Still his was the minority view. Although Muir's writing swept more Americans into the countercurrent, most adhered to the traditional values represented by Kilbourne. Most believed in Manifest Destiny and the Land of Gilpin. Most heard the call westward toward growth and development. The cattle kingdom, the Wild West, the "Marlboro Man" mentality still ruled the nation's vision of the landscape. John Muir later set a second milestone on that alternate path. Before he did, though, change in American landscape thinking would pass another important marker.

CHAPTER 10

A Tale of Two Parks

[Yellowstone] is a name that conjures up something of
geysers and bears, canyons and great waterfalls, and it
carries with it a romantic aura that cannot quite be defined.
Perhaps that is because The Yellowstone is, in part, a product
of our daydreaming.

> Aubrey L. Haines, *The Yellowstone Story:*
> *A History of Our First National Park*

On March 1, 1872, President Ulysses S. Grant signed an act designating a 2-million-acre tract in northwestern Wyoming as the world's first national park. The creation of Yellowstone National Park could not have occurred fifty years earlier, for the prevailing landscape values then were still too hostile toward wilderness. The thought of protecting a piece of wilderness for recreational use and enjoyment would have been nonsensical in James Kilbourne's day. The creation of the park marked the slight softening in the nation's landscape values. It also reflected the blend of landscape illusions with reality that colored our thinking in the nineteenth century. To most Americans, the landscape, especially the western landscape, remained a place known by fact and fantasy. Thus the founding of Yellowstone was a tale of two parks.

As legislation to establish Yellowstone was drafted in the months immediately before Grant put pen to paper, exploration of the region had barely begun. Since the return of the Lewis and Clark expedition through the region in 1806, the headwaters of the Yellowstone River remained little known to all but local traders and trappers, mostly illiterate men, so their knowledge rarely reached the public and was usually buried with them.

When stories did filter outside their small number, the public was skeptical. Trappers like John Colter, Joe Meek, and especially Jim Bridger were notorious liars, well known for their tall tales, making the distinction between fact and fiction difficult. As Nathaniel Langford wrote in an 1871 *Scribner's Monthly* article following his survey of Yellowstone, "Old mountaineers and trappers are great romancers. I have met many, but never one who was not fond of practicing upon the credulity of those who listened to his adventures. Bridger . . . has been so much in the habit of embellishing his Indian adventures, that they are received by all who know him with many grains of allowance. This want of faith will account for the skepticism with which the oft-repeated stories of the wonders of the Upper Yellowstone were received."[1]

To trappers like Bridger, storytelling was a form of entertainment rather than a means of communicating precise geographical knowledge. In addition, the clever trapper hesitated to divulge such information for fear of revealing valued lands. It was also a time when much of the eastern public romanticized stories of the Wild West and its colorful people, making trappers like Bridger folk heroes, even with all their idiosyncrasies. The public wanted to believe the incredible stories of Yellowstone because these stories fulfilled their romantic fantasies, but were cautious because of the limited credibility of the source.

In 1869 three hearty adventurers overcame the local fear of Indian attack and the lack of an available protective cavalry escort to explore the Yellowstone region to see firsthand if the lingering stories of its many fabulous wonders were true. David E. Folsom, Charles W. Cook, and William Peterson, although none a seasoned explorer accustomed to the hardships of the remote backwoods, set out for the region despite ominous warnings from friends. For several weeks in late September the trio wandered throughout Yellowstone, encountering only small bands of curious but friendly Sheepeater Indians. Upon first glance into Yellowstone's Grand Canyon, Folsom wrote, "Language is inadequate to convey a just conception of the awful grandeur and sublimity of this masterpiece of nature's handiwork." Later, while breaking camp at Yellowstone Lake, he wrote with remarkable clairvoyance, "This inland sea . . . is a scene of transcendent beauty which has been viewed by few white men, and we felt glad to have looked upon it before its primeval solitude should be broken by the crowds of pleasure seekers which at no distant day will throng its shores."[2]

Unfortunately the Folsom trio, like the trappers, provided only questionable accounts of the area and rudimentary updates to existing maps. Both the *New York Tribune* and *Scribner's Monthly* rejected their accounts as "unreliable material."[3] However, their stories of Yellowstone's spectacular geysers, magnificent waterfalls, and breathtaking canyons were enough to stimulate several acquaintances to plan another expedition the following summer. Among those who heard Folsom in Helena, Montana, that fall was Nathaniel P. Langford. A prominent local official, Langford played a key role in organizing the next expedition, conducted in September 1870.

Langford's interest combined curiosity about the rumored wonders with strong business interests in the Northern Pacific Railroad (NPRR) and the overall commercial development of the Montana region, both of which would likely benefit from tourism. As efforts to mount the new expedition congealed, fear of hostile Indians again surfaced and redirected the character of the planned expedition away from a purely civilian one. The expedition evolved into a nineteen-member party led by Henry D. Washburn, a former major general in the Union Army, accompanied by a small military escort commanded by second lieutenant Gustavus Doane. Other members of the expedition included prominent men of irreproachable character. Two novice artists also went along, Private Charles Moore and Henry Trumball, a civilian.

Following much the same route as the Folsom trio, the Langford-Washburn-Doane expedition again had no trouble with the native Sheep-eater Indians, and confirmed at last the "beautiful, picturesque, magnificent, grand, sublime, awful, terrible" Grand Canyon of Yellowstone and the region's other "wonders" and "curiosities" described by Folsom and the trappers.[4] Truth was found to underlie many of the trappers' tales.

On the night of September 19, while sitting around a pleasant campfire, Langford, Cornelius Hedges, and others in the group pondered Yellowstone's future. During that conversation, some speculate, the idea of preserving the region for public use was first discussed, and with it, the "national park idea" was born.[5] Most of the group shared the belief that private claims around the spectacular wonders should be permitted because of their potential tourism value. According to Langford's diary of the expedition, Hedges "did not approve of any of these plans—[he felt] that there ought to be no private ownership of any portion of that region, but that the whole ought to be set apart as a great National Park." Langford continued, "[This] suggestion met with an instantaneous and favorable response from all—except one . . . and quickly became the main theme of later conversations."[6]

The park Hedges and Langford envisioned was a small one, consisting of only enough land around each of the geysers and the rims of the canyons to maintain public access to see the wonders and curiosities.

Otherwise, they feared, private entrepreneurs might buy the land and charge tourists exorbitant fees to see the sights. Their interest was not in wilderness preservation or environmental protection. To Langford and Hedges, Yellowstone's attraction lay in its novelties and natural anomalies as scenery that should be preserved for public spectacle.

Like Folsom's group, the Langford party spent the winter following its return spreading the Yellowstone story. Initial newspaper accounts of the expedition remained skeptical. The *Rocky Mountain News* of Denver titled its article on the expedition "Montana Romance" and warned the reader that the account would "draw somewhat on the powers of credulity." The tone of the *New York Times* article, although enthusiastic, suggested the story was the "realization of a child's fairy tale."[7] Yet it was through Langford's articles, "The Wonders of Yellowstone," which appeared in the May and June 1871 issues of *Scribner's Monthly,* that a curious nation first received a detailed description of the region.

The tone of Langford's articles reflected popular landscape attitudes of the days in which nature's unfamiliar features were perceived with childlike fascination. Consequently, Yellowstone's wonders were ripe for exaggeration and for the creation of a popular set of illusions, many of which still haunt its current image. "The Wonders of Yellowstone" quickly spread a fanciful image both to *Scribner's* large readership as well as to Congress. Prior to the 1872 congressional vote establishing the national park, copies of Langford's articles were placed on every legislator's desk, along with actual Yellowstone photographs taken by William Henry Jackson during an expedition in 1871.

From his opening paragraph, Langford chronicled the expedition's adventures in a romantic Buck Rogers–like fashion. The article was illustrated with woodcuts made by artist Thomas Moran based on Private Moore's and Henry Trumball's drawings and Langford's notes and descriptions. Together the story and illustrations created a vivid Yellowstone image. With threatening language similar to Hawthorne's, Langford described Yellowstone's geysers and gas vents: "We suddenly came upon a hideous-looking glen filled with the sulphurous vapor emitted from six or eight boiling springs of great size and activity. . . . The springs themselves were as diabolical in appearance as the witches' cauldron in *Macbeth,* and needed but the presence of Hecate and her weird band to realize that horrible creation of poetic fancy."[8]

Yellowstone's Grand Canyon was also a subject for Langford's dramatic pen: "The brain reels as we gaze into this profound and solemn solitude. . . . You feel the absence of sound, the oppression of absolute silence. If you could only hear that gurgling river, if you could see a living tree in the depth beneath you, if a bird would fly past . . . you would rise from your prostrate condition and thank God that He had permitted you to gaze, unharmed, upon this majestic display of natural architecture."[9]

Despite Langford's reverence for Yellowstone's natural features, his concern centered on their protection as curiosities from private exploitation. His comment about the Devil's Slide encapsulated those concerns: "In future years, when the wonders of Yellowstone are incorporated into the family of fashionable resorts, there will be few of its attractions surpassing in interest this marvelous freak of the elements."[10] Langford's landscape values had softened sufficiently, compared to those typified by Kilbourne, to let him value the preservation of wilderness; yet he still saw wilderness, unfamiliar nature, in foreboding terms. And although he found value in such places, that value remained a commodity for human use. Wilderness, in this case, was to be set aside as scenery at which the public could gaze in amazement. In contrast, Emerson was concurrently extolling the transcendental value of nature, and Muir was glorifying wilderness as the clearest reflection of God's grand plan.

As Langford compiled his extensive notes over the winter of 1870–71, he also presented several very popular public lectures in Helena, New York, and Washington, D.C. The lectures were part of his business arrangements with the NPRR to provide railroad publicity through the promotion of the Yellowstone region. Among those who heard him was Ferdinand V. Hayden, head of the U.S. Geological and Geographical Survey of the Territories that had been under way in the northern Rockies since 1867. Hayden's interest in Yellowstone stretched back to his participation in an earlier, unsuccessful expedition to the region in 1860. With his interest again sparked by Langford's fanciful descriptions, he quickly redirected the focus of his survey from southern Wyoming northward to Yellowstone.

The 1871 Hayden expedition, like the King, Wheeler, and Powell surveys, was a major scientific venture funded by Congress. The large entourage included surveyors, meteorologists, geologists, botanists, zoologists, and other scientists. Also joining the group to provide a better record of the scenery were photographer William Henry Jackson and painter Thomas Moran, the latter joining the expedition on behalf of the NPRR.

Hayden's expedition provided the first definitive descriptions and illustrations of Yellowstone. Those resulted in sufficient congressional support to set Yellowstone aside in early 1872. Other contributing factors included public interest stirred by the trappers' tales and Langford's accounts; political pressure from the NPRR based on its commercial interests; and regional political interests seeking further territorial recognition.

The congressional action, however, was not based on a thorough understanding of Yellowstone's natural condition, for only small portions of the region had been surveyed, and little of its physical character or human history had been studied. Rather, as Langford and Hedges proposed, Congress simply sought to protect the geysers, canyons, and waterfalls from private exploitation. On February 27, 1872, the U.S. House of Representatives Committee on the Public Lands reported on the proposed bill:

> If this bill fails to become a law this session, the vandals who are now waiting to enter this wonderland will, in a single season, despoil, beyond recovery, these remarkable curiosities which have required all the cunning skill of nature thousands of years to prepare.
>
> We have already shown that no portion of this tract can ever be made available for agricultural or mining purposes. Even if the altitude and the climate would permit the country to be made available, not over fifty square miles of the entire area could ever be settled. . . .
>
> The withdrawal of this tract, therefore, from sale or settlement takes nothing from the value of the public domain, and is no pecuniary loss to the Government, but will be regarded by the entire civilized world as a step of progress and an honor to Congress and the nation.[11]

And congressional concern was only for the few acres immediately adjacent those curiosities. The remaining land, the vast majority of the 2 million acres, was included because other wonders were suspected to exist in the outlying, unexplored acreage, vacant lands also felt to have no other commercial value and that could, therefore, easily be lumped into the protected zone.

When President Grant signed An Act to set apart a certain Tract of Land lying near the Head-waters of the Yellowstone River as a public Park [sic], he inadvertently triggered a long-running debate over the park's dual purpose. The act states:

> [The Yellowstone region] is hereby reserved and withdrawn from settlement, occupancy, or sale . . . and set apart as a public park or pleasuring-ground for the benefit and enjoyment of the people. . . . Regulations shall provide for the preservation, from injury or spoliation, of all timber, mineral

deposits, natural curiosities, or wonders, within said park, and their reten-
tion in their natural condition. The secretary [of the interior] may in his dis-
cretion, grant leases for building purposes . . . of small parcels of ground, at
such places . . . as shall require the erection of buildings for the accommodation of visitors. . . . He shall provide against the wanton destruction of the
fish and game found within . . . , and against their capture or destruction for
the purposes of merchandise or profit.[12]

The law left unanswered how a balance was to be achieved between
a "public park or pleasuring-ground" and the "preservation from injury
or spoliation" of its natural resources in "their natural condition." It also
left unresolved how that "natural condition" was to be defined, since perception of the region was already a blend of fact and fantasy.

On April 8, 1903, President Theodore Roosevelt arrived at the Cin-
nabar terminus of the NPRR's Yellowstone spur to begin his third visit
to the park. Accompanied for much of his two-week tour by noted naturalist John Burroughs, Roosevelt explored the back country, despite
areas of deep snow, in his indomitable "roughing it" fashion. To his
sharp hunter's eye, Yellowstone was again displayed in full glory as its
range and forests teemed with elk, antelope, cougar, and deer. Roosevelt's was a leading voice for a growing national sentiment in which the
untrammeled wilderness, thought to be on the verge of extinction as the
nation rapidly settled its hinterlands, was perceived nostalgically and
romantically.

Wilderness, Roosevelt believed, promoted "that vigorous manliness
for the lack of which in a nation, as in an individual, the possession of
no other qualities can possibly atone." To counter the apparent loss of
those essential traits, he felt, "we need a greater and not less development of the fundamental frontier virtues."[13] Those, he thought, were best
developed in contact with wilderness. An advocate of Turner's Frontier
Thesis, Roosevelt believed that wilderness conquest, marked by the nation's westward-marching frontier, greatly contributed to the development of our national values and institutions. As the frontier approached
extinction at the Pacific Coast, he sought to preserve frontier values by
preserving wilderness bastions like Yellowstone.

Was it really such a bastion? Was it still a pristine remnant of the
wilderness that once covered the continent? Was Yellowstone a place
untrammeled by humans, where the primeval environment could yet be
experienced in order to reconnect to our frontier heritage and rekindle
our frontier values? That was then the common belief, and one, I believe, still prevalent today.

But illusion clouded reality in 1900, and clouds it now a century later. Despite the act to preserve Yellowstone signed by President Grant a generation earlier, Roosevelt's beloved Yellowstone differed from that seen by the first Euro-American, John Colter, in 1807, and it differed from the place that later amazed Langford. For although the law established the goal of maintaining the park's natural condition, that condition was difficult to define, and the goal difficult to achieve.

Like much of the upper Midwest, the Yellowstone region had been in a constant state of flux since the retreat of the last glacier, and it had been inhabited by people continuously over that period. The Early Hunters were the first inhabitants of Yellowstone, appearing about eleven thousand years ago when the region was still cool and moist in the aftermath of the glacier. They lived in small groups and roamed the region hunting mammoths and ancient bison until the megafauna went extinct about eight thousand years ago as the climate gradually warmed and dried. Although the Early Hunters disappeared as well, the human presence in Yellowstone continued as a new group occupied the region.

The Foragers too were a nomadic people, but they roamed within a more limited area as they searched constantly for edible plants and animals. The Foragers' marginal survival tended to limit their movements to seasonal migrations from mountain hunting grounds in the summer, where the drought-plagued effects of the warmer, drier climate were less severe, to the valleys for winter protection.

About two thousand years ago, the Yellowstone climate moderated toward current conditions, triggering a wide range of human and environmental adaptations. As wildlife diversity increased, the Foragers grew more dependent on hunting and began to use communal hunting techniques. Those advances enabled a more secure lifestyle and the emergence of the Late Hunters. The direct forerunners of the Sheepeater Indians encountered by Langford, the Late Hunters roamed the Yellowstone plateau in increasing numbers and began to use the bow and arrow, as well as to cook with vessels and grind vegetable foods.

By the time John Colter saw the Yellowstone plateau, Indians had long been environmental agents modifying the landscape to better suit their wants and needs. The development of the Indian lifestyle and the landscape evolved progressively in an intertwined cycle of cause and effect, change and adaptation. Early white explorers thought the Yellowstone plateau a primeval place touched only by the hand of God; instead, it was a place also shaped by the hand of man.

Fire was the primary tool the Yellowstone Indians used to shape their

Eden. Over time, it emerged as an integral component of the land-scape's "natural" disturbance regime. As we are only now beginning to fully understand, fire plays a critical role in maintaining many ecosystems, especially through its ability to arrest or reverse plant succession. The Indians in Ohio and Yellowstone learned that lesson long ago.

Lieutenant Doane noted the Indians' use of fire as the Langford expedition neared Yellowstone: "From this camp was seen the smoke of fires on the mountains in front, while Indian signs became more numerous and distinct." He later speculated that "the great plateau had been recently burned off to drive away the game, and the woods were on fire in every direction."[14] While it is more likely the Indians were using fire to entrap game as had Scoouwa on the Sandusky Plains, the numerous parklike grasslands the expedition encountered and the thick tangles of berries, briars, and brambles noted by early explorers along the streams were promoted by the fires he described.

Fire was only one tool the Indians used to reshape the land. Advances in hunting techniques, particularly the use of buffalo jumps and traps, and their use in conjunction with fire, combined to make the Indian a significant influence on the Yellowstone plateau's game populations. Those hunting practices gave a survival advantage to species in which they had no interest or were less able to successfully hunt.

Therefore the Yellowstone Roosevelt admired in 1903 was not a last bastion of primitive wilderness untrammeled by humans, for the Indians had long been reshaping it. Nor was it even the same as that described by Langford, for the Indians had been evicted from the park by the 1880s and so no longer manipulated its character. In addition, the Yellowstone of 1903 differed from the Yellowstone of 1872, despite the provisions of the 1872 act, because it had also been radically reshaped by its new white landlords to better suit their desires.

Under the act, a Congress struggling with severe budget problems made no financial provision for the new park's management. Recognizing the potential problem, it assumed that the anticipated revenues from visitation and concession fees would be adequate for the secretary of the interior to pay for proper park management. In effect, Congress accepted the park because it thought the park would be financially self-sufficient.

However, much of Yellowstone's recreational attraction and, in turn, its commercial attraction, was contingent on its easy accessibility by train from eastern cities. This fueled the NPRR's interest. Unfortunately, development of the requisite rail access was delayed six years because of NPRR financial problems. Consequently, when Nathaniel Langford accepted the appointment, without salary, as Yellowstone's first superintendent, he found that "no horde of tourists clamored at the gates of Wonderland, no worthy concessioners [sic] bid for privileges, no funds came forth from franchise fees, and—because of the state of penury imposed upon the Park by its promoters and the Congress—there was no progress."15 Yellowstone was a national park only on paper. Without proper management, its wildlife fell easy prey to the poacher.

During the 1870s and 1880s, the spread of settlement on the Great Plains decimated wildlife throughout the West. The great herds of buffalo, elk, and other species were driven to virtual extinction within a fifty-year span. Without proper protection and enforcement, Yellowstone was not spared that carnage. Poachers hunted there, despite its designation as a national park, with the same immunity they enjoyed on the Plains, killing elk, bighorn sheep, deer, antelope, and moose by the thousands. By the mid-1880s whites had substantially depleted Yellowstone's normal game populations. To stop the poaching and to bring order to the frontier chaos, the U.S. Cavalry was dispatched in 1886 to rescue the park. The cavalry managed the park off and on until 1918, two years after the National Park Service was established in 1916. The cavalry's presence in Yellowstone left indelible marks on both the park and the Park Service.

With proper military efficiency, the cavalry quickly controlled the poaching and began active programs to reestablish game. Troopers fed elk and antelope. Patrols were sent out to chase herds into the park. Garbage was put out for bears to feed on. And a seven-foot-high fence several miles long was constructed to keep in the animals. The cavalry also began intermittent efforts to control predators to encourage the growth of the preferred populations on which they fed. Results were nearly instantaneous. Without the Indians, with fewer natural predators, and with wildlife migrating into the park for sanctuary from the unrestrained carnage outside, game populations exploded.

The cavalry even began to introduce exotic fish into about half of the park's streams. Acting Superintendent Captain F. A. Boutelle, a fishing enthusiast, wrote in 1891, "Besides the beautiful Shoshone and other smaller lakes there are hundreds of miles of as fine streams as any in existence without a fish of any kind. . . . I hope . . . to see all of these waters so stocked that the pleasure-seeker in the Park can enjoy fine fishing within a few rods of any hotel or camp."16

He set out to achieve his dream, seeding fish species from around the world into many of the park's lakes and streams, while native Yellowstone cutthroat and grayling were exchanged with hatcheries, fishery exhibits, and aquariums across the country. Many attempts to introduce the exotic fish were dismal failures. Others succeeded too well. In some places brown and rainbow trout drove out the native cutthroat and grayling. The yellow perch put in Goose Lake colonized so well they had to be poisoned with derris root to control their numbers. But like Boutelle, most visitors enjoyed the abundance of fish and the fabulous angling it offered. That was what they came to enjoy; that was what they came to expect: Yellowstone as a wonderland of natural curiosities and what Teddy Roosevelt called a "natural breeding-ground and nursery for those stately and beautiful haunters of the wilds," all present in a pristine, primeval wilderness untrammeled by humans.[17] That is what we still expect today. This intoxicating vision, seen by Langford and Roosevelt, became Yellowstone's reality for many Americans. While the wilderness condition of Yellowstone was changed, and in its place a landscape manipulated by its white managers created, many preferred the new landscape to the original.

With only small refinements, the landscape illusion still persists, accompanied by a set of expectations about Yellowstone and, perhaps, the environment as a whole. Many of us expect to see spectacular natural wonders and an abundance of friendly wildlife passively grazing outside the lodge window in our national parks or jovially begging for handouts outside our car windows. But few of us are prepared to seek out or appreciate nature's sublime and infinitely subtle variations as Muir advocated. Much like Emerson and Thoreau, many Americans expect a modicum of life's luxuries in a safe, comfortable setting, free of risk from the animals or the elements. And we expect this from a place we perceive as pristine wilderness.[18]

Accompanying such expectations, the basic illusion developed ominous implications. Throughout our history, we have segregated wildlife into good and bad species, with large predators such as the grizzly bear, coyote, cougar, wolf, and wolverine sentenced to elimination because they posed a threat, often imagined, to humans or our commercial interests. Similarly, fire came to be perceived as a threat to our interests, so we sent Smokey the Bear to prevent it.

Yet landscape values changed, slowly shifting, softening incrementally. Transcendentalism, although advocated only by a small group of devotees, marked this change. John Muir's more widespread popularity marked its further movement. The creation of Yellowstone, whatever its motivations, marked still another milestone. These markers, each a bit farther forward than those preceding, demarcated the advancing philosophical boundary of the nation's landscape awareness and thought, just as the frontier boundary drawn by the Census Bureau traced the inexorably westward movement of our physical boundary. Those incremental advances culminated at the conclusion of the century in the first great national environmental debate, which focused the public's changing attitudes toward nature like a magnifying glass focuses a beam of light. The debate sought to define our stewardship values and find the greater of two environmental goods. It was perhaps the nineteenth century's most important marker in the progression. And at its center were two protagonists, once master and acolyte, then friend and compatriot, and finally bitter philosophical enemies—one a forester and politician, a patrician from the East Coast named Gifford Pinchot, and the other a singular writer from San Francisco named John Muir.

CHAPTER 11

The Greater Good

Through all the wonderful, eventful centuries since Christ's
time . . . God has cared for these trees, saved them from
drought, disease, avalanches, and a thousand straining,
leveling tempests and floods; but he cannot save them from
fools—only Uncle Sam can do that.

John Muir, *Our National Parks*

Gaunt and scraggly, John Muir never looked the role of a national figure who influenced presidents and the politically powerful or the cap-
tains of industry and the social elite. He always looked more like a
homeless wanderer. His role as a national figure started auspiciously
when he published his first article in the nation's leading newspaper. On
December 5, 1871, shortly after Emerson's visit to Yosemite, Muir's first
piece, "Yosemite's Glaciers," appeared in Horace Greeley's *New York
Tribune*. The article was not a colorful piece of philosophical prose but
a significant scientific explanation of how glaciers carved Yosemite's valleys. Muir's explanation, based on physical evidence he gathered while
wandering the park, countered the prevailing wisdom proposed by the
California state geologist, Josiah Dwight Whitney, after whom Mount
Whitney was named. Whitney believed the park was formed by "random 'cataclysmic' shiftings and heavings of the Earth's crust."[1] Whitney's brilliant former assistant, Clarence King, thought much the same
even though he found remnant traces of glaciers in the valley; however,
he did not make the connection between their presence and the forma-
tion of the valley.

Muir intuitively made that connection, then sought physical evidence
to support his hypothesis. Following Lyell's concept of uniformitarian-
ism, he found that evidence sprinkled about the valley in glacial scor-
ing, glacial mud (finely ground-up rock), and small residual glaciers in
the upper reaches of the Merced River basin. Scientists were skeptical
though, so he gathered additional empirical evidence using stakes planted

in snowfields to measure glacial movement (about an inch per day). His hypothesis was soon corroborated and widely accepted.

The article led to academic and scientific renown—not the type Muir particularly sought or approved of, since he viewed the formal scientific community with disdain, but it created other writing opportunities that led to his broader public appeal. He soon published articles on Yosemite's glaciers in the *Overland Monthly,* a California-based literary magazine; excerpts were reprinted in the *American Journal of Science* and the *Proceedings of the American Association for the Advancement of Science.* The *Overland* articles provided more opportunity for his creative writing talents to emerge and more opportunity to tell his personal philosophy. Other articles on Yosemite soon appeared in the *Tribune* and the *Boston Evening Transcript.* By the autumn of 1872, with his writing career under way and with his personal problems with James and Elvira Hutchins at a boil, he moved to the Russian Hill district in San Francisco, not to return to his Yosemite home for two years and never again to be there for other than brief visits.

Always uncomfortable in the crowds and civility of urban life, Muir spent the next few years preaching the virtues of nature and wilderness. The more converts he won, the more environmental souls he saved, the more he also inadvertently created an army of heretics—converts meant visitors to the remote, untouched wilderness he so eloquently extolled. Ironically, the furry "hoofed-locusts" he once shepherded during his first days in Yosemite were replaced with a new hoofed locust of his own persuasion. He was once again a shepherd tending a flock of nature seekers.

At the same time, other, less sensitive forms of human intrusion encroached upon his temples as stockmen and lumbermen also threatened his wilderness sanctuaries. By the 1870s those conflicts created a debate regarding the proper use of the nation's dwindling wildland, and with it gave birth to the American conservation movement. As Thoreau and Muir proposed, and as Turner and Roosevelt would proclaim, wildlands were valuable not just as a potential source of natural resources but as important psychological and cultural resources. By the end of the nineteenth century, American values had softened sufficiently to let people see the dilemma. What was the best use of the scarce wilderness resource? For many people wilderness no longer posed a physical or psychological threat, nor was it so prevalent that it could be used wastefully. The issue became one of finding the greater good between conflicting, competing interests and defining our stewardship responsibilities.

Muir remained settled in San Francisco, successfully spreading his gospel and swirling in the currents of the conservation movement until he married in 1880. Children quickly followed, Wanda in 1881 and Helen in 1886, by which time the Muirs had built a home on his father-in-law's prosperous fruit farm in Martinez, fifteen miles north of Oakland. Married life and fatherhood apparently gave Muir a long-sought sense of emotional security. He no longer rambled as freely and extensively. Perhaps his family obligations, his sporadic writing, and his extensive reading now satisfied his environmental soul.[2]

As the conservation debate gradually intensified during the last quarter of the 1800s, Muir's personal philosophy continued to evolve, shaped by provocative people and ideas. Like Henry George, a political economist living in San Francisco, Muir began to question the nature of private land ownership because of the widespread abuses that stemmed from monopolies and a laissez-faire, free-market system that often rewarded greed and short-term return over environmental or social responsibility. Like the Grange, to which he was introduced by his father-in-law, Muir became concerned about the relationship between labor and management, such as that between the farmer who produces a crop and the distributor who sells it. And like the influential English art-critic-turned-social-philosopher John Ruskin, he began to think that environmental degradation resulted not so much from individual greed as from fundamental flaws in the nation's economic system. Muir had read some of Ruskin's works while living in Yosemite, but thought little of his romantic notions of scenery. Ruskin, he felt, needed a good dose of Yosemite to sharpen his landscape vision, as did Emerson and Thoreau. However, when he reexamined Ruskin's work in the mid-1880s, he found valuable ideas, including Ruskin's criticism of cities and technology and his call to reward landowners who maintained their property in "conditions of natural grace."

These influences led Muir to assume a more proactive philosophical stance toward the preservation of the wilderness landscapes he so cherished. He began to advocate more overt governmental intervention to offset private abuse of the public domain. Others were beginning to take the same stance as they, too, reacted against the wanton degradation caused by most of the logging, mining, grazing, and farming that encroached farther and farther into the nation's dwindling wilderness. Those were the founders of the American conservation movement.

Joining Muir were newly formed groups of outdoorsmen who sought to protect the components of wilderness on which their outdoor interests were based; in so doing, they were also sympathetic to the protection of wilderness in general. The first such group was likely the Appalachian Mountain Club, formed in the 1870s by a group of Boston academics and hikers who wanted to promote and protect their interests in hiking the Appalachian wilderness. The Audubon Society, organized by George Bird Grinnell in 1886, was interested in protecting birds and their eggs. The Boone and Crockett Club, a group of big game hunters from New York City established in 1887 by Grinnell and Theodore Roosevelt, aimed to protect wilderness as a means of protecting quarry. Club membership was limited to one hundred men who had killed "in fair chase" at least three species of North American big game.

On the West Coast, a group of professors from Berkeley and Stanford met in the office of a San Francisco lawyer on June 4, 1892, to discuss the organization of an alpine club modeled after the Appalachian Mountain Club. Their version would be dedicated to "exploring, enjoying, and rendering accessible the mountain regions of the Pacific Coast," and it would "enlist the support of the people and the government in preserving the forests and other features of the Sierra Nevada Mountains."[3] Those twenty-seven men formed the Sierra Club and elected John Muir its first president, a position he held for twenty-two years until his death in 1914.

Those who conceived American conservationism were mostly people of great distinction and social position. And most, like the leaders of transcendentalism, were among the liberal, intellectual, wealthy elite of the East Coast: Charles Sprague Sargent, George Bird Grinnell, and Robert Underwood Johnson, to name a few. Sargent, a Boston blue blood, was the director of the Arnold Arboretum at Harvard University and a great friend and supporter of Muir. Following in the footsteps of Emerson and the transcendentalists, he was welcome in the fashionable parlors of Beacon Hill and eastern high society. Sargent was a key matchmaker in the movement, instrumental in connecting Muir with those of like minds in the eastern establishment.

Grinnell was also from a wealthy, well-connected family. Omnipresent in conservation causes, George was editor and publisher of *Forest and Stream,* the leading sportsmen's magazine of the day. His activism likely stemmed from the influence of the man for whom he named his organization of birders. The Grinnells had had a long relationship with the Audubons. George grew up in Audubon Park in upper Manhattan, the property once owned by its namesake. He hunted muskrats as a boy with Audubon's grandson along the Harlem River in the 1860s, and even attended a school run by the painter's widow.

Robert Underwood Johnson was not a member of the eastern elite by birth. He grew up in a small rural community in Indiana. Like Muir, he was a transplanted midwesterner who was raised in a conservative Scotch Presbyterian (and Quaker) home. Much of Johnson's effect in the conservation movement, like Grinnell's, stemmed from his position as an editor of *Century,* the nation's leading literary magazine and successor to *Scribner's Monthly* (*Scribner's* had published many of Muir's best early works in the 1870s while Johnson was a junior editor). He used the magazine as a forum to promote his proconservation views. By the time Johnson assumed an associate editorship of *Century* in the 1880s, the magazine was at its pinnacle, paying the highest fees for authors of the caliber of Henry James and Mark Twain. The magazine sought to promote American literature and culture as separate and distinct from those of Europe, and even to reform national politics. It rarely shied from social or political crusades. Since the emerging conservation debate was pure Americana, the magazine was naturally attracted to it and Muir's voice.

The first conservation battles fought by those founders usually concerned the conflict between conserving the beautiful scenery of a wilderness area for recreational viewing and pleasure-seeking and the commercial exploitation of its resources. One of the first battles was fought over Niagara Falls. The conflict was apparent there by the early 1800s as crass commercialism ringed what many considered the nation's greatest natural spectacle. Alexis de Tocqueville saw this conflict during his visit to the falls in 1831. He warned its grandeur was about to be lost and prophesied a sawmill at its base within ten years. Few paid heed as calls for exploitation drowned those for conservation. The state of New York, in response to a campaign led by Frederick Law Olmsted, and Charles Eliot Norton of Harvard, finally set aside a four hundred–acre park around Niagara Falls in 1885. Canada followed suit on its side about a year later.

The American reserve soon proved too small to protect the falls from the effects of nearby hydropower and industrial development. On visiting the falls in 1906, English author H. G. Wells commented, "Its spectacular effect, its magnificent and humbling size and splendor, were long since destroyed beyond recovery by the hotels, the factories, the powerhouses, the bridges and tramways and hoardings that arose about it." Yet like most commentators of the day he concluded, "It seems altogether well that all the froth and hurry of Niagara at last, all of it, dying into hungry canals of intake, should rise again in light and power."[4] Public preference still valued development over conservation.

The Adirondacks were the focus of another of the nation's early conservation debates. Like Niagara Falls, the Adirondack region in upstate New York had long been valued as a place for passive outdoor recreation, where people marveled at its spectacular scenery. Perhaps no other American wilderness was so often written about or was so widely known by the mid-1800s, in part because of its proximity to the eastern cities. By that time too the expansion of settlement and resource exploitation threatened the remnant, islandlike wilderness adrift in the sea of eastern civilization. That threat prompted calls for its protection.

While most of the calls came for aesthetic and recreational reasons, the actual proposals for its protection were politically justified on more utilitarian needs—to protect the timber as a valuable resource and, drawing on the revelations in *Man and Nature,* to protect the forest as a steady source of water for the region's rivers and canals. Big business joined forces with the conservationists in support of protection in the 1870s and early 1880s as water levels in the Erie Canal and Hudson River were purported to be dropping. The supposed crisis made for strange bedfellows as Grinnell's *Forest and Stream* and Greeley's *Tribune* both supported the proposal.

On May 15, 1885, New York State set aside 715,000 acres as wild forestlands in establishing the Adirondack Forest Reserve. As with Niagara, though, the issue was far from over. In 1892 the state revised the reserve and made it a state park of more than 3 million acres in response to growing support for aesthetic values. The new park had a dual purpose: for recreation available to everyone for health and pleasure, and as forestland to ensure a future supply of timber and to preserve the headwaters of the chief rivers of the state.

Fearing that designation as a park might make it susceptible to abuses, the state constitution was revised in 1894 to permanently preserve the wilderness. Strongly influenced by Sargent and his supporters, the state declared an area nearly the size of Connecticut as "forever wild." While the heart of the matter remained utilitarian—protection of the watershed for commercial reasons such as supplying municipal water—aesthetic considerations, reasons less utilitarian, also gained recognition.

By the end of the 1880s even Muir's beloved Yosemite, despite its protection in 1864 as a state preserve, was threatened by internal mismanagement and uncontrolled logging and grazing about its borders. The temple itself was now defiled with dilapidated tourist facilities, saloons, a pigsty, butcher shops, cropland, vegetable stalls, and lumberyards. Yet Muir's pen was mute as family and farm life occupied his attention. That silence was soon shattered by events that thrust him back into the center of the American conservation debate where he remained the rest of his life.

In the spring of 1889 Robert Underwood Johnson, while visiting California to gather material for articles on the state's early settlers, arranged to meet Muir at Johnson's hotel in San Francisco in hopes of prodding the gentleman farmer out of his self-imposed silence. Muir too had a purpose in agreeing to the meeting. He wanted to enlist *Century*'s support in the defense of Yosemite and other threatened California wildernesses.

That was the first time the two met face-to-face. As was Muir's way, he talked. Johnson politely listened. The meeting went well, though, and several days later Johnson visited the Muirs in Martinez. Again the visit was pleasant but no agreement was reached; however, Johnson, unlike Emerson, accepted Muir's invitation to accompany him to the high altar—to go camping in Yosemite. There, despite the human vandalism they saw about the preserve, Yosemite's inherent wonder once again worked its magic on both men as they hiked the beautiful meadows of wildflowers and the thundering canyons of the Tuolumne River and slept on beds of spruce needles beneath a blanket of stars, "the biggest stars I have ever seen," recalled Johnson. He noted, "The first impression is wonder that there should be so much in so small a space . . . the total effect is overwhelming so that you don't take it in at first, but are surprised anew at each new look."[5]

While sitting around the evening campfire along the upper reaches of the Tuolumne River in early June, the two talked about Yosemite's future, as Langford and Hedges had talked around a campfire in Yellowstone. Johnson asked Muir about Yosemite's fate. Muir responded by describing the devastating effects of overgrazing and mismanagement, to which Johnson replied, "Obviously the thing to do is to make Yosemite National Park around the Valley on the plan of Yellowstone."[6] Enclose the state preserve by a much larger national park, he proposed. Muir agreed wholeheartedly, and a deal was struck. Muir would write two articles for *Century* and champion the cause. Johnson would lobby friends in Washington and write editorials.

The ensuing campaign established the prototype for many conservation debates to come. A loose collection of part-time volunteers motivated mostly by philosophical and recreational interests challenged the economic interests of those whose income was tied to the issue. John P. Irish, a powerful political boss and editor of an Oakland newspaper, led the formidable opposition; Irish was also secretary and treasurer of the Yosemite Board of Commissioners, who oversaw the reserve for the state. The board was widely known to favor commercial interests (Irish later was an agent for the local cattleman's association).

To counter the vested commercial interests of the Yosemite concessionaires—the lumbermen, cattlemen, and sheepherders, and Leland Stanford's all-powerful Southern Pacific Railroad that dominated California politics and controlled rail access to the park—the conservationists sought to bypass local politics by taking the issue to a more receptive national audience. They quickly enlisted the support of sympathetic people such as Olmsted and California Senator George Hearst. Eastern newspapers such as the *New York Evening Post* joined the call. By March 1890, Congressman William Vandever from Los Angeles introduced legislation proposing a new Yosemite reserve of about 185,000 acres. Key areas Muir wanted included, however, were overlooked.

Muir's articles finally appeared in *Century*'s August and September 1890 issues, as the bill neared its hearings before the House Committee on Public Lands. His call to protect a much larger area than Vandever proposed was carried to the magazine's two hundred thousand readers, many of whom were among the nation's most influential and powerful people. His arguments for protection drew on Marsh and described the many utilitarian benefits that would result from protecting the forest, the soil, and the water. He suggested the area was worthless for mining and agriculture, as Langford had successfully argued in the Yellowstone debate. For perhaps the first time in a conservation debate, Muir also argued that Yosemite should be protected solely as a wilderness, to remain untouched for no utilitarian reason other than its own preservation.

The conservationists knew public support across the country still might not be sufficient to offset local commercial opposition. With Irish and the Southern Pacific opposing the plan, congressional approval appeared unlikely. Realizing Leland Stanford held the key to the entire debate, Johnson called on the godfather of California politics only weeks after leaving Muir in Yosemite. Leland was noncommittal, but fate soon interceded. Over the winter, Stanford and his partner, Collis Huntington, had a fight to the finish for control of the company. Huntington won and replaced Stanford as president. He also promised to reform and reduce the railroad's political meddling. Quietly, then, the corporate giant threw its support behind the Vandever bill, which had been amended to include the additional lands Muir wanted in the new park. On September 30, 1890, the bill passed both chambers of Congress with little discussion. The next day, with President Benjamin Harrison's signature, nearly a million acres in Yosemite became the nation's second national park.

The stunning victory left Muir with little time to celebrate, as words once again flowed from his pen in response to another major battle, one at the heart of the conservation debate. The new battle addressed the conflict between the competing interests for control of the nation's remaining forest wilderness.

By the last quarter of the 1800s the vast majority of the most valued public land was forested, and most of that was in the West in places like Yellowstone and Yosemite. Such mountainous landscapes usually possessed the spectacular scenery and natural wonders that were most sought after as public pleasuring grounds. By this time too their commercial value as sources of timber or minerals was becoming apparent. In contrast, the grassland or scrub growth that covered the remainder of the public domain on the Plains was of much less interest. The positive value assigned to wilderness—whether by the consortium of conservation-oriented philosophies espoused by Thoreau and Muir, or Turner and Roosevelt, or by those intent on the continued use of wilderness for economic exploitation and development—was assigned to forested wilderness, not desert or prairie wilderness. Nor were conservation and stewardship values focused on the commonplace landscapes of typical daily life, whether farmland, suburb, or city in the East or West. Instead, the conservation movement arose mostly from a philosophical battle of wills regarding the use of the nation's remaining forested wilderness, the battleground of which was a proposal to set aside our national forests.

The battle's first skirmish began quietly in 1876, when Dr. Franklin B. Hough was hired by the Department of Agriculture to collect statistics and distribute information on the nation's remaining forests. The federal action was taken in response to pressure principally from the American Association for the Advancement of Science (AAAS), and from Hough himself, who was the director of the National Census and a disciple of George Perkins Marsh. Hough was instructed to develop a report detailing the nation's current and projected forest production and consumption. The first volume of his three-part report, published in 1877 and distributed to twenty-five thousand readers, broadened awareness of the prevailing plunder of the nation's forests, to little result.

Abuse of the nation's forests continued unchecked throughout the 1870s, as did general abuse of the public land laws and land distribution programs. By 1880 Dr. Bernard E. Fernow, a German-trained forester, became the head of Hough's small organization, which was renamed the Division of Forestry.[7] Ironically, though, during his tenure as chief (1880–98), Fernow questioned the federal government's role in large-scale forest management. Under his skeptical management, the division received little money (its annual budget was around $28,500) and held little power, so it was of minimal consequence. Fernow was politically unable or philosophically unwilling to implement the scientific forest management practices he had learned in Germany. His division functioned mostly as an information bureau that tacitly tolerated continued clear-cutting of the nation's timberland to satisfy short-term supply as cheaply as possible. The division had little effect on reducing the environmental consequences of logging, or on clarifying the long-term implications of that production. Public and private forestry had not yet incorporated the lessons offered by Marsh's *Man and Nature.* American forestry was neither a profession nor a science.

The origins of federal forest management began with concerned citizens outside the government, especially Charles Sprague Sargent. Inspired by *Man and Nature,* Sargent in 1880 surveyed on behalf of the government the forestland remaining in the public domain. The lack of modern management and the environmental abuses he found shocked him and prompted him to call for federal reform and regulation pending a more complete study by experts. The fledgling American Forestry Association joined his call. Few listened, however, and little progress toward either the reform or the study was made (*Century* would finally endorse Sargent's proposals nearly nine years later, in 1889, as Muir and Johnson headed into the high country of Yosemite).

Over a decade passed following Hough's 1877 report and Sargent's 1880 survey before the next rumbles in the gathering storm over the nation's forests occurred. Calls for land reform from the AAAS, the American Forestry Association, the American Forestry Congress, and others found a receptive ear in President Benjamin Harrison in the early 1890s. On March 3, 1891, just before adjournment, Congress passed a bill repealing the widely abused Timber Culture Act and the preemption laws. Attached to the bill was a little-noticed rider proposed by William Hallett Phillips, an obscure Washington lawyer, socialite, and member of the Boone and Crockett Club. The innocuous proposal was supported by other wealthy, influential people, mostly easterners, including Sargent and Johnson, and John Muir. The rider rode through on the bill's coat-tails without comment. Its enactment, though, precipitated the storm over the nation's forests.

The new amendment empowered the president to withdraw forest-land from the public domain and set it aside as a "forest reserve" (later called a national forest). President Harrison quickly seized the opportunity and, at the behest of the secretary of the interior, John W. Noble, over the next two years designated fifteen reserves totaling 13 million acres. As it had with the establishment of Yellowstone, Congress left much unsaid in the new Forest Reserve Act. It failed to specify the function of the reserves the chief executive could so easily create with the stroke of a pen. Were they to protect timber and water resources as in the Adirondacks, or scenic and recreational resources as in Yosemite and Yellowstone, or wilderness in general?

Muir argued that the intention was clear since the law followed so closely on the Yosemite debate. The forest reserves were to be like national parks and remain uncut and undeveloped. Secretary Noble generally agreed, as did his successor, Hoke Smith, who dutifully banned sheep-grazing in the reserves, despite opposition from stockmen, after Muir and Johnson met with him in 1893 to discuss the matter. The conservationists found a willing audience among key government administrators, often their peerage who shared their sympathies. Others with vested commercial and political interests received only a deaf ear. These conservation decisions were not based on popular public opinion or open debate, but on backroom politics, albeit refined.

Regardless of interpretation, the new reserves existed only on paper for, as had happened at Yellowstone, no one was assigned to enforce the rules or regulations. Use of the forest reserves remained a free-for-all. Clear-cutting and exploitation continued unabated despite Noble and Smith, Johnson and Muir. By the mid-1890s the chaotic situation became unacceptable to the conservationists.

In the aftermath of New York's constitutional guarantee that the Adirondacks would be "forever wild," Johnson's *Century* organized a symposium in 1894 to examine a Sargent proposal that called for forestry instruction at West Point and for the army to protect forest reserves, as in Yellowstone and Yosemite National Parks. The symposium attracted many of the nation's leading conservationists, most of whom agreed with the Sargent proposal.

One visionary twenty-nine-year-old consulting forester did not, and he voiced a powerful dissenting opinion. There, Gifford Pinchot struck a mortal blow through the heart of the unified conservation movement, forever splitting it into two factions. That split was another milestone in the progression of American landscape values and stewardship thinking. The difference between the two factions was no longer between a good and bad; instead it had become a difference in the definition of the greater good. Both factions loved the land and valued wilderness, only they did so from very different perspectives.

Pinchot, along with several others from the Division of Forestry, proposed that the reserves should be managed not by the army but by "a forest service, a commission of scientifically trained men" (which he would presumably head).[8] While the idea appeared reasonable, the guiding philosophy cut clear through the movement. He proposed that forest reserves be used not just as a onetime supply of a single resource; instead, they should serve as a source of multiple resources managed to extract their maximum sustained yield. Pinchot wanted to generate the greatest possible benefit to the greatest number of people, advocating the utilitarian philosophy of John Stuart Mill and Jeremy Bentham. In his autobiography, *Breaking New Ground,* he wrote:

> Forestry is Tree Farming. Forestry is handling trees so that one crop follows another. . . . Trees may be grown as a crop just as corn may be grown as a crop. The farmer gets crop after crop of corn, oats, wheat, cotton, tobacco, and hay from his farm. The forester gets crop after crop of logs, cordwood, shingles, poles, or railroad ties from his forest, and even some return from regulated grazing. . . . The purpose of Forestry, then, is to make the forest produce the largest possible amount of whatever crop or service will be most useful, and keep on producing it for generation after generation of men and trees. . . . It is true indeed that the forest, rightly handled—given the chance—is, next to the earth itself, the most useful servant of man.[9]

Thus the forests were not to be preserved untouched as Muir proposed; rather, they were to be the source of valuable natural resources managed scientifically to squeeze out every ounce of benefit. Nor were they only to be pleasuring grounds for hikers and campers like the national parks. Instead, the forests were to be harvested annually like a crop, and every resource possible was to be extracted for the maximum benefit of society. They were to yield an annual income like the interest on a savings account, and every penny of interest was to be spent. With enlightened management, he preached, natural resource "investments" would yield even greater rates of return than nature alone would produce. Like Marsh, Pinchot wanted his treasured forests to be used wisely, not blindly or wastefully. Pinchot believed a properly managed forest that supplied many resources and recreational opportunities satisfied more human needs. That was the better use of the prized possession. It achieved the greater good.

The philosophies of John Muir and Gifford Pinchot were, ironically, at opposite ends of the environmental spectrum even though both arose from the same intention, to better husband, to better steward the beloved forests. Pinchot's consumption-based conservation philosophy of maximum sustained yield quickly assumed the mantle of the conservation movement and was soon applied to virtually all of the public domain, whether wilderness or not, whether forest or not. Applicable to the everyday landscape, it encapsulated the public's prevailing values, just as Turner's Frontier Thesis had. Muir's philosophy of minimum possible use became the cornerstone of the preservation movement and remained focused primarily on the dwindling supply of wilderness—initially forested wilderness but later other wilderness landscapes.

Pinchot's appearance at the symposium and his proposals were not totally out of the blue. Although young and self-employed, Pinchot was a wealthy, well-connected easterner like Grinnell and Sargent. And like so many others in the movement, his conservationist inclinations grew out of the opportunities afforded by his economic and social station.

Pinchot's forestry career and his conservation philosophy grew, as Char Miller described in "The Greening of Gifford Pinchot," in large part from his father's quest for personal penance. James Pinchot, like his own father (Gifford's grandfather), Cyril, was a successful speculator in the timber industry. For two generations the Pinchots had profited from New York, Pennsylvania, Michigan, and Wisconsin timberland. As was common in the industry during much of the 1800s, they bought land, clear-cut the forest, then floated the logs tied together in giant rafts down rivers swollen with spring rain to eastern markets. The profits were reinvested in more virgin land, creating a vicious cycle of land exploitation and degradation that left giant swaths of the landscape denuded of forest cover and exposed to the erosive effects of wind and water. That was the short-sighted cycle George Perkins Marsh decried in *Man and Nature*.

In the 1850s, though, James Pinchot left the timber industry and moved to New York City, where he became a successful distributor of domestic and commercial furnishings. His business benefited from the prosperity and economic boom that resulted from rapidly expanding industrialization after the Civil War. His wife, Mary Eno, was also wealthy, adding to the family fortune. Mary was the daughter of Amos Eno, a successful New York City land developer who owned prime properties such as the Fifth Avenue, the city's grand hotel, and other choice real estate along principal avenues like Broadway. Gifford was born in 1865 into a family of great wealth and social position.

By the 1870s James retired, even though he was only in his midforties. And as Miller proposed, he was gripped by a growing sense of shame and guilt over the devastation wrought by his, and his father's, timber speculation. James saw young Gifford as the vehicle for his redemption and sought to direct Gifford along a path that would atone for the landscape sins of the father and grandfather.

Miller suggested that the guilt was fostered in part by the European landscapes James saw while on holidays in the latter half of the 1800s. James fell in love with the picturesque English countryside of quaint villages scattered about a bucolic landscape of green pastures, fields, and forests. In contrast, the abused land around his home in Milford, Connecticut, land devastated by the same shortsightedness he practiced throughout New England and the Midwest, stood in stark testimony to the misuse and arrogance documented by Marsh and later implied by Jean Giono. And as Giono created Elzéard Bouffier to replant trees to restore the Provence landscape to its former health, James sent his son Gifford to restore the American landscape.

Gifford's turning point came during his senior year at Yale University as he struggled to choose a career. Amos Eno pressured his grandson to join his business. James, impressed by the apparent harmony between people and nature in the English landscape, and bothered by guilt over his landscape malevolence, encouraged the type of modern forestry practiced in Europe as the profession for his son. The father prevailed, and an idealistic Gifford chose forestry as a form of public service instead of continuing the family quest for mammon.[10]

After graduation in 1889 young Pinchot went to Europe in search of forestry training, since no U.S. schools taught the subject. There, on the advice of eminent foresters, he enrolled at the *L'Ecole nationale forestier* in Nancy, France. In the classroom and in the French national forests, he learned how to manage the forest as a crop that could yield a continuous supply of resources as had been practiced out of necessity for decades on manorial estates and royal forests.

In contrast, the American attitude toward the forests that covered the eastern third of the continent had been since settlement one of all-out war. Here the goal was the removal of the forest to make room for agriculture and settlement, and later to feed the Industrial Revolution's insatiable hunger for fuel and material, especially the ferocious appetite of the railroads. Americans were understandably slower to adopt concepts of forest management because they considered the supply as limitless. When Pinchot returned to America in 1890 to begin his career, he was perhaps the first American formally trained in modern forestry.

While studying European forestry Pinchot also acquired the seed of a later belief that natural resources belonged to all the people and should therefore be used for public benefit to achieve social equity—they should serve the greatest good for the greatest number, rather than serve only the interests of the social and economic elite. That recognition was accentuated during a tour of a Prussian forest. Pinchot wrote in his autobiography, "I shall never forget the old peasant who rose to his feet from his stone-breaking, as the Oberfoerster came striding along, and stood silent, head bent, cap in hands, while the official stalked by without the slightest sign that he knew the peasant was on earth."[11] The feudal roots of the European landscape remained recognizable. This utilitarian belief would be a hallmark of the Progressive Movement he championed in the early 1900s.

When young Pinchot first returned from Europe he was uncertain about his future. Again benefiting from family contacts, he sought the advice of leading conservationists and government officials, including Sargent, Fernow, Charles Schurz, the secretary of the interior to whom John Wesley Powell had delivered his *Report on the Lands of the Arid Region,* even Vice President Levi P. Morton. Through those contacts he was employed early in 1891 as a consulting forester by Phelps, Dodge, and Company, a large mining and timber company. Phelps sent the novice to examine its holdings in Pennsylvania. With that work quickly completed, he then accompanied Fernow for several weeks as a guest on an inspection tour of the hardwood forests in the southern Mississippi River basin in Arkansas. By March, Phelps sent him to Arizona to survey company lands for reforestation. Before returning east, he toured California, visiting Yosemite and the Mariposa sequoia grove, then continued up the Cascades through the old-growth forests of the Pacific Northwest. On return, he praised the way John P. Irish and the state commission managed the park while, at the same time, he dismissed *Century*'s editorial criticism of the commission. He felt the greater good was achieved by the full exploitation of the park's resources, since the measure he used to calculate benefit relied primarily on tangible economic benefits, not intangible aesthetic and recreational measures.

By late 1891 he returned to New York in search of work. Once again family connections supplied the answer. There his father helped arrange meetings between Gifford, George Vanderbilt, and Frederick Law Olmsted.[12] As a result, Vanderbilt hired Pinchot, at the behest of Olmsted, to manage the extensive forests at Vanderbilt's sprawling Biltmore estate, perhaps America's closest cousin to the grand European baronial estates. Olmsted, whose firm had been working on the innovative master plan for the seven thousand–acre grounds since 1888, wanted the estate to have a model farm, a great arboretum, and a vast game preserve. He intended the estate to be, as Pinchot recalled, "the first example of practical forest management in the United States." Olmsted had recognized the value of sound forest management from his days of setting aside Yosemite and the Mariposa sequoia grove as a state preserve. Energetic and idealistic, Pinchot made the most of the opportunity to test his theories on Vanderbilt's forests, which expanded to nearly two hundred thousand acres; today, that land forms the core of Pisgah National Forest. Pinchot later credited Olmsted with making Biltmore the "nest egg for practical Forestry in America."[13]

Pinchot's landmark work at Biltmore left him time for consulting work, so he opened an office in New York in late 1893. His work with privately owned forests steadily grew, as did his influence in the expanding forestry and conservation movements. During the debate on the Adirondacks at the New York constitutional convention, he opposed the "forever wild" clause because it would prohibit lumbering. He soon realized that the greatest opportunity and the most desperate need for his skills rested in managing the hundreds of millions of acres in the public domain. That understanding led him to abandon his private consulting practice in 1898 when asked to replace Bernard Fernow as the head of the Forestry Division in the U.S. Department of Agriculture. There his professional ideals, his political ambitions, and his social agenda blossomed.

When Pinchot countered Sargent's notion that the army should manage the national forests at the 1894 *Century* symposium, and proposed instead that the national forests be managed by professionally trained foresters, his viewpoint arose from his family heritage, his European forestry education, and his sense of social responsibility. So did his redefinition of conservation to mean maximum sustained yield for the benefit of society. Still, he joined the rest in supporting Sargent's proposal to form a government commission to propose management guidelines based on a more detailed study of the nation's remaining forests than the one he (Sargent) conducted in 1880.

The commission was eventually established in 1896 following several years of political maneuvering. After efforts to have Congress establish and fund the body stalled, Sargent, Johnson, Pinchot, and other proponents persuaded Hoke Smith to ask the National Academy of Science to organize it (much as the NAS conducted the study of the geological surveys); that would make the twenty-five thousand dollar requisite funding far easier to obtain from Congress. The strategy worked, and a six-person commission set out to survey the nation's forests that summer. Commission members included Sargent, who headed the group, William Brewer of Yale University, Alexander Agassiz of Harvard University, General H. L. Abbott, an engineer, Arnold Hague of the U.S. Geological Service, and Gifford Pinchot. John Muir joined the group as an ex officio member, much to Pinchot's delight.

Pinchot had met Muir the year before the *Century* symposium and had presented himself somewhat as a neophyte before the revered master. Pinchot sought and Muir offered advice and encouragement. They became friends, though Pinchot's outlook on nature was probably too gentle, too romantic, and removed, like that of the transcendentalists, for Muir's taste. Pinchot and Muir shared a love of the land and a desire to conserve it; unfortunately, their concepts of conservation differed profoundly. While those differences had surfaced before the commission convened, they were not yet divisive. The tour fostered their friendship. All of that changed when the commission set about preparing its final report.

Sargent and Muir felt the report should identify forest wilderness areas that needed preservation, and they hoped the report would lead the government to set aside additional lands using the existing Forest Reserve Act. Pinchot and Hague felt the report should lay the basis for enlightened forest management open to carefully conducted exploitation under the control of a civilian, professionally trained forest service. The Sargent-Muir faction won as the report limited resource exploitation. The report also recommended the army be given the responsibility to patrol the reserves as it had at Yellowstone and Yosemite. On February 22, 1897, just before leaving office, President Grover Cleveland added thirteen new reserves totaling 21.4 million acres. Like President Harrison before him, President Cleveland made no mention of utilitarian purposes. The apparent preservation victory didn't last.

Sargent, without the consent of the other commission members, had persuaded a sympathetic president to make those designations without further consultation or debate. Sargent's action outraged the opposing commissioners, and the president's action outraged Congress and commercial interests all across the country. The issue, far from settled by Sargent and President Cleveland, was carried over to the new McKinley administration. And once again, the president and the secretary of the interior were persuaded by personal pleas, from Sargent and Johnson, not to succumb to mounting political pressure to rescind the new reserves. But the debate by then had spread to Congress, which sought legislation to resolve the issue. Muir, writing in *Atlantic Monthly* and *Harper's Weekly*, championed the preservation cause: "Much is said on questions of this kind about 'the greatest good for the greatest number,' but the greatest number is too often found to be number one. It is never the greatest number in the common meaning of the term that make the greatest noise and stir on questions mixed with money. . . . Complaints are made in the name of poor settlers and miners, while the wealthy corporations are kept carefully hidden in the background. . . . Let right, commendable industry be fostered, but as to these Goths and Vandals of the wilderness, who are spreading black death in the fairest woods God ever made, let the government up and at 'em."[14]

On June 4, 1897, Congress finally spoke when it passed the Forest Management Act and clarified the purpose of the reserves: "to furnish a continuous supply of timber for the use and necessities of the citizens of the United States."[15] Responding to the bulk of popular opinion from western legislators and lumber, grazing, and mining interests, Congress rejected Muir's preservation philosophy in favor of Pinchot's conservation philosophy. National forests were not set aside to preserve wilderness, Congress declared. They were set aside to protect utilitarian needs. The nation's traditional landscape values prevailed. The land was still regarded as a commodity to be used for economic growth and development. Preservation remained a philosophy held by a small minority comprised mostly of elite easterners. Feeling the national forests were a lost cause, Muir thereafter devoted his attention to the care of the national parks, directing his efforts toward the establishment of new ones where his preservation philosophy might still apply.

Pinchot was rewarded several months later with the appointment as head of the forestry division, where he replaced his lackadaisical colleague, Bernard Fernow, who had resigned when asked by Congress to justify the division's existence. Politically astute and aggressive, Pinchot built the floundering division from a staff of approximately 125 persons with little authority to roughly 1,500 people who managed nearly 150 million acres of federal forest. Working closely with President Roosevelt,

Pinchot became the government's most influential land manager, molding his division over his tenure (1898–1910) to reflect his personal philosophy while spreading his vision over every aspect of federal land policy. In 1905 his division, renamed the U.S. Forest Service, managed all the national forests, by which time his conservation philosophy of maximum use and sustained yield for the greatest good to the greatest number, dominated the government's approach to the entire public domain, as it does today. Pinchot's philosophy, not Muir's, defined the nation's stewardship attitudes at the turn of the century.

The philosophical battle lines between Pinchot and Muir, between conservation and preservation, were firmly drawn by 1900 as the two leaders parted ways, personally and philosophically. But the battles over Niagara Falls, the Adirondacks, Yosemite, and the national forests were merely preparatory skirmishes contested by the economically and politically powerful. They were not all-out philosophical warfare, for the American public was yet to be called to war. Now the two estranged leaders battled for the environmental soul of the country, a battle that took the war into American homes beyond those with vested philosophical, political, or commercial interests. It's appropriate, then, that their ultimate battle was waged over Muir's sanctum sanctorum—part of his Yosemite temple, the Hetch Hetchy Valley.

The Hetch Hetchy Valley, through which the Tuolumne River flowed, was just north of the valley through which the Merced River flowed, the heart of Yosemite. The two steep-walled valleys were cousins, if not sisters, in appearance. Both were spectacular, glacier-carved granitic valleys. And both were within the boundary and protection of the national park after Muir successfully argued for Hetch Hetchy's inclusion in 1890. Yet Yosemite's proximity to San Francisco posed a threat to the park greater than that posed by Muir's ever-growing herd of two-legged locusts and the logging and mining industries. By the late 1800s, the quest for a new form of resource extraction generated by the city's growth challenged the park.

Yosemite lies about 150 miles east-southeast of San Francisco, which sits on a dry peninsula surrounded by salt water. Since its settlement, the city by the bay had suffered a chronic freshwater supply problem. As the city boomed in the late 1800s, the problem posed a potential obstacle to future growth—San Francisco needed a major new source of municipal water. When city engineers looked for likely sources around the

region in the 1880s, they quickly focused on the steep-walled glacial valleys in the Sierra Nevada, which were ideal for the construction of a
reservoir.

Perfect, city officials thought. Dam the Tuolumne River and pipe the
pristine water to San Francisco. The dam could also generate hydropower, and the reservoir could be used for recreation. To the preservationists, the plan was sacrilege. Fortunately for them, the Hetch Hetchy
Valley, protected as part of Yosemite National Park, was safe from such
a blasphemous proposal. Proponents of the dam needed a loophole in
the law to allow them to proceed.

To create that loophole, San Francisco Mayor James D. Phelan and
others quietly persuaded Congress to pass a bill in February 1901 permitting water conduits "for domestic, public, or other beneficial uses"
to cross national parks.[16] No one noticed. Muir's Sierra Club never
learned of it until after the fact. Phelan then applied to Secretary of the
Interior Ethan Hitchcock in 1903 for permission to dam Hetch Hetchy.
Hitchcock refused because, like Muir, he believed that the national parks
should be free of such utilitarian uses. Undaunted, Phelan tried again two
years later and, again, Hitchcock said no. By that time Pinchot had become embroiled in the issue in support of Phelan's application; but so
long as Hitchcock, long a thorn in Pinchot's philosophical side, remained
in office, little could be done. Hitchcock soon left office and the new secretary, James Garfield, was Pinchot's friend.

In the interim another event shaped the issue. On April 18, 1906, a
devastating earthquake shook San Francisco and fires consumed much
of the city. The catastrophe ignited public support for additional water
supplies (although salt water puts out fire as effectively as freshwater).
Private business interests and other politics also fanned the flames of public interest. In November, Pinchot encouraged the city to reapply to the
new secretary. The next summer, while the Sierra Club blithely camped
in Yosemite, Secretary Garfield held a hearing in San Francisco to consider the city's reapplication. Proponents presented the case for the dam.
No one from the Sierra Club or any other group was there to speak
against it. Review of the application was well under way before the opposition became aware.

Muir and Johnson spearheaded the opposition. Acting quickly, they
went straight to the president, whom they considered a friend and sympathetic supporter. President Roosevelt had a long personal involvement in conservation, and an equally long friendship with both lobby-

ists (he had accompanied Muir in 1903 on one of Muir's pilgrimages into Yosemite). But Roosevelt's philosophy was more compatible with that of Pinchot, whom he had rapidly promoted and anointed as his chief conservation advisor. Roosevelt's conservation stance, as revealed in a 1903 speech on the goal of forestry, was clear: "The object is not to preserve forests because they are beautiful—though that is good in itself—not to preserve them because they are refuges for the wild creatures of the wilderness—though that too is good in itself—but the primary object of forest policy . . . is the making of prosperous homes, is part of the traditional policy of homemaking in our country."[17]

Roosevelt saw value in leaving some areas untouched so long as other important utilitarian uses were not sacrificed. As much as Roosevelt admired Muir and believed in the value of wilderness, he ultimately prized it for its tangible usefulness to people, whether for resource extraction, as big game habitat, or as a wellspring of Turneresque American values.

When Muir and Johnson met with Roosevelt, the president was uninvolved in the issue. Reacting noncommittally to their plea, he sought more information by asking Garfield to study alternative ways to solve the city's problem. Muir wrote Roosevelt after the meeting, stating that, as soon as the facts became clear, 90 percent of the city residents would oppose it. He felt national opinion would be equally strong, citing the outcry over the mismanagement of Niagara Falls.

Garfield offered no viable alternatives, although the report's conclusion was probably shaped by powerful business and political undercurrents in San Francisco—the economic and political stakes in the Bay Area were high. Pinchot wrote the president, torn between the two sides, "I fully sympathize with the desire of Mr. Johnson and Mr. Muir to protect Yosemite National Park, but I believe that the highest possible use which could be made of it would be to supply pure water to a great center of population."[18] Roosevelt felt his hands were tied by the report and announced his support for the city's application unless public opinion was against it. He explained to Muir: "So far everyone that has appeared has been for it and I have been in the disagreeable position of seeming to interfere with the development of the State for the sake of keeping a valley, which apparently hardly anyone wanted to have kept, under national control. . . . (P.S.: How I do wish I were again with you camping out under those great sequoias or in the snow under the silver firs!)."[19]

Muir and Johnson feverishly rallied broader opposition to the dam in late 1907 and early 1908 to give Roosevelt a politically acceptable way out of the dilemma. Both sides bombarded the administration with expert opinion and engaged in a battle of words fought through the mass media. That spring Muir even suggested as a compromise that San Francisco use Lake Eleanor as a reservoir instead of Hetch Hetchy. The natural lake was only several miles north of Hetch Hetchy and within the park. Its use would cause less damage than damming the valley, and might better match the city's stated water supply criteria. It would also be less expensive.

But stronger, more subtle forces were at work. Hetch Hetchy was a pawn in a larger power game played with criteria hidden from the public—a struggle to control the city's electric power and, with it, much of San Francisco's future development. The Hetch Hetchy supporters hardly noticed Muir's offer. In May 1908, with Roosevelt's begrudging blessing, Garfield used the right-of-way loophole to approve the application, subject to congressional confirmation.

The challenge now facing Muir and the other opponents was daunting. They had to bombard Congress with sufficient public pressure to reject the proposal despite support from the president, Pinchot, key administrators, powerful business interests, and a seven-to-one polarity in support of the proposal from the citizens of San Francisco. Muir's boast that nine out of ten local residents opposed the dam was badly mistaken. The only way the preservationists could win the battle was to take the debate to a larger audience, a national audience free to decide on philosophical grounds instead of vested economic, political, and practical interests.

With the issue in the hands of Congress, the next hope to stop the application came during its review in committee. Each side unleashed its best speakers and strongest arguments. The preservationists mobilized a grassroots army of volunteers. Thousands of outdoor enthusiasts across the country wrote letters in opposition to the dam. A flurry of passionate articles by Muir and others appeared in leading newspapers and magazines nationwide. The tactic worked. As the congressional debate progressed, popular opinion swelled to stop the dam, weakening even Roosevelt's support. In his final State of the Union address on December 8, 1908, Roosevelt noted that Yellowstone and Yosemite "should be kept as a great national playground. In both, all wild things should be protected and the scenery kept wholly unmarred."[20]

The vote in the House Committee on Public Lands in 1909 was eight in favor, seven opposed. The application was approved, but by so slim a margin the resolution was pigeonholed and further action in the House postponed. When the corresponding Senate committee later failed to forward the application to the floor for consideration, the resolution was withdrawn until the next session of Congress and a new administration took office. Roosevelt was out. William Howard Taft was in.

Taft was less opinionated on such matters than his predecessor, although as a strict legal constructionist, he was sympathetic to Johnson's argument that approval of the dam would violate the intent of the law that established Yosemite as a national park in 1890. The new administration also had political reason to retaliate against Pinchot, who had harshly criticized the administration, in particular the new secretary of the interior, Richard A. Ballinger, on other conservation policies. The preservationists hoped to convince Ballinger to reject Pinchot's pet project and revoke Garfield's permit approval, killing the issue before congressional reconsideration.

In the fall of 1909 the preservationist's cause looked even brighter when Muir took President Taft on a tour of Yosemite in October, although the stout president would not camp with him. Later Muir escorted Ballinger around Hetch Hetchy. When Ballinger asked the city to show just cause why he should not revoke the permit, the opponent's case seemed won.

In response, the proponents of the dam, with their backs to the wall, stalled for time. The Taft administration, in no rush to make a decision on the contentious issue, granted the city extension after extension, postponing the hearing two and a half years, until November 1912. Over the thirty months the war of words to win public opinion continued unabated. During the delay Walter Fisher replaced Ballinger as secretary of the interior. He too was disinclined to move the permit forward; so the dam's proponents again stalled for the next political administration to take office on March 4, 1913.

Woodrow Wilson's road to the presidency had passed prominently through California, and San Francisco. In reward, he appointed Franklin K. Lane, a former San Francisco city attorney under Mayor Phelan and an active Wilson campaigner, as the new secretary of the interior. The Hetch Hetchy proponents' patience finally paid off. A new permit for the dam was quickly approved by Lane and passed to Congress for approval. House committee hearings began in late June. By the end of August, the bill reached the House floor and on September 3 was approved 183 to 43, with 203 representatives not voting. No one from a western state voted against it.

Muir, Johnson, and the preservationists mounted one last-ditch effort to stop the bill in the Senate, to no avail. With a supportive president and a contingent of southern Democrats aligned with their leader, the bill passed, unceremoniously, at 11:57 P.M., on December 6, 1913. The vote was 43 in favor, 25 opposed; 29 failed to vote. All the fighting, all the rhetoric, all the accusations, had been heard again and again. All the passion had been exhausted. Congress finally approved the damming of Hetch Hetchy after a decade-long battle that left both sides spent. On December 19, President Wilson, ignoring desperate pleas for a veto from the preservationists, signed the bill into law.

Who won the war? What was the greater good? The answer depended on one's landscape values. Muir concluded his book *The Yosemite* by stating,

> These temple destroyers, devotees of ravaging commercialism, seem to have a perfect contempt for Nature, and, instead of lifting their eyes to the God of the mountains, lift them to the Almighty Dollar.
>
> Dam Hetch Hetchy! As well dam for water-tanks the people's cathedrals and churches, for no holier temple has ever been consecrated by the heart of man.[21]

In contrast, Pinchot wrote: "As to my attitude regarding the proposed use of Hetch Hetchy by the city of San Francisco[,] . . . I am fully persuaded that . . . the injury . . . by substituting a lake for the present swampy floor of the valley . . . is altogether unimportant compared with the benefits to be derived from its use as a reservoir."[22]

Water began to flow to San Francisco from the Hetch Hetchy reservoir impounded by the expensive new O'Shaughnessy Dam after its completion in 1934 (at a cost of over $100 million, it was nearly twice the estimate). During the twenty years it took to design and construct the project, the city of Oakland tapped its own water source in the Sierra foothills at much less cost than San Francisco's new supply. Although the law approving Hetch Hetchy specified all power generated by the dam was to be sold by public agencies, every kilowatt was initially controlled by the Pacific Gas and Electric Company, then a private company. Although the conservationists claimed the reservoir's beautiful shimmering water would actually improve the valley's spectacular scenery, effectively countering preservationists claims it would despoil the view, Hetch Hetchy's shoreline initially proved to be a muddy collar of decaying vegetation as the water level fluctuated. And although the conservationists claimed the reservoir would provide a wonderful recreational resource, to protect the water quality no boating was ever allowed nor were roads and cabins built on the steep valley walls. The Hetch Hetchy reservoir stood in silent isolation, tucked safely away in Yosemite National Park.

In December 1914, about a year after the devastating decision, John Muir contracted pneumonia and died suddenly. Muir's principal protagonist, Gifford Pinchot, had been fired by President Taft for insubordination in 1910 as a result of the Ballinger-Pinchot affair, but his career had just begun as a public official and advocate for a Progressive conservation agenda in which natural resources were a means to help remedy social inequities. Ironically, in the 1920s Pinchot began to question some of the tenets he instilled in the Forest Service. He challenged the agency's allegiance to the timber industry and emphasis on production at the expense of other, less economic benefits like recreation, aesthetics, and spiritual refreshment. Forests, Pinchot came to believe, were not abstract commodities or a crop. Instead, they were "a living society of living beings, with many of the qualities of societies of men."[23] Yet he held onto the belief that forests are resources that could yield multiple benefits for all people, resources that could advance human well-being and social equity. His views of nature remained essentially the same ones that Muir had contested in the Hetch Hetchy debate. Before his death in 1947, Pinchot concluded *Breaking New Ground* with:

> I believe in free enterprise—freedom for the common man to think and work and rise to the limit of his ability, with due regard to the rights of others. But in what Concentrated Wealth means by free enterprise—freedom to use and abuse the common man—I do not believe. I object to the law of the jungle.
>
> The earth, I repeat, belongs of right to all its people, and not to a minority, insignificant in numbers but tremendous in wealth and power. The public good must come first.
>
> The rightful use and purpose of our natural resources is to make all the people strong and well, able and wise, well-taught, well-fed, well-clothed, well-housed, full of knowledge and initiative, with equal opportunity for all and special privilege for none.[24]

The true meaning of the Hetch Hetchy battle, as Nash suggested in *Wilderness and the American Mind,* was not about winners and losers, or about the differences between conservation and preservation philosophies. The battle's significance was that it ever occurred at all. The essence of Hetch Hetchy was expressed by Missouri Senator James A. Reed as the Senate debate neared the final vote. Bewildered by the proceedings, he wondered why "the Senate goes into profound debate, the country is thrown into a condition of hysteria" merely over a piece of wilderness.

And the intensity of the hysteria seemed to Reed to increase with one's distance from the place, so "when we get as far east as New England the opposition has become a frenzy." To Reed, it was all "much ado about little."[25] Perhaps fifty years earlier, the damming of Hetch Hetchy would not have raised an eyebrow, but American landscape values had softened over the 1800s. The traditional hostility had dissipated. By the start of the twentieth century we became concerned about the nature of that use. Environmental hostilities and stewardship values were called into question. The land for many Americans was no longer infinite, and its wealth of resources no longer inexhaustible. The question became one of determining the greater good among conflicting interests. The Hetch Hetchy debate as a national cause célèbre carried those questions into many American homes and would be the model for many future land use debates.

Still, most of the values remained little changed. By the beginning of the 1900s, most people continued to believe humans existed separate from and were external to the environment; therefore, human actions are unnatural in the sense of being outside nature. The role of the Indian, or contemporary society, in shaping the landscape was largely ignored. The cause-and-effect relation between people and the environment, the recognition of people as agents of landscape disturbance, was yet to be fully realized. Most Americans continued to see the environment as a commodity, whether for farming and resource extraction, or for scenic beauty and pleasure, or for spiritual enlightenment and refreshment. Land was property, something to be bought and sold, something to be used for human purposes. Most people also perceived the landscape more than ever in rational, scientific terms with little historical, emotional, or cultural connection. Pinchot's philosophy gave substance to those values. Marsh's message, Muir's message, and Powell's message had not yet been fully heard.

· Time Travel ·

Summer vacation began at night when I was young. The drive to the shore took fourteen hours on Route 40, the old National Road, and the Pennsylvania Turnpike. We loaded the blue-and-white Oldsmobile after Dad got home from work, and set off after sundown to avoid as much traffic as possible. Even in the darkness I sensed change in the landscape before sunrise.

I could smell it through the open car windows as livestock, then smokestacks, replaced row crops. I could feel it in the rise and fall of the roadway as we left flat farmland and entered the Appalachians. And I could see it in the sequence of small towns we passed through so intimately on the narrow two-lane road. Community spacing and

form differed; housing styles did too. Those were subtle differences
I sensed, without realizing their causes or interconnections. I didn't
know we had crossed from one of the nation's great physiographic
and sociocultural regions to another—from the corn belt of the Midwest to the nation's Atlantic origins.

Visually, it's a subtle change, not pronounced like that from the
Great Plains to the Front Range of the Rockies. It's an equally sharp
physical change, though. Over the stretch of several miles, the highway ascends slightly from vast till plain to gently rolling foothills
of the Appalachian Plateau. As motorists cross that narrow band,
they leave behind the landscape of national expansion; ahead is the
landscape of colonization. Behind is the interior of America; ahead
is coastal America. Behind is the West; ahead is the East. Political,
economic, social, and cultural differences arising two hundred years
ago still differentiate these landscapes today.

I've driven to the shore as an adult. Interstate 70 has replaced the
National Road. The drive is now much different: a far less intimate
way of experiencing landscape; far faster and more convenient. The
previous visceral sense of landscape change has been lost. Interstate
travel disconnects you from the surrounding landscape, fundamentally
changing the travel experience. Travel once meant experiencing landscape change. Interstate travel is time travel that simply transports you
from point A to point B with little sense of place in between. The drive
through any mountain range illustrates the point. The landscape the
local roads transect is often a world apart from that of the interstate
only moments away. Although a series of long grades, the interstate
is far flatter and faster than the steep, winding side roads. The interstates' standardized grades and curve radii flatten and smooth the
landscape much more than the local roads. To test this, exit in a
mountain range and drive any direction for five minutes. Try, for
example, the Stinking Creek exit on I-75 in Tennessee or the Idaho
Springs exit on I-70 in Colorado.

The same disconnection occurs in flat landscapes, too. I-70 lulls
the motorist nearly to sleep as it crosses the broad, unbroken expanse
of the Great Plains in Kansas. The small towns of Grinnell, Grainfield,
Park, and Quinter rush by in a blur of uninterrupted motion as the
highway bypasses each hamlet, smoothly bridges each stream, and
diverts each county road overhead.

That motion compels us forward. Speed and convenience overwhelm interest in the local landscape. We avoid "Daisy's Roadside

Diner" (good home cookin') or the "Blue Swan Motor Court" motel (kitchenettes, free TV). McDonald's and the Holiday Inn beckon. Each is about the same, regardless of its location. This sameness is one of their most attractive features—uniformity, familiarity. Those in Orlando look like those in Topeka or Timbuktu. So do the road signs, the service stations, and the convenience stores, even the mowed strip and planting style along the right-of-way. The interstate landscape is remarkably similar despite the vast variation in the surrounding landscape. It's symptomatic of our limited landscape literacy, and symbolic of our amazing mobility. We can now cross an entire continent, and reach every major city, without exiting the androgynous interstate corridor.

The same effect also occurs in cities. To see a city, to know it, don't ride the subway or drive the freeways. Walk it, ride the bus, or drive the streets instead. Each change in conveyance changes our sense of landscape. How we move about profoundly affects how we see landscape, more than the route, perhaps more than any physical factor. We mentally organize and structure our landscape—what we expect to see, what we look for, and what we experience—around the mode of travel. A simple demonstration makes the point. Take a short route you know well via one mode of transit, car for example, and record what you see the next time you travel it. Then retrace your path by foot, bicycle, or bus and record those observations. The diaries will differ dramatically.

Landscapes are designed in response. As the mode of travel has changed, the design of the surrounding landscape has changed to reflect the traveler's needs and interests. When people walked, or rode the trolley, storefronts with large display windows lined the sidewalks, often cluttered with wares, forming a narrow corridor for Main Street. Displays were complex, signs were small, and stores inconspicuous, appealing to passersby who had the opportunity to record the complicated message.

Advent of the car changed that message. Faster movement meant the message had to be simpler and seen from a greater distance. Decisions confronted the driver in rapid sequence: What is this store? Should I stop? Where do I park? The landscape responded. Buildings pulled back from the street to provide parking and easy access. Structures assumed a more descriptive character to communicate the interior business. Signs grew immense and bright to attract attention. The strip with its red, barnlike steak house was born, complete with split rail fence along the curb (no sidewalks since the landscape was no longer designed for pedestrians) and a small pasture of plastic cows. High-speed commerce tailored to the car dominated. Freeways advanced the style.

Advent of modern mass marketing, especially television, changed the landscape again. Provision for the car remained paramount—easy access, convenient parking—so buildings still stood in isolation. Advertising, though, meant the traveler could recognize a store by a logo learned from the media. Architectural styles, signage, and landscaping could assume a less literal form and communicate a new set of messages not limited to product.

What landscape experiences do we miss as we pass by in such a rush? What need have we for intimate travel on slower side roads? What do we lose on the interstates? We lose those landscape details I sensed as a child. We lose the richness and diversity of landscape. Travel becomes transportation through time rather than landscape. In return, perhaps we gain the opportunity to see large landscape patterns, regional differences that were once obscured by detail and the slow pace that made the differences imperceptible.

Mode of travel has shaped each era's perception of landscape. Imagine the differences between the pioneers' vision of landscape as they crossed the continent on foot, horseback, or wagon, versus that of the interstate or airplane traveler, differences arising not from the landscape itself (which has surely changed), but from the speed and convenience of conveyance. Will our "travel" by the Internet again change the landscape and how we perceive it? Time will tell.

CHAPTER 12

Widening the Circle

The great fault of all hitherto has been that they believed
themselves to have to deal only with the relations of man
to man. A man is ethical only when life, as such, is sacred
to him, that of plants and animals as that of fellow men.

Albert Schweitzer

The gradual softening and shifting of American landscape values in the nineteenth century continues, driven by many of the same forces evident in the 1800s. Today we enjoy greater prosperity and affluence than people of James Kilbourne's day, and technology affords us far more comfort than he enjoyed. These advances have given us an even greater sense of physical security from the environmental threats that plagued our predecessors in the last century, just as they enjoyed more security than those who preceded them. More people now recognize the inherent value of wilderness as little-touched lands rapidly disappear from the most remote recesses of the continent. We continue to romanticize nature and the way it was "conquered." We perpetuate landscape myths and illusions such as the Frontier Thesis and the suburban Arcadian myth. And we replay the debate between preservation and conservation begun by Hetch Hetchy.

But new forces over the last century also shape our landscape values and our landscape. Public input, politics, and the courts now shape them in a way much different than in the past. While politics has always affected our landscape, the struggle for influence during the nation's early history centered in the corridors of power. Jefferson and Hamilton struggled to shape the national landscape with policies and programs created in the capital. There it courted little public debate. Even most of Muir's and Pinchot's philosophical battles were waged in the private parlors of the American political, economic, and social elite. Only in Hetch Hetchy did the debate reach Main Street.

Today the environmental debate reaches much farther into everyday American life. Millions of people participate in decision-making for land use issues, be they local, regional, or national, in ways and numbers unimaginable a century ago. Recent growth in environmental groups illustrates the trend. The Sierra Club, Muir's small group of hikers and wilderness enthusiasts in San Francisco, and the Audubon Society, Grinnell's small bevy of birders in New York, each has more than a half million members nationwide.

The growth of environmental concern as a grassroots issue politicized the landscape. Public opinion yielded political pressure. Washington responded to public pressure by enacting new environmental laws. The landscape became a political hot potato at all levels of government, and government regulation of private property grew far more pervasive. The result is a proliferation of rules and regulations that affect everything from the height of fences in backyards to the location of suburban shopping malls to the quality of air and water.

The courts have greatly influenced the American landscape too. Unprecedented judicial activism has promoted the popularizing and politicizing of the landscape by giving substance to such public debates. The rules and regulations that dictate everything from that backyard fence to the levels of chemicals in the air we breathe or the water we drink are outgrowths of that activism and reflect changing judicial interpretation of the Constitution and congressional intent.

The emergence of these new forces has occurred within the context of a deepening devotion to science. We see the landscape more than ever through a thick scientific lens and a more sophisticated ecological understanding. Whether the tint in that lens is a cold, impersonal hue as it was for Pinchot, or a warm, reassuring one as it was for Muir, depends on one's attitude. For some, greater scientific understanding reinforces the traditional Judeo-Christian landscape values typified by James Kilbourne and the enlightened thinking of Thomas Jefferson. For others, it nurtures alternative philosophies. Some have abandoned the traditional anthropocentric valuation of nature in favor of a biocentric view. Others, like the romantics reacting against the Enlightenment, call for emotional and intuitive perspectives that rely on distinctly nonscientific ways of knowing—folklore, mysticism, and superstition—in opposition to the dominance of our rational, analytical perspective.

The first major marker of these twentieth-century changes was signified by the life and writing of Aldo Leopold, a disciple of Pinchot. To set that marker, Leopold ran counter to the prevailing landscape thinking and shed Pinchot's popular utilitarian philosophy. Once free of those constraints, he redefined conservationism and redirected American landscape values. For Leopold, as for John Muir, wilderness became a necessary salve for the human soul. And for Leopold, as for Muir, science sharpened his view of nature's wonder—warming and enriching, not demeaning and disconnecting.

Aldo Leopold's most lasting effect on American thought stems from his book, *A Sand County Almanac*. Published in 1949, a year after his death, Leopold's classic called for people to adopt a "land ethic" in which we perceive ourselves as integral parts of nature, a viewpoint that values nature from a biocentric orientation, much like the values typified by Jonathan Alder. Leopold's land ethic widened the sphere of elements to which we grant rights. Where before we granted rights primarily to people, Leopold proposed that rights be extended to include nature. This extension and the stewardship obligations it implies are central to the environmental debate today and mark this century's most notable milestones in the movement of our landscape values. "There are some who can live without wild things, and some who cannot," began Leopold. "These essays are the delights and dilemmas of one who cannot." He continued:

> Like winds and sunsets, wild things were taken for granted until progress began to do away with them. Now we face the question whether a still higher "standard of living" is worth its cost in things natural, wild, and free. For us of the minority, the opportunity to see geese is more important than television, and the chance to find a pasque-flower is a right as inalienable as free speech.
>
> These wild things, I admit, had little human value until mechanization assured us of a good breakfast, and until science disclosed the drama of where they come from and how they live. The whole conflict thus boils down to a question of degree. We of the minority see a law of diminishing returns in progress; our opponents do not.

Later he noted:

> Conservation is getting nowhere because it is incompatible with our Abrahamic concept of land. We abuse land because we regard it as a commodity belonging to us. When we see land as a commodity to which we belong, we may begin to use it with love and respect. There is no other way for land to survive the impact of mechanized man, nor for us to reap from it the esthetic harvest it is capable, under science, of contributing to culture.
>
> That land is a community is the basic concept of ecology, but that land is to be loved and respected is an extension of ethics. That land yields a cultural harvest is a fact long known, but latterly often forgotten.[1]

Leopold did not always hold such beliefs. He came to them only as a mature adult after discarding the carefree, consumptive attitudes of adolescence. Aldo Leopold was born in 1887 in Burlington, Iowa, a small town overlooking the Mississippi River. His parents were outdoor enthusiasts, his father a prosperous furniture maker. As a youngster, Aldo hunted quail and partridge in the forested western bluffs overlooking the river, and ducks in the marshy bottomlands along the opposite riverbank. Outdoor activities like camping, hunting, and birding dominated his childhood.

His formal education took him east to the Lawrenceville School in New Jersey (a private preparatory school), then to Yale, where he graduated in 1908. Traditional curricula, though, left him dissatisfied, as they had Muir and Powell. While book and laboratory learning were wonderful, they could not completely fulfill the yearnings of one who truly loved the land. Landscapes are best learned by firsthand experience.

Forestry, Leopold felt, came close, so he stayed at Yale to study in its forestry school established in 1900 by an endowment from the Pinchot family. As the nation's first forestry program, Yale supplied Pinchot's new Forest Service with most of its professional staff. Leopold was no exception. He completed the yearlong curriculum steeped in Pinchot's utilitarian conservation philosophy in 1909, then took a Forest Service position in the Southwest District. There he managed wildlife in the remote Arizona and New Mexico territories (wildlife management was not yet a separate profession and was then a part of forestry).

The territories were wild landscapes of desolate scrubland and desert, deep canyons, high plains, and forested mountains. They were the landscapes of Powell's *Arid Region* report. Some settlers had arrived by the early 1900s, and were ranching, recreating, and, where nature permitted, logging and mining, but statehood did not reach the region until 1912. Jefferson's hand joined Pinchot's in shaping the remote lands. Powell's hand might have had a small role by then as well.

At the time, Pinchot's guiding philosophy of scientific management for maximum sustained yield meant primarily consumptive forms of resource use, not passive forms, such as hiking and camping, or the maintenance of aesthetic resources such as scenery. Wildlife management meant the promotion of species with commercial or recreational value. Species that preyed on or competed with livestock or preferred game animals, or species that posed a nuisance, or species that simply got in the way were eliminated. Sound management dictated the elimination of such threats to key commercial "crops" harvested from the national forests. As Pinchot preached, "The purpose of Forestry, then, is to make the forest produce the largest possible amount of whatever crop or service will be most useful, and keep on producing it for generation after generation of men and trees."[2] By the early 1900s that philosophy governed management of the national forests and grazing lands in the public domain.[3] The "management" of the wolf and coyote illustrated the approach.

The melancholy howl of the wolf had long haunted the wilds of North America. Into the hearts and minds of many pioneers it also cast fear unlike the stirrings of any other creature. Perhaps its nocturnal cry most awakened the settlers' dread of the dark, unknown wilderness. Perhaps the native carnivores most symbolized the aliens' environmental antipathy. Wolves were the antithesis of the preferred pastoral world, savage predators that mercilessly preyed upon the tame, the defenseless, the innocent. They were truly wild creatures, demons in the dark, without hope of redemption. Settlers feverishly waged a holy war to eradicate them. Before the crusade roughly 2 million wolves inhabited North America. By 1900 only several thousand remained in the West. Twenty years later they were virtually extinct in the lower forty-eight, as were many other predators, including mountain lions and grizzly bears.

The war against the coyote followed suit, although this animal was perhaps more despised than feared. Waged mostly in this century, its soldiers came armed with new weaponry such as airplanes from which sharpshooters fired high-powered rifles. The arsenal included cyanide guns and baits such as sheep and deer carcasses laced with deadly poisons. The weaponry enabled the soldiers to slaughter the enemy by the millions.

Aldo Leopold, like most in the Forest Service and the general public, was a willing warrior in the crusades. When placed in charge of fish, game, and recreation for his district, he took up the crusade and eagerly waged war on the hated varmints. He also set about stocking streams for fishing and the range for hunting, as had the cavalry in Yellowstone. The concerted government action proved phenomenally successful. The deer population exploded on the Kaibab plateau in northern Arizona where Leopold worked. In 1906 the population was around four thousand. After less than twenty years of all-out warfare on predators, the number reached one hundred thousand—a hunter's paradise and proof positive that the prevailing land management philosophy was sound. Through scientific management, nature could yield more resources for human benefit.

The management effort proved too successful, though, for it soon began to yield other, unexpected, results. Over the winters of 1924–25 and 1925–26, nearly sixty thousand deer died of malnutrition. By 1939 the population was reduced by starvation and hunting to ten thousand. Without the natural check on the herd provided by predators culling the weak, the number of deer had skyrocketed far above the land's natural carrying capacity. The deer had simply eaten themselves out of house and home, overgrazing, overbrowsing, and "highlining" trees. The delicate balance between deer and environment was unwittingly upset, and it toppled like a house of cards. Scientific management was fine in principle, provided one possessed sufficient understanding of the landscape. Unfortunately the Kaibab disaster showed that such understanding was lacking.

Recognition of the shortcoming spread slowly. Leopold eventually learned the lesson, as did others in and out of the Forest Service. Yet old concepts and deep-rooted environmental conceits gave way begrudgingly. For hundreds of years we had segregated and classified wildlife into "good" and "bad" species based on *our* purposes and *our* values. Most considered the behavior of predators as contrary to human sensitivities or economic interests, even though for the animal it was no less natural than the behavior of other more endearing creatures like deer, rabbits, and robins. "Bad" animals were targeted for elimination with little understanding of their place in the landscape. Even today we struggle to overcome such conceits.

For Leopold and many others who grew up carrying a shotgun, schooled in the Pinchot philosophy, the Kaibab experience only suggested the need for more complete scientific understanding. They still believed in the principle of controlling predators. They still categorized wildlife as good and bad and granted few rights to animals. Some of these beliefs remained little changed when Leopold published *Game Management* in 1933. The landmark book pioneered many of our modern wildlife management practices and gave birth to the field as a profession separate from forestry. Yet underlying its more enlightened understanding of the role of predators, the book perceived wildlife as commodities to be managed from a Pinchotesque philosophy.

Another experience in the 1920s more fundamentally redirected his thinking, an experience much like John Muir's epiphany triggered by the white orchid he found in the dismal Canadian swamp. As Leopold later recalled,

> [Several colleagues and I] were eating lunch on a high rimrock, at the foot of which a turbulent river elbowed its way. We saw what we thought was a doe fording the torrent, her breast awash in white water. When she climbed the bank toward us and shook out her tail, we realized our error: it was a wolf. A half-dozen others, evidently grown pups, sprang from the willows and all joined in a welcoming mêlée of wagging tails and playful maulings. What was literally a pile of wolves writhed and tumbled in the center of an open flat at the foot of our rimrock.
>
> In those days we had never heard of passing up a chance to kill a wolf. In a second we were pumping lead into the pack, but with more excitement than accuracy: how to aim a steep downhill shot is always confusing. When our rifles were empty, the old wolf was down, and a pup was dragging a leg into impassable slide-rocks.
>
> We reached the old wolf in time to watch a fierce green fire dying in her eyes. I realized then, and have known ever since, that there was something new to me in those eyes—something known only to her and to the mountain. I was young then, and full of trigger-itch; I thought that because fewer wolves meant more deer, that no wolves would mean hunters' paradise. But after seeing the green fire die, I sensed that neither the wolf nor the mountain agreed with such a view.[4]

Leopold began to sense that the wolf had as much right to be as he. As the green fire was extinguished, Leopold began to abandon the traditional Kilbourne-like landscape values and adopt in their place values more compatible with those represented by Alder. The experience planted the seed of his land ethic. But additional nutrients would be needed for its germination; a few he found in the ongoing debate about wilderness.

By the second decade of the twentieth century some in the Forest Service began to question the degree of the agency's consumption-based management orientation. The dissenters advocated other nonmaterial uses for the national forests, such as for passive recreation and scenery, in part a reaction to the success enjoyed by the national parks as public vacation meccas attracting the newly mobile public eager for outdoor recreation in the scenic splendor of nature. Not to be outdone, the Forest Service simply expanded its notion of multiple use and began to promote the national forests for similar uses.

In 1917, with annual visitation to the national forests reaching nearly 3 million, the Service commissioned Frank Waugh, a landscape architect, to study the recreational-use potential of the lands it managed.

Waugh reported that the wild beauty of forests possessed a recreational value that should be given equal consideration with economic criteria in determining management practices. Soon some in the Forest Service, including Leopold, discussed the possibility of keeping sections of the forests "wild" for passive use. Those discussions led Leopold to Denver to explore the idea with Forest Service staff from that region. There he met Arthur Carhart, a young landscape architect who served as that district's "recreation engineer." Carhart's ideas had a crystallizing effect on Leopold's thoughts about wilderness preservation and helped to trigger the designation of the first wilderness areas in the national forests.

Carhart was hired by the Forest Service early in 1919. His first assignment was to hike into Trappers Lake, an isolated section of the White River National Forest in northwestern Colorado, to survey a road around the lake in preparation for a subdivision of about one hundred summer homes, two commercial sites, and a public marina he had been assigned to design. Such extensive recreational developments were becoming popular in the Forest Service as a means of competing with those in the national parks. Carhart spent the summer working at the remote lake, fifty miles from the nearest town or conveniences. Upon his return he had a different type of development in mind.

As an alternative to providing cabins that catered to the well-to-do and special interest groups, Carhart proposed that the area immediately around the lake remain roadless and free of structures. His plan called for a large public campground to be located at the lake's outlet, where it would be less intrusive in the pristine landscape, and where the access road could be screened by vegetation. With echoes of Olmsted, Langford, and Muir, Carhart proposed that the area be held in perpetuity in the public trust because of its irreplaceable beauty. His passionate report persuaded his supervisor to study the matter further; he suggested that Carhart meet with Leopold, the assistant forester at Gila National Forest in New Mexico. They met on December 6.

In a memorandum documenting the meeting, Carhart recommended that the Forest Service group undisturbed lands into four management categories reflecting the land's scenic quality and suitability for recreational development, with one prohibiting structures in superlative areas. Several days later, Carhart wrote Leopold to say, "There is a limit to the number of lands of shore line on the lakes; there is a limit to the number of lakes in existence; there is a limit to the mountainous areas of the world, and in each one of these situations there are portions of natural scenic beauty which are God made, and the beauties of which of a right should be the property of all people."[5]

Yet "wild," "preservation," and "undisturbed" meant different things to different people, just as it had in Muir's day. Pinchot's utilitarian conservation philosophy saturated Forest Service thinking, even that of wilderness preservation advocates like Carhart and Leopold. Where today we might define wilderness preservation as the total prohibition of human disturbance in a landscape, to Leopold, Carhart, and others around the 1920s, it meant instead the exclusion of the extensive tourist-oriented recreational development then becoming popular in the national parks and forests. Less extensive use, such as unpaved access roads and campgrounds, together with various resource uses, including grazing and hunting, might be permitted. Muir's concept of more restrictive wilderness preservation had not yet been accepted. Still, the concept of wilderness as a place of reduced resource use, a concept in which passive recreational use based on aesthetic considerations was coequal with more intensive uses, gained government support.

Carhart's alternative plan for Trappers Lake was approved about three years after its proposal, making the lake the nation's first designated wilderness area. A month later Carhart resigned from the Forest Service, frustrated with bureaucratic red tape and disillusioned with institutional obstinacy.

Leopold returned to his Albuquerque office after the Denver meeting invigorated and ready to push for wilderness preservation. In a *Journal of Forestry* article in 1921, he took Carhart's thinking one step farther and gave a more definitive shape to the emerging concept by defining wilderness as "a continuous stretch of country preserved in its natural state, open to lawful hunting and fishing, big enough to absorb a two weeks' pack trip, and kept devoid of roads, artificial trails, cottages, or other works of man."[6] Such landscapes, he realized, did not appeal to the majority who wanted more extensive recreational development; but for the minority who preferred more primitive recreation, nothing less sufficed. He closed the article calling for the undeveloped portion of Gila National Forest to be set aside as a wilderness preserve.

The article proved a powerful stimulus in the district. Several months later Leopold was instructed to survey the forest for such purposes and to formulate the governing policies. On June 3, 1924, 574,000 acres in Gila National Forest were designated primarily for wilderness recreation. The designation had the support of local sporting associations who favored the plan because it protected hunting areas from the intrusion of intensive recreational development. It also had the support of local cattle ranchers who would be permitted to graze their herds on range protected from tourist disruption. While the reserve had no paved roads or tourist hotels, cattle roamed freely, and game was managed (meaning the predators eliminated) to promote deer populations for hunters.

That same year, Leopold transferred within the Service from the Southwest District office to Madison, Wisconsin, where he became the associate director of the Forest Products Laboratory. Leopold's tenure in the Southwest District had not always been happy. He even left the Service in January 1918, disgruntled with its staunch Pinchotesque land management philosophy and the slow pace of change, but returned about eighteen months later as change appeared likely. Perhaps the new position with the lab provided a better opportunity to further his interests in wildlife conservation and wilderness preservation. Perhaps it was an escape from the same bureaucratic entanglements that frustrated Carhart. Perhaps it was simply time for a change.

He worked at the lab until 1928, when he left the Service to survey game populations in the upper Midwest, a study supported by the Sporting Arms and Ammunition Manufacturers' Institute. Despite his new sponsor's bias toward hunting, Leopold was objective in his work and remained influential in advancing concepts of game management and conservation by writing articles for professional journals and outdoor magazines and by public speaking. He advocated a more comprehensive, scientific approach to game management and conservation, and stressed the importance of understanding the life cycle of each species and the intricate interrelations among species in determining management plans. He urged greater understanding of habitat requirements and the effects of hunting and use of land surrounding the habitat. Yet he still advocated predator control and viewed wildlife as a product, a commodity, a crop. Those themes formed the heart of *Game Management*. With the book's release in 1933, he accepted a chair in game management created for him in the Agricultural Economics Department at the University of Wisconsin. There he continued his intellectual wanderings down the path of wilderness preservation until his death fifteen years later.

The Trappers Lake and Gila National Forest wilderness preserves set an important precedent for wilderness preservation on public land. They also fueled the wilderness debate begun by Muir and Pinchot, a debate that continued throughout the 1920s in the Forest Service, in environmental interest groups like the Sierra Club, and in the public sector. Wanderers like Leopold sought philosophical bases to justify wilderness preservation, and they looked for practical arguments to counter the common utilitarian concerns. As they built their case, wilderness preservation evolved. That evolution continues today, a principal product of and contributor to the definition of new stewardship values.

Leopold found justifications in many sources, including Thoreau, Muir, Darwin, Lyell, Liberty Hyde Bailey, Albert Schweitzer, and, for a while, Frederick Jackson Turner. They were the crucial nutrients that fertilized his latent land ethic. They enabled the seeds of thought that led to his eventual rejection of Pinchot's conservation philosophy to germinate.

Like Thoreau and Muir, Leopold acquired an aesthetic appreciation of wilderness and a reverence for pristine nature. From Darwin and Lyell, he found a way to dismantle the ideological barrier between humans and the environment erected by our Judeo-Christian heritage and, in its place, to perceive a sense of community, a oneness, among all living things. From Bailey, a horticulture professor at Cornell University and leading advocate of scientific-based agriculture and land management, he acquired a George Perkins Marsh belief that the earth was a divine gift; therefore, wasteful abuse and mismanagement of the land were not only economically foolish but also morally wrong.

From Schweitzer, Leopold acquired the belief that all life is equally deserving of respect, wolf and human alike. In 1915 the Alsace-German theologian, who had become a medical missionary, was on the deck of a small steamer slowly moving up the Ogowe River in French equatorial Africa when it passed quietly through a herd of hippos, leaving them undisturbed in the river behind. The boat's peaceful passing triggered a revelation. He wrote, "There flashed upon my mind, unforeseen and unsought, the phrase, 'Reverence for life.'" From this he constructed a philosophy in which every living thing possessed an equal will-to-live; hence human conduct must be governed by giving "to every will-to-live the same reverence for life that he gives to his own."[7]

Leopold found in Schweitzer the belief that rights should be extended beyond humans and granted to all living things, similar to many Native American and Eastern philosophies. "Ethics thus consists in this," Schweitzer said, "that I experience the necessity of practising the same reverence for life toward all will-to-live, as toward my own. Therein I have already the needed fundamental principle of morality. It is *good* to maintain and cherish life; it is *evil* to destroy and check life."[8]

These arguments helped Leopold justify wilderness preservation. Yet

other wilderness advocates, such as John Muir, had proposed similar justifications only to be overwhelmed by the utilitarian concerns of people
like Gifford Pinchot. The question of the greater good was a philosophical stumbling block even though American values had softened since the
Hetch Hetchy controversy a generation earlier. Leopold found the needed
counterargument in Frederick Jackson Turner's Frontier Thesis. There he
found a popular way to sell wilderness and to counter the persuasive
utilitarian arguments of conservationists. His implication was clear: preserve wilderness to preserve America and American values. He wrote,
"I am glad I shall never be young without wild country to be young in.
Of what avail are forty freedoms without a blank spot on the map?"[9]
While he later largely abandoned his Turneresque justification in favor
of more fundamental arguments for the extension of rights to all nature
as Schweitzer and others suggested, it nonetheless proved an effective
counter to Pinchot-like positions in the debates during the 1920s, 1930s,
and 1940s—and perhaps today.

Leopold felt the full effect of those critical nutrients on his farm.
There his land ethic took physical and philosophical root. By the 1930s
Leopold had bought a 120-acre abandoned farm in the sand counties of
central Wisconsin, coincidentally just thirty miles west of John Muir's
boyhood farm in Marquette County. There he spent weekends with his
family working to rehabilitate the land laid waste by the previous owner's greed and ignorance. A bootlegger who hated the farm, the owner
had stripped the land of its fertility, burned the farmhouse, and abandoned the property like an orphan, leaving it to the county as the only
compensation for delinquent taxes. Over the years Leopold guided the
land's recovery and, in so doing, found that the fertile ground nurtured
his incipient thoughts. He set about planting trees, as Elzéard Bouffier
had done, restoring the land bit by bit to health. The farm provided the
primary setting for *A Sand County Almanac*.

The book begins with a delightful collection of eloquent essays organized around a year in the life of that beautiful landscape. Wandering the
corridors of time, he showed fellow time-travelers clues about the landscape's legacy overlooked by all but the most sharp-sighted. His land
ethic, though implied in the sketches of seasonal change, fully blossomed
in the several essays that followed. In them he redefined conservation
from Pinchot's utilitarian interpretation of maximum consumption to a
less economic one in which limits and restrictions are placed on our environmental use. This revised interpretation is the most common con-
notation today: "Conservation is a state of harmony between men and

land. By land is meant all of the things on, over, or in the earth. Harmony with land is like harmony with a friend; you cannot cherish his right hand and chop off his left. . . . The land is one organism. Its parts, like our own parts, compete with each other and co-operate with each other. The competitions are as much a part of the inner workings as the co-operations. You can regulate them—cautiously—but not abolish them." Finally he made his call for a land ethic:

> There is as yet no ethic dealing with man's relation to land and to animals and plants which grow upon it. Land . . . is still property. The land-relation is still strictly economic, entailing privileges but not obligations.
>
> The extension of ethics to . . . [the land] is, if I read the evidence correctly, an evolutionary possibility and an ecological necessity. . . .
>
> All ethics so far evolved rest upon a single premise: that the individual is a member of a community of interdependent parts. His instincts prompt him to compete for his place in the community, but his ethics prompt him also to co-operate (perhaps in order that there may be a place to compete for).
>
> The land ethic simply enlarges the boundaries of the community to include soils, waters, plants, and animals, or collectively: the land. . . .
>
> In short, a land ethic changes the role of *Homo sapiens* from conqueror of the land-community to plain member and citizen of it. It implies respect for his fellow-members, and also respect for the community as such.[10]

Aldo Leopold died fighting a brushfire on a neighbor's farm in 1948. The manuscript for *A Sand County Almanac* was his philosophical last will and testament, and quickly became the conservation movement's bible, a movement he was instrumental in redirecting to one less consumption-based. Like Marsh, Powell, and Muir before him, Leopold's message was too far forward in advance of popular opinion and thought. While his land ethic philosophically widened the circle in environmental ethics, most Americans in the early post–World War II years adhered to Pinchot's concept of conservation. Recently, though, Leopold's ethic was the focus of a pivotal environmental debate. That debate, the most crucial since Muir and Pinchot waged philosophical war over Hetch Hetchy, was litigated before the Supreme Court like many landscape issues today, not argued in the court of public opinion or the corridors of Congress.

CHAPTER 13

Litigating the Land

There is hardly a political question in the United States which
does not sooner or later turn into a judicial one.

Alexis de Tocqueville, *Democracy in America*

Leopold's assertion that "there is as yet no ethic dealing with man's
relation to land and to animals and plants which grow upon it" was
probably overstated. Schweitzer proposed such an ethic, as had others
throughout Western history. Leopold was correct in suggesting that in
the mid-1900s American society as a whole had yet to recognize such an
ethic either philosophically or practically. Although small segments had,
and some small precedents had been established, overall rights remained
extended little beyond people.

In *The Rights of Nature*, Roderick Nash outlined the sphere of entities granted rights in America, at least in principle if not practice, and he
noted that the sphere has gradually expanded. Those originally granted
full rights, the small inner circle, included mostly white, male landowners. More and more groups have been embraced since, including slaves
in 1863 (Emancipation Proclamation), women in 1920 (Nineteenth
Amendment), Native Americans in 1924 (Indian Citizenship Act), and
African Americans and other minorities in 1957 (Civil Rights Act).
Rights were extended even to nonhuman entities with the 1973 Endangered Species Act.

As a society, we have begun to wrestle with Leopold's land ethic and
grant limited rights to nature, rights that impose on us obligations toward nature which we exercise through self-restraint. This expansion,
Nash suggested, is a by-product of our sociocultural development. He
and others speculate we will extend them even farther in the future to

include additional animals, plants, nonliving elements, and eventually entire biotic communities. At the moment, we dabble at the fringes of such a momentous legal and philosophical step as if testing the water with our collective toe.

That extension, though, was recently tested in a landmark court case, a battle whose outcome might have established for the first time the legal rights of nature as proposed by Leopold. The battleground lay in the same region over which so many of Muir's battles were fought—the Sierra Nevada, just south of his beloved high temple, Yosemite—and concerned the use of lands for which he and Pinchot previously struggled: our national parks and forests. This battle, fought over the legal doctrine of "standing," began quietly in 1965 and ended abruptly in April 1972. Such a frontal assault has not happened since; instead, small skirmishes have eaten away at the margins, slowly eroding the prevailing separation between humans and environment as our landscape values continue to change.

The spectacular Mineral King Valley had been a part of Sequoia National Forest since 1926, and it had been designated a National Game Refuge by special act of Congress that same year. Although once extensively mined, by the early 1960s the valley was used primarily for passive recreation such as hiking and camping. Because of its remoteness and pristine condition, save for the remaining scars from the mining, the Forest Service had been interested since the 1940s in expanding the valley's recreational use. With mounting public demand for ski facilities in the region, the Service in 1965 issued a prospectus soliciting bids from private developers for the construction and operation of a ski resort and summer recreation facilities. Six bids were received, and an offer made by Walt Disney Enterprises was accepted. Disney received a three-year permit to begin planning and design for the Mineral King Resort. The Forest Service approved the plan in January 1969.

The plan called for construction of a full-service resort that would include motels, restaurants, swimming pools, ski trails, lift lines, a cog-assisted railway, parking lots, and support facilities sufficient to accommodate fourteen thousand visitors daily. The resort, to be located on eighty acres of the valley floor, was to be granted a thirty-year use permit by the Forest Service. The ski facilities on the mountain would be operated under a revocable special use permit. To provide access to the beautiful valley, the State of California would construct a twenty-mile-long highway across a portion of Sequoia National Park. A high-voltage power line to the resort would also be routed across the park.

To construct the essential arteries, a special permit similar to the type that triggered the Hetch Hetchy controversy was required from the secretary of the interior.

The Sierra Club followed the issue with great concern and provided input when possible in opposition to the Forest Service's plan.[1] When the plan proceeded despite its objections, the club filed suit in the U.S. District Court for Northern California in June 1969. In the suit the club sought a declaratory judgment ruling that the project violated various federal laws and regulations; consequently, it requested temporary and permanent injunctions restraining federal officials from granting the required permits. Among many concerns, the Sierra Club claimed that construction of the highway and power line across Sequoia National Park would violate the purpose of the park set forth in its creation. While the club's complaint raised fundamental procedural questions, nuts-and-bolts-type issues, and questions similar to those debated seventy-five years earlier by Muir and Pinchot, the heart of the case was a critical legal issue little known beyond the bar: the issue of standing.

Standing in this context refers to the legal right to bring a lawsuit, to come before the bar and be heard in a court of law. Historically, one was granted standing provided that certain conditions be met. Principal among them, for our purposes, was the requirement that the plaintiff be the injured party. Only in certain instances could a plaintiff who was not the aggrieved party bring a case on behalf of another unable to speak for himself or herself: minors, for example, could be represented by their parents or legal guardians, and a person could represent a corporation or ship in a court case. The standard test was whether the plaintiff had suffered "injury in fact."

This phrase was historically interpreted by the courts to mean direct bodily or economic injury. In the mid-1950s, though, the Supreme Court began, at first indirectly, to ease the test for standing by expanding the concept of injury. That expansion, particularly as applied to environmental issues, would reach a peak with Court decisions handed down in the early 1970s. During this time, the courts were beginning to recognize that softer injuries, such as those offensive to one's senses or aesthetic values, could be sufficient to constitute legitimate legal injury; thus the courts were, in a way, extending rights to new groups of people who suffered harm that previously had not been judged worthy of legal protection. That extension opened the door for new regulations, including aesthetic-related architectural standards and many environmental standards.

In the Mineral King case, known as *Sierra Club v. Morton,* the Sierra
Club claimed "a special interest in the conservation and sound maintenance of the national parks, game refuges, and forests of the country"
as the basis of its standing.[2] By not claiming that any specific member
of the club would be injured by the action, the Sierra Club hoped to establish a legal precedent. They wanted the court to recognize that third
parties could challenge federal actions that disturbed the environment
in a manner the plaintiff found objectionable, even though no one suffered physical or economic injury.

After two days of hearings, the district court upheld the Sierra Club's
petition. The respondents (the government) appealed. The Ninth District Court of Appeals reversed the lower court's decision, and the plaintiffs appealed the case to the Supreme Court. In reversing the district
court, the Court of Appeals wrote that the plaintiffs made "no allega-
tion in the complaint that members of the Sierra Club would be affected by the actions of [the respondents] other than the fact that the
actions are personally displeasing and distasteful to them." The court
concluded, "We do not believe such club concern without a showing of
more direct interest can constitute standing in the legal sense sufficient
to challenge the exercise of responsibilities on behalf of all the citizens
by two cabinet level officials of the government acting under Congressional and Constitutional authority."[3]

Although the Sierra Club petition did not raise the idea, the case
could be interpreted from another perspective, that of extending rights
to nature itself, as Leopold proposed. Christopher Stone, a law professor at the University of Southern California, explored that alternative interpretation in an influential article for the *Southern California Law Review.* The article was later expanded into the book *Should Trees Have
Standing? Toward Legal Rights for Natural Objects.* From that alternative interpretation, the case could have been an attempt to have the
courts recognize what society had yet to accept: that nature possessed
the right to exist in and of itself, separate from human valuation, a right
that could be violated or injured sufficiently to warrant legal protection.

If the court agreed, it could grant standing to the Sierra Club, which
had a legitimate interest in the protection of nature, permitting it to
speak on behalf of the Sierra, which, like a minor, corporation, or ship,
was unable to speak for itself. Hence, from the alternative interpretation the club sought standing to bring a suit on behalf of the Mineral
King Valley and Sequoia National Park because those magnificent landscapes would be injured by the Disney development.

If the court accepted the alternative, or even if it accepted the Sierra Club's stated position, the ramifications would be significant practically and philosophically. If either were upheld, American landscape values and land management practices might be affected. After all, groups like the Sierra Club or Audubon Society could routinely challenge proposals that called for landscape disturbance on the grounds that their membership would suffer a generic injury or, alternatively, because nature itself would be injured. How would the court weigh the injury in balance to the benefit? And to whom would the injury and benefit be measured: people or nature?

The Mineral King case was America's most significant environmental debate since Hetch Hetchy. Victory for the Sierra Club would hasten change in our landscape values and mark another milestone in the shift away from our historical view of nature primarily as property. Because those implications seem so profound, it's difficult today to believe the Court would take such a step. Why did the Sierra Club try it? And why had the courts been easing the test for standing? The explanation for both actions involved the gradual softening of American landscape values and the broadening of sociocultural values. By the late 1960s those changes indicated the nation might be at the brink of such recognition.

Remember, the post–World War II period in the 1950s and 1960s was a time of tremendous optimism about the future and the role that science and technology would play in shaping it; however, that optimism was tempered by a growing ecological awareness stimulated by books like *A Sand County Almanac* and *Silent Spring*. It was a period of phenomenal growth in grassroots environmental activism, as indicated by the rapid growth in advocacy groups like the Sierra Club. And it was a period of tremendous political activism, as was evident from popular causes like the Civil Rights movement and the anti–Vietnam War movement. Political administrations like those of Presidents John F. Kennedy and Lyndon Johnson reflected and contributed to that milieu, as did Congress. Administrative policy and federal legislation were affected too.

None of this was lost on the courts. Judges, particularly at the appellate level, at times have significant latitude in their interpretation of the Constitution, legislation, and legal precedent. As people, they can choose to let their interpretations reflect their personal beliefs as well as the weight of popular opinion—in our society the law is a living element that reflects our changing attitudes and values; it is not static and stayed. Consequently, during the 1950s, 1960s, and early 1970s, many of the most influential courts, including the Supreme Court, gradually modified the judicial environment. Their reasons varied—judicial philosophy, social responsibility, political theory, legislative precedent, environmental ethics, even popular opinion.

One important modification can be seen in the context of a conceptual model that envisions many forms of judicial argument, including environmental issues, as occurring within an adversarial context; as such, the courts function as the mediator, or referee, to fairly enforce the governing rules defined by law and administrative regulations. The easing of the test for standing assumes special significance in this context.

As the environmental movement evolved in the post–World War II era, it often became a confrontation between two combatants, much like the Hetch Hetchy controversy had been. One combatant was the party proposing some type of environmental disturbance—the proponent. This party was typically the government or a large corporation or private land developer that had received government approval. The other combatant was the party seeking to prevent the action—the opponent. This was often a local nonprofit environmental advocacy group or someone seeking to protect personal interests. Eventually many of these battles were waged in the courts, with the opponent challenging the proponent mostly on authority of the Administrative Procedures Act (APA) or, later, an environmental law like the National Environmental Policy Act.

The APA is an old law that entitles a citizen to challenge a government action that adversely affects him or her.[4] Often, in order to prevail, the plaintiff must prove that the government action was arbitrary or capricious, a tough standard of proof. If successful, the plaintiff might receive an injunction—again depending on the situation—halting the government action until the government reconsidered; but the court would not reverse the action per se, it would only require the action to be reconsidered.

To bring a suit based on the APA, the plaintiff had to establish standing by way of the traditional "injury in fact" test. Once the case began, the plaintiff waged legal war with the government in a David-versus-Goliath confrontation. The problem that some began to detect in this method of resolving environmental disputes was the disparity between the two combatants. While the opponent often was composed of few people who possessed limited financial and technical resources, the proponent was large, well endowed, and driven by an institutional bias toward development. The legal safeguards and means for redress available to the opponents were perceived by some as limited in comparison to the institutional incentives and protections given the proponents.

In the sociopolitical climate of the day, some perceived the playing field to be tilted unfairly in favor of development and environmental disturbance over environmental quality and protection. To remedy this, at least implicitly, the Supreme Court led a gradual easing in the test for standing by broadening the definition of "injury." An early milestone in the changing judicial environment occurred in the landmark 1954 Supreme Court decision for the *Berman v. Parker* case.[5] Although the case involved a specific urban renewal project, it had much broader significance for general land use and environmental issues. For the first time the Court recognized aesthetic and other intangible concerns as falling within the sphere of legitimate government interest. Justice William O. Douglas, writing for the majority, stated, "The concept of the public welfare is broad and inclusive. . . . The values it represents are spiritual as well as physical, aesthetic as well as monetary."[6] In expanding the sphere of government interest, the decision indirectly loosened the test for standing, since people could now, by definition, seek legal redress for suffering injury to the newly recognized concerns.

By 1970 the Court handed down three decisions in rapid succession that further redefined the boundaries of standing. The decisions recognized that the test was "injury in fact, economic or otherwise," explicitly adding aesthetic and less direct or nonphysical harm to the test.[7] One effect was the opening of the courtroom door to more people to bring suits against proposed environmental actions. The expansion balanced the playing field within our adversarial framework by enabling many Davids to confront a Goliath. The goal was to find the "truth" or the "public good."

The Court did not act alone. Its action was in concert with Congress (which passed new environmental laws) and the administration (which promulgated new environmental regulations). When the Court permitted more people to bring suits via an eased test for standing, it added more combatants on the side of environmental protection. When Congress enacted new environmental legislation in the 1960s and 1970s, it created stricter standards for environmental quality while it signaled the judiciary that a better balance between development and environmental quality was sought. That signal encouraged the courts to take a hard line on environmental protection when the opportunity arose. And when the administrations promulgated new environmental rules and regulations, they established strict standards and procedures for the courts to uphold.

The rash of innovative land use and pollution control laws passed by Congress revolutionized the federal role in environmental management, especially by establishing national standards for pollutants. Among the many early laws were ones addressing water resources (1965), air quality (1967), wild and scenic rivers (1968), and solid waste disposal (1970); more followed in the 1970s.

Perhaps the most significant Act was the National Environmental Policy Act of 1969 (NEPA).[8] Signed into law by President Richard Nixon on January 1, 1970, NEPA significantly changed government's environmental decision-making, as well as decision-making in other programmatic areas. In so doing, it served as a model for similar legislation worldwide.

NEPA was very much a product of the times. Many legislators felt the political pressure resulting from the period's frantic environmental activism was so intense that to vote against environmental legislation was tantamount to political suicide. The resulting legislation was often piecemeal, a knee-jerk reaction to the latest or loudest environmental issue, since political and legislative agendas were often set by others. Congress was reacting, not anticipating. The situation was complicated by the institution's inexperience in environmental matters, because historically the body gave little informed attention to environmental issues, and most legislators were lawyers, not scientists or environmental experts. In response, Congress sought a unified statement of national environmental policy as a means of providing guidance and coherence to this rapidly growing concern. NEPA was the result, yet its eventual effects went well beyond those anticipated by Congress.

The roots of NEPA can be traced to legislative proposals made almost a decade earlier, though these proposals had received little serious consideration because the political climate was calm and cool to the issue—the hot winds of public environmental opinion were just beginning to blow. By 1968 the Washington climate had changed. In June a House subcommittee reported that fragmented federal decision-making and environmental neglect had contributed to the nation's environmental problems. The Senate followed suit. Its Committee on Interior and Insular Affairs, chaired by Senator Henry Jackson, sponsored a joint Senate-House colloquium to review national environmental policy, which resulted in an influential report, *Congressional White Paper on a National Policy for the Environment*.

The immediate legislative history of NEPA began in early 1969 with two draft bills, one introduced in the House (H.R. 6750) on February 17, and one introduced in the Senate (S. 1075) the next day. Both were modest bills, and neither committed the government to much beyond the conduct of environmental research and a philosophical statement about the importance of the environment. The unambitious nature of the bills may have been purposefully designed to avoid premature congressional policy arguments and unnecessary turf squabbles over jurisdiction between competing committees. The watery bills were promptly directed to the sponsors' desired committees for hearings and development.

Hearings in Jackson's Senate committee began amid the blustery political winds blowing about Washington that spring. Scores of scientists detailed the extent of the nation's environmental deterioration. Public policy and resource policy experts outlined the federal government's contribution to the deterioration by its actions or indifference; many advocated a unified policy that would guide future federal action. By June the Jackson bill included a more elaborate philosophical statement of national policy and a provision that called for the issuance of a "finding" for proposed agency actions. The finding, to be issued by the responsible federal official, would describe the environmental impacts anticipated from a proposed action. The provision committed the government for the first time to a change in administrative procedure—it was an action-forcing provision. Other considerations in the bill recognized the need for greater environmental understanding via government-sponsored research, and the need for Congress and the public to be better informed on the state of the environment. The bill was passed unanimously by the committee and forwarded to the floor. The full Senate passed the bill by unanimous consent on July 10.

The companion bill in the House also moved rapidly through committee, with seven days of hearings held during the hot summer. On September 23 the House approved the bill, little changed from the original, by a resounding 372 to 15 vote. But one amendment stated that "nothing in this Act shall increase, decrease, or change any responsibility of any Federal official or agency," apparently negating the Senate's action-forcing provision.[9]

When the two bills reached a joint House-Senate conference committee to resolve the differences, the political maneuvering and haggling intensified. Political horse-trading and compromises ensued. One compromise rephrased the requirement for a "finding" to a "detailed statement." Another struck a balance between Senator Jackson's desire that the action-forcing provision apply to all agencies and Congressman Wayne

Aspinall's insistence that the act leave all agency mandates unchanged. The result was a requirement that all agencies comply with the action-forcing provision "to the fullest extent possible." As the final bill was hammered out, the rapidly approaching conclusion of the term added urgency to complete the work. On December 17 the committee reached agreement, and several days later both chambers agreed to the conference report.[10] So it was done. Perhaps in a bit of a rush. Perhaps with less thoughtful consideration and debate than one might wish. But President Nixon inaugurated the new decade with the new law.

In its five short pages NEPA put forth three main provisions. The law begins with the much-debated philosophical statement of national environmental policy, a "motherhood-and-apple-pie" statement about our newfound national concern for and love of the environment: "The Congress, recognizing the profound impact of man's activity on the interrelations of all components of the natural environment . . . declares that it is the continuing policy of the Federal government . . . to create and maintain conditions under which man and nature can exist in productive harmony, and fulfill the social, economic, and other requirements of present and future generations of Americans."[11]

Truly amazing language for a nation with the landscape legacy we share. Apparently our landscape values had changed significantly since Kilbourne's day. The law committed the government to improve and coordinate all federal plans, functions, programs, and resources so that the nation might (1) fulfill the responsibilities of each generation as trustee of the environment for succeeding generations; (2) assure safe, healthful, productive, and aesthetically and culturally pleasing surroundings; (3) attain the widest range of beneficial uses of the environment without degradation, risk to health or safety, or other unintended consequences; (4) preserve historic, cultural, and natural aspects of our national heritage, and maintain an environment that supports diversity of individual choice; (5) achieve a balance between population and resource use that permits high standards of living and a wide sharing of life's amenities; and (6) enhance the quality of renewable resources and approach the maximum attainable recycling of depletable resources.[12]

NEPA's second purpose was to create a body, called the Council of Environmental Quality (CEQ), to assist the government in environmental affairs and oversee the law's action-forcing provision. Housed in the Executive Office of the president, CEQ was to consist of three people appointed by the president, with one acting as chair. Though functioning as the government's environmental watchdog, CEQ was not granted any enforcement power. Beyond its oversight responsibility, CEQ would sponsor environmental research and serve as a fact finder and source of policy recommendations on behalf of the president and Congress. These it would accomplish mainly by assisting the president in the preparation of an environmental quality report that the chief executive would present annually to Congress.

As a final stipulation, the law included a two-part action-forcing provision. The first applied to all government activities. It stated that the policies, regulations, and public laws of the United States would be administered in accordance with the policies set forth in the act. In addition, all agencies of the federal government would take an interdisciplinary approach to ensure that the natural and social sciences and the environmental design arts were integrated into planning and decision-making that might affect the environment. The agencies would also identify and develop methods with CEQ that would ensure that unquantified environmental amenities and values were given consideration in decision-making along with economic and technical ones.

The second part applied only to those federal actions that would significantly affect the quality of the environment. For them, a detailed statement on the anticipated environmental impacts would be made by the responsible official. Before making that statement, the official would obtain the comments of any federal agency that had jurisdiction by law or special expertise related to the proposed action and its consequences. Copies of the statement together with the views of interested federal, state, and local agencies would be made available to the president and the public.[13]

That was it, short and sweet. What did Congress really intend NEPA to accomplish? Did Congress really envision the profound change in national environmental policy it mandated by the law's flowery language? As an indication of Congress's commitment to NEPA, consider how much funding it allocated for implementation, including funding for research: three hundred thousand dollars for the remainder of 1970, then seven hundred thousand dollars for fiscal 1971, then $1 million each year thereafter.[14]

Many managers greeted the new law and its vague requirements with reticence. NEPA appeared to stand between agency proposals and implementation. Managers did not welcome the intrusion on their authority, nor the additional bureaucratic hurdle, nor the second-guessing of other federal, state, and local agencies, and the public. After all, in the past they often had to justify proposals to Congress to get funding.

Some felt that running that twisted gauntlet was sufficient scrutiny to
protect public interest and environmental quality. Many interpreted the
"detailed statement" and other language in the law loosely, and prepared
only brief explanations of proposed projects. To many, the document
was perceived as a perfunctory exercise to justify what was already decided. That was where the courts intervened.

Triggered by a progression of a hundred court cases in the 1970s
brought under NEPA, and the APA, the courts forced the government
into a behavior not fully envisioned by Congress when it created the
law. As plaintiffs gained legal access to the bar using the new law and
tests for standing, the courts held the government agencies to a rigorous
interpretation of NEPA's intent and forced them to prepare more detailed statements, which came to be known as environmental impact
statements (EISs). The courts based their stance on the law's flowery language in the philosophical statement of national policy, and on the legislative mandate suggested by the new environmental laws passed by a
Congress ruffled by the winds of public opinion now blowing around the
country. Some justices, like William O. Douglas, may have been motivated also by personal judicial and environmental philosophies.

People challenged those brief statements on the grounds that they
violated NEPA's requirement that the EIS be a "detailed statement" that
documented impacts "to the fullest extent possible." Who was to say
what constituted "detailed" or "the fullest extent possible"? The courts.
And time after time they agreed with the plaintiffs, forcing the respondents to correct the substantive deficiencies in the document by rewrit-
ing the EIS until the court ruled it was satisfactory. EISs were also challenged on their timing, as well as their content or completeness. Plaintiffs
claimed perfunctory EISs violated the procedure specified by the law because the agency had already decided to proceed with the project before
the EIS was complete; consequently, the timing of the decision contradicted NEPA's intent that the EIS precede the decision and be an integral part of decision-making.

Among the law's vague phrases interpreted by the courts to shape the
EIS were: "to the fullest extent possible," "utilize a systematic, interdisciplinary approach," "include on every recommendation or report on
proposals for legislation and other major Federal actions significantly
affecting the quality of the human environment," "ensure that presently
unquantified environmental amenities and values may be given appropriate consideration in the decisionmaking along with economic and
technical considerations," "a detailed statement," and "copies of such

statement and the comments and views of the appropriate Federal, State, and local agencies . . . shall be made available to . . . the public. . . . "

What did Congress mean by such statements? How was an agency to respond? Given the statement of national environmental policy that began the law, the law's legislative history, and the sociopolitical climate, the courts took a strict interpretation and held the agencies' feet to the fire that Congress inadvertently ignited. The courts seized the opportunity left by the law's vague language to define the details of the EIS, and they interceded to fill the procedural and substantive vacuums left by Congress. The courts gave body to the EIS skeleton. In so doing they opened up much of federal decision-making to greater public scrutiny.

Yet NEPA said nothing about the decision the government could make after completion of the EIS process. An EIS would not, for example, require the Bureau of Reclamation *not* to dam the Grand Canyon, even if the environmental impacts were significant. NEPA only required the director to consider the consequences. Nor could the court require that a decision be made in light of data in the EIS. It could only grant injunctive relief to the plaintiffs to prohibit the agency from proceeding with the project if the EIS had substantive or procedural errors, in which case the court could only force the government to go back and make the necessary corrections.

Instead, NEPA's profound effect on decision-making occurred indirectly, when it required full public disclosure during the EIS process. Public scrutiny and easier access to the courts created uncertainty in project planning as a result of the potential for a lawsuit. They also created the threat of political embarrassment if an agency's final decision contradicted data in the EIS. Those ominous threats can be just as effective as a decision requirement or the strict standards set by environmental laws and regulations. Uncertainty is often the Achilles's heel in development because it characterizes as too speculative otherwise reliable long-term planning and financial forecasting. Hence, people planning major projects often prefer an alternative course of action that avoids potential lawsuits and satisfies environmental regulations, even if the costs are higher. In addition, uncertainty is often less affected by the changing winds of Washington politics than by specific standards, which can be circumvented by loopholes or diluted by lax enforcement. Uncertainty in project planning and the threat of political embarrassment are powerful motivations for change in administrative and corporate behavior.

While the courts could shape NEPA's EIS requirements, they could not enforce the other components of the law as directly or strictly. The other action-forcing requirements lacked the clear basis for judicial action inherent in the EIS. The courts could only require that procedures be followed, and that substantive concerns be a part of the decision-making. They could not dictate the outcome of that consideration. Nor did the courts recognize any substantive basis to NEPA's statement of national environmental policy. Although the statement was noble, it was merely a goal, and, as such, the courts were incapable of compelling administrative or legislative compliance. Except for the EIS, NEPA had few teeth.[15]

With all this swirling about, the Sierra Club initiated the Mineral King suit in the summer of 1969. Perhaps the time was right after all to attack head-on the questions posed by Leopold's land ethic. Perhaps it was not so crazy to think the Supreme Court might take such a step. As the case proceeded in the early 1970s, Congress continued to pass environmental protection laws, such as NEPA, which the courts interpreted strictly. The boom in environmental activism, marked by the first Earth Day on April 22, 1970, coupled with NEPA's statement of environmental policy and the proliferation of environmental laws, produced an air of optimism, even anticipation. When given an opportunity to amend its petition to claim injury to its members, the Sierra Club held steadfast.

Finally on April 19, 1972, two years after NEPA was implemented, the Supreme Court made its decision. In the end, while sympathetic to the club's concern for the park's impending injury due to the Disney development, the Court rejected the club's request for standing since it did not claim injury to its members. By a vote of four to three (Justices Lewis Powell and William Rehnquist, newly appointed, did not participate), the Court narrowly rejected the Sierra Club's argument. Although the majority decision never mentioned the expansion of rights to nature nor couched the central legal issue as such, preferring instead to keep it more narrowly defined as a classic question of standing, Justice Douglas's dissenting minority opinion took up the cause and cited Professor Stone's law review article (Justices Harry Blackmun and William Brennan dissented as well, although not entirely for the same reasons).

Douglas's dissent is in many circles more famous than the majority opinion. With compelling eloquence he advocated the alternative interpretation of the case's central legal issue and echoed Leopold's call for a land ethic. He wrote, "The critical question of 'standing' would be simplified and also put neatly in focus if we fashioned a federal rule that allowed environmental issues to be litigated . . . in the name of the inanimate object about to be dispoiled [sic], defaced, or invaded by roads and bulldozers and where injury is the subject of public outrage. . . . This suit would therefore be more properly labeled as *Mineral King v. Morton*." He continued:

> Those who hike [Mineral King], fish it, hunt it, camp in it, or frequent it, or visit it merely to sit in solitude and wonderment are legitimate spokesmen for it, whether they may be few or many. Those who have that intimate relation with the inanimate object about to be injured, polluted, or otherwise despoiled are its legitimate spokesmen. . . .
>
> The voice of the inanimate object, therefore, should not be stilled. That does not mean that the judiciary takes over the managerial functions from the federal agency. It merely means that before these priceless bits of Americana (such as a valley, an alpine meadow, a river, or a lake) are forever lost or are so transformed as to be reduced to the eventual rubble of our urban environment, the voice of the existing beneficiaries of these environmental wonders should be heard.[16]

But Douglas, Blackmun, and Brennan lost. The majority chose not to interpret the issue so broadly nor to expand the concept of standing any further (although Congress included provisions in later environmental laws such as the Clean Air and Clean Water Acts to allow suits to be brought by citizens who would not otherwise meet the direct injury test).[17] As a nation we came tantalizingly close to Leopold's land ethic but hesitated at the brink. Perhaps our landscape values had not yet softened sufficiently. Perhaps there remained more philosophical groundwork to be laid in order to form a solid basis for such a step forward. Or perhaps it would be a step beyond our best interests. Might it extend rights too far and thus not be an inevitable step or progression, as some suggest? These are questions we now debate. John Muir would probably be pleased that the debate he helped start continues. James Kilbourne and the early pioneers would probably be amazed.

While our landscape values have progressed, less change is noticeable in our treatment of the land. From a practical standpoint, we act in the same way that Pinchot proposed. Stewardship, then, is both a philosophy in which we believe and a behavior manifested through our daily actions. It is both what we say (a noun defined by rhetoric) and what we do (a verb). It is an abstract ideal and a tangible outcome in the real world. Rarely are the two definitions the same—our actions rarely match our rhetoric.

Although the Sierra Club lost the Mineral King case, the resort was never built. As in many cases, the costs of long delays due to the lawsuit helped dissuade Disney from proceeding with the project, and in 1978

Congress transferred the Mineral King Valley to Sequoia National Park. In the end, was the playing field well balanced? Did the "system" work? Was the public good achieved? Perhaps, perhaps not.

The American landscape today is largely a legal, litigated landscape, a landscape shaped more and more by rule and regulation based on emotion and subjectivity as well as reason and objectivity; science and law, philosophy and ethics respond to each. The American landscape is one for which legislation is often passed on the basis of moral commitment, but, as Oliver Wendell Holmes suggested, only rarely is a moral basis created by that legislation.[18] The implications of this interplay are seen in the story of our national strip mine reclamation law described next.

CHAPTER 14

The Emotional Landscape

The mountaineer can present no enigma to a world which
is interested enough to look with sympathy into the forces
which have made him.

> Harry Caudill, *Night Comes to the Cumberlands:
> A Biography of a Depressed Area*

Most Americans have vivid mental images of surface coal mining. ravaged Appalachian Mountains, orange-tinted streams clogged with sediments, and the abject poverty hidden in the back hollows of the region. It is a profoundly emotional issue as demonstrated by the following excerpt from a 1967 Department of the Interior report titled *Surface Mining and Our Environment:*

> Our derelict acreage is made up of tens of thousands of separate patches. In some regions they are often close together. Where one acre in ten is laid waste, the whole landscape is disfigured. The face of the earth is riddled with abandoned mineral workings packed with subsidence, gashed with quarries, littered with disused plant structures and piled high with stark and sterile banks of dross and debris, and spoil and slag. Their very existence fosters slovenliness and vandalism, invites the squatter's shack and engenders a "derelict land mentality" that can never be eradicated until the mess itself has been cleaned up. Dereliction, indeed, breeds a brutish insensibility, bordering on positive antagonism, to the life and loveliness of the natural landscape it has supplanted. It debases as well as disgraces our civilization.[1]

Yet few of us have firsthand knowledge of the industry. Instead, our understanding of its operation and its effects on the physical and social landscapes has been shaped mostly by the media.

On August 3, 1977, President Jimmy Carter signed into law the Surface Mining Control and Reclamation Act of 1977.[2] For the first time, minimum national standards regulated the operation and reclamation of surface coal mines and the rehabilitation of abandoned mines. At their core was a standard that required operations to return the site to its pre-mined condition, to take it "back to approximate original contour."

At first glance, "back to contour" is a logical method of reclamation: simply put the land back the way it was before mining it, so in several years one scarcely knows it was ever mined. The needed resource is extracted, leaving little trace of our action. But emotion, not the rational consideration of alternatives supported by factual evidence, led Congress to establish that standard in the same manner that our perceptions of strip mining are also most often emotionally based. Consequently, the story of this law tells us much about our general landscape values and regulatory process.

President Carter's action culminated a long and turbulent legislative dance dating back to the 1940s. Just before World War II, Congressman Everett Dirksen proposed a law to establish reclamation standards on all surface mining "as may be necessary to make the contour of the land approximately the same as before the mining operation began."[3] The country's attention was focused on other matters, though. After the war the issue reappeared. In 1949 Congressman Brooks Hays introduced a bill that called for "Federal and State cooperation in restoring mined lands."[4] His effort provoked little interest. Between 1949 and the early 1960s, numerous bills were introduced and many hearings conducted, to little avail.

Concerns about strip mining ignited, however, as environmental awareness and activism heightened nationwide in the 1960s and early 1970s. Harry Caudill's startling 1962 biography of Appalachian strip mining, *Night Comes to the Cumberlands,* was a key spark. Caudill's vivid description lifted the veil of ignorance that had cloaked one of the nation's least understood regions, and America was shocked by its first glimpse into the little-known land. National attention focused on the environmental and social problems created by the highwalls, benches, and out-slopes of the traditional "drill, shoot, and shove" mining method on the steep mountainous slopes of central Appalachia, problems exacerbated by the years of neglect of abandoned mines.

Such mining spread rapidly as a result of technical improvements in mining capabilities and changing coal markets. Unfortunately the surge in supply was largely uncontrolled and environmentally reckless. Local outcry triggered some Appalachian states, such as Kentucky and West Virginia, to regulate the industry in an attempt to keep the overburden, the material removed to expose the coal, as close to its original position as possible. Meanwhile, West Virginia Congressman Ken Hechler led a group proposing federal legislation in the 1960s to abolish all strip mining. More moderate views found outlets in legislators like Ohio Senator Frank Lausche, who proposed several bills to require federal study of the problem. Reacting to public pressure, in part triggered by Caudill's provocative book, the congressional moderates won a provision requiring a detailed study of surface mining in the act creating the Appalachian Regional Commission in 1965.[5] The resulting 1967 report, *Surface Mining and Our Environment,* added fuel to the rising fire.

The colorful, emotionally charged report examined the status of all surface mining. Pages of large, clear, compelling photographs of "soiled air and eroded land, gouged and torn landscapes and piles of barren mine spoil, scarred and ripped hillsides, sediment choked streams and polluted rivers, and a habitat . . . destroyed by acid mine water" set the tone.[6] The combination of the report's vivid graphics, evocative adjectives, and allusions made a persuasive presentation.

Beyond the emotionalism, the report presented a wealth of practical information. Charts, tables, and diagrams described the surface-mining industry. That pragmatism carried over to the report's conclusion that basic principles of sound resource management dictate that we eliminate unnecessary damage from future mining and begin to repair damage from past mining. It went on to recommend that the federal government establish minimum standards and reclamation requirements for mining within the public domain, standards the states should also enforce on private property. In the absence of satisfactory state compliance, the federal government should intervene with standards to control or eliminate water pollution, control soil erosion, eliminate health and safety hazards, conserve natural resources, and preserve and restore natural beauty. Those goals would be accomplished by making issuance of a mining permit contingent on the operator's submission of an acceptable mining and reclamation plan with time limits imposed for the completion of reclamation, the posting of a performance bond sufficient to cover the anticipated cost of reclamation, and the imposition of penalties for noncompliance. Programs were also recommended to rehabilitate abandoned mine lands and for research.[7]

Surface Mining and Our Environment had an enormous influence on Congress even though many of its regulatory concepts had been drawn from existing state procedures. Consideration of these concepts at the national level was a significant step and set the broad outline for all subsequent federal action on strip mining. The report still influences how strip-mining issues are considered.

Within a year, new congressional hearings were held. Vivid photographs and heartfelt testimony by locals, supported by factual data, again set the tone of the advocates' presentations. Senator Henry Jackson's Committee on Interior and Insular Affairs received testimony regarding three proposed laws to regulate the industry. Of those, Jackson's bill (S. 3132) received greatest attention and was identical to the report's recommendations. The other bills differed by limiting regulation to coal mining alone and in procedures for abandoned mine land reclamation. Industry opposition and insufficient political support in Congress brought the proposed bills to a quiet halt. Even so, the likelihood of a federal law regulating surface mining took a giant step forward.

Like the report, the 1968 hearings established several trends. First, the general outlines of federal intervention in the surface-mining industry proposed in the report gained widespread congressional acceptance, as federal interest in the problem and recognition of a need for action increased. Second, Congress chose not to dictate reclamation standards. It deferred their development, permitting each state in consultation with the federal agency to oversee the law's implementation. Third, reclamation was interpreted to mean simply the return of the mined site to a useful condition. Restoration of the land to its original condition was specifically ruled undesirable.

Efforts to regulate or abolish strip mining intensified in the wake of the 1968 hearings. By the beginning of the new decade, as Nixon signed NEPA, the environmental movement was in full swing. The media flooded the country with reports of pending environmental catastrophes. Much of the debate bordered on hysteria, and its participants often resembled evangelical preachers. Capitalizing on that atmosphere, abolitionists led by Congressman Hechler, and other more moderate activists, mounted a grassroots lobbying campaign to expose the public in key congressional districts to the effects of unregulated coal strip mining in Appalachia. Firsthand accounts of its horrors aroused sentimental support from local media and environmental groups, which often produced political pressure on the local congressional delegation as the issue struck a national nerve.

By 1971, armed with resolve bolstered by an upsurge in popular political support for legislative action, Congress again focused on the issue with three dozen new bills and extensive hearings in the House and Senate. Thousands of pages of testimony were received. The bulk concerned the scope of the law (whether it should apply to all surface mining or only surface coal mining), the jurisdiction (whether the regulations and enforcement should be handled by federal or state governments), the treatment of abandoned mine lands (to what extent they should be re-claimed), and other procedural and administrative issues. For the most part the bills resembled the recommendations made in *Surface Mining and Our Environment,* and the development of reclamation standards was typically deferred to each state. Congress appeared hesitant to ad-dress the complex details of strip-mining regulation because of its lack of familiarity with such intricacies and the recent emergence of the issue as a cause célèbre.

Amid that flurry of activity, the cornerstone of the federal law was laid. House Bill H.R. 6482, proposed by Ohio Congressman Wayne Hays, established "back to approximate original contour" as one of its fun-damental reclamation standards. Although it represented a profound change in the concept of reclamation as earlier defined by the 1967 De-partment of the Interior report, it received little attention in the hear-ings. What led to the silent acceptance of the standard that today shapes much of our landscape? What were the roots of "back to approximate original contour"? Emotions, not sound data.

The back-to-contour standard was included because it seemed !ike a good idea to its authors, according to their knowledge of strip mining, the trends in state reclamation laws, and their personal aesthetic values.[8] Congressman Wayne Hays owned a farm in the rolling foothills of east-ern Ohio (Belmont County), where the economy was based on farming and coal mining. Those mines occurred on the softer, more gently roll-ing slopes of the Appalachian foothills and offered reclamation oppor-tunities different from those feasible on the steeper mountainsides in eastern Kentucky, West Virginia, and other regions in central Appalachia. Given the contour farming and agricultural practices common around his home, the reclamation of strip mines to approximate original con-tour seemed logical. At the same time, Ohio and other states tried to keep overburden closer to its original placement by revising their recla-mation laws to reduce landslides, erosion, stream siltation, and general environmental damage. It was reasonable that backfilling to the land's original configuration would follow; and in the early 1970s it did.

Proponents of back-to-original-contour also reacted aesthetically to the hideous scarring of mountains resulting from unregulated strip min-ing. Most participants in the legislative process thought such mines were ugly. Therefore, they reasoned, the best way to preserve the landscape's beauty was to return it to its original condition. Aesthetic concerns were commonly thought to even outweigh unemotional cost-benefit analysis as a factor in the regulatory debate. The prognostication proved right. Hence members assumed it offered the least environmentally damaging and most aesthetically acceptable mining-reclamation method, and they assumed it was economically viable. On the surface the assumptions seemed sound. Beneath, they had little factual basis.

Recognizing the trend toward a law based on back-to-contour, and the dangerous lack of information on its consequences, the Tennessee Valley Authority (TVA) in 1971 began the first major test of that mining and reclamation approach on steep Appalachian slopes. Its objectives were to determine costs, assess impacts, and eliminate uncertainties about the technique. Unfortunately TVA results from the Massengale Mountain experimental mine were not published until late 1975, several years after Congress became irrevocably committed to the standard. In the interim, were there sufficient reasons for Congress to question its assumptions and either delay the specification of back-to-contour as the cornerstone of the law or select another, better proven standard? Politically, no. Public pressure for regulation was swelling. Most participants were willing to accept the assumptions at face value. The standard was simply too reasonable.

Yet ample evidence warranted caution. Significant questions had been raised about the environmental consequences of back-to-contour on steep Appalachian slopes. There, it would require unconsolidated rubble to be placed on slopes up to 25 percent. Such steep, barren slopes would be difficult to stabilize before revegetation and thus subject to sheet erosion and landslides. Other reclamation standards offered proven methods that lessened those problems. Backfilling the highwall to its original contour was also questioned in relation to long-term land use. Returning to contour would inhibit the future mining of any coal remaining in the mountain, and the opportunity to create an economically viable postmine land use on the site would be lost. The most telling argument against back-to-contour, though, was its cost. In 1971 it was widely believed the standard would substantially increase the cost of strip-mined coal, placing it at a severe cost disadvantage in relation to deep-mined coal. Serious economic questions could have been raised about the standard, but few were asked. The standard did not become economically acceptable until coal prices quadrupled during the 1973 oil embargo.[9]

Testimony during the 1971 hearings overlooked these issues and focused instead on facts and figures describing the amounts of land disturbed by strip mining, current and projected coal production, and typical reclamation costs of less stringent state standards. A handful of case studies were frequently cited as proof that strip mines could be economically and technically reclaimed, even though the case studies differed significantly from the situations and sites typical of most Appalachian mines, and even though the mining and reclamation techniques of some differed from those being considered.

The lack of more applicable data was, again, understandable because it simply did not exist. States had required reclamation for only a dozen years, and neither the mining industry nor the states had initiated significant reclamation research programs. Recognizing the urgent need, Senator Jackson requested the CEQ to conduct a fact-finding study. Results were published in March 1973.[10] So, while Congress recognized a need for strip mine regulation and began its development, it did so without data on which to base decisions. Whether by accident or insight, however, it refrained from establishing standards beyond the one pertaining to original contour.

Lacking detailed data, the 1971 testimony remained emotionally charged with many personal accounts by witnesses. Congressman Hays's account of a strip miner's attitude toward family burial grounds, a major part of Appalachian culture, was typical: "Even the burial plot of President Nixon's great-great-grandfather and his family, on the old Milhous farm in my county, is threatened. . . . I was talking to a shovel operator the other day and I said, 'You know, there are some old family burial plots in this county and a lot of them are unmarked. Did you ever run into them?' He said, 'Oh yes, I turned up six of them.' I said, 'What did you do with them?' He said, 'I put them in the bottom of the [strip mine] pit and I covered them up real fast.'"[11]

The Hays bill, H.R. 6482, became the "markup" (major) bill in the House and was passed 265 to 75 on October 11, 1972, despite opposition from the American Mining Congress, the National Coal Association, and much of the surface-mining industry. In the other chamber, Senator Jackson's bill, S. 630, served as the markup bill. Although unanimously reported out of committee, the legislative session ended before the bill reached a floor vote.

Attention returned quickly to the strip mine issue during the next congress. H.R. 6482 was soon reintroduced in the House, and S. 425, replacing S. 630, was introduced by Senator Jackson in the Senate. Busy committee dockets prevented the scheduling of hearings and markup sessions until late in the session. S. 425 was eventually passed by both chambers on December 16, 1974, and forwarded to President Gerald Ford, who pocket vetoed the legislation over the holidays.

In the two years between the passage of H.R. 6482 in the House and the passage of S. 425 by both chambers, the law changed dramatically. H.R. 6482 totaled seventeen pages, while S. 425 contained seventy-three. Congress's earlier hesitancy for detail was clearly overcome. The new detail in S. 425 tightened environmental protection standards and addressed water quality, revegetation, blasting, and other concerns, as Congress became more familiar with environmental issues. Still, reclamation remained based on the original contour standard. More than ever it was the focus of the law.

The ninety-fourth Congress tried again to enact the strip mine legislation, passing a bill in May 1975 little changed from S. 425. President Ford again vetoed the attempt, citing primarily economic reasons for his action. Federal strip mine legislation was dead until a change in administration and a change in the political climate.

With President Carter's entry into the White House, those changes occurred. The White House soon sent word to Congress that any strip mine bill would be signed. Congress resurrected the idea as H.R. 2 (the lower the number of a bill, the higher its priority). Years of effort were finally rewarded when the president signed the bill into law on August 3, 1977.

After a decade of debate, the federal government intervened in the surface coal mining industry and established minimum reclamation standards nationwide, even though many states had enacted their own reclamation standards years earlier, and even though each state had always possessed the ability to regulate the industry within its borders as it saw fit. Yet the public, speaking through political pressure on Congress, wanted new standards, even though few Americans had ever seen a strip mine and fewer still were affected by mine reclamation. We passed a law regulating the industry and setting a standard for reclamation based on emotion and aesthetics, not sound ecology, economics, land use planning, or resource use. Had the 1973 oil embargo not occurred and its higher price driven up the price of coal, the costs of that standard might have effectively killed the surface-mining industry. Perhaps that was an unspoken reason many proponents of the bill so quickly latched onto back-to-contour despite its many uncertainties.

Back-to-contour was set during the frenzied passion of the environmental movement, a time of great concern about environmental quality and the legacy we would leave our children. However, other techniques that reduced the amount of backfilling on steep slopes, techniques that reduced erosion and landslides more completely while enabling more complete and rapid revegetation, were the more ecologically and functionally sound. But with them the land would not look as "natural" afterward. They would leave little scars for future generations to see.

The regulation of strip mining reminds us that landscape is a personal perception shaped by many forces, including science and emotion, and landscape perception is part and parcel of one's stewardship values. We scar the land all the time, for numerous reasons: we mine forest and mineral resources; we cut ski trails into the mountains of our national forests; we dam rivers to create reservoirs for flood control, electric power, and pleasure boating; we stretch utility corridors and highways across the landscape; we even clothe the land with a fabric of farms dotted with cities and suburbs. Many of these land uses create impacts as profound and long-lasting as those of strip mines, but people feel they are more acceptable than the scars left by strip mining.

While it might seem that our landscape values and perceptions have evolved significantly since James Kilbourne started to reshape central Ohio into his image of paradise, from another perspective those values are remarkably constant. Despite the profound changes in our society, our technology, and our environmental understanding, and despite the upsurge in grassroots environmental activism, the proliferation of environmental protection laws, the statement of national environmental policy in NEPA, sympathetic judicial activism, even the adoption by some of alternative landscape values such as deep ecology and ecofeminism, little has changed at the core. While we may seriously consider Leopold's land ethic and our environmental obligations to future generations, while we may extend limited rights to endangered species and other select elements of nature, and while we may appreciate nature more biocentrically, most Americans still cling, like Kilbourne and Pinchot, to the belief that humans exist separate from and have dominion over nature. Most still believe nature is a commodity we are free to buy and sell to satisfy our wants and needs. We may recognize the need to better husband its resources as Pinchot proposed, and not to squander our patrimony as Marsh admonished, and the overt environmental hostility depicted by Hawthorne may be largely dissipated, yet at the core our landscape values are still dominated by our European traditions and Judeo-Christian heritage. Perhaps our technology and scientific understanding, our prosperity and some aspects of our culture, have advanced faster than those values.

Nowhere is this more evident than in our devotion to the Arcadian myth and its physical expression in the suburban landscape. This romantic landscape is one of our most distinctive and significant—a truly American landscape. Another is the rural landscape of the national grid established by the Land Ordinance, the Homestead Act, and their many relatives. A third is the business landscape of our urban cores. As a son of suburbia, I view it as a tourist, a temporary visitor. Urban America is merely a place to me. I see the middle landscape of suburbia far more sharply than places like Manhattan. I will leave its canyons of towering corporate cathedrals to other guides more familiar with the territory. Instead I will look next at suburbia and the character of our current auto-oriented landscape.

· Forgotten Sensations ·

I crossed the bayou late at night on a New Orleans bus bound for Houston. The ride was hypnotic after two days of train travel from Boston. The bus was dimly lit. Little traffic or roadside development offered additional light. I could see only the faint outlines of the landscape. Far off in the distance, a burning flame served as a beacon to mark each petrochemical plant we passed. For part of the trip, the bus crossed a narrow causeway, the water's primordial blackness beneath lending it a deep, viscous feel. Windows were slid open to allow the heavy, humid air inside. The landscape had a distinct smell. I was shaken when the bus arrived in Houston at dawn, my landscape senses stirred by the trip.

The landscape around Laramie, Wyoming, aroused similar sensations. While the Great Plains always affect me inexplicably, here the effect was magnified. Laramie lies on the high plains at the edge of the mountains. The far-off peaks emphasized the landscape's unexpected flatness and fertility. Here I felt the earth's massiveness and sensed my mercurial presence on its surface as never before. The sky and quality of light contrasted with that solidity. They possessed a fragile crystalline clarity I found sympathetic and attractive at a deep emotional level. Like me, they were ephemeral, while the earth was eternal and unyielding.

The long, low mesas of the Hopi nation elicited other landscape sensations. The sacred mesas lay within a complex mosaic of Sonoran desert, high plains, mountains, and alpine meadows, canyons, and arroyos. The spicy fragrances of mesquite bush, greasewood, cedar, sage, creosote, juniper, and pine scented the crisp air. It was a rocky, horizontal landscape of sparse vegetation and earthy smells, a place of burnt umber and sharp textures juxtaposed with sensuously rounded forms. Massive, billowing clouds, seemingly ready to fall, loomed above in an endless electric blue sky.

I was overcome by a sense of the landscape's timelessness, by its contrasts in color and light, form and texture. It was a place of impenetrable stillness and solitude. Yet I sensed a human presence reaching back millennia that had acquired a great wisdom about and harmony with the land. That wisdom and harmony were expressed in simple yet subtly complex adaptations, as if a whisper on the wind rather than blatant technological change, as if a boisterous yell in a crowd, wrought on the landscapes familiar to me. Here I felt truly in another land, not in the sense that it was different physiographically or culturally from my midwestern home. It certainly was. Yet so were the bayou and high plains, which did not prompt this response. Nor have other Native American landscapes I've visited in the East, Midwest, Great Plains, and Southwest. Here I felt strangely out of place, alien, as if the land were full of forces beyond my experience, forces I was unable to interpret or understand. I lacked the language to decipher their message.

Those experiences meant little to me at the time. Each elicited an emotional response I could not explain. We've all had such reactions, whether to exterior or interior, urban or suburban, rural or wild landscapes. Nor are they a function of our familiarity with the landscape, or the timing of the exposure. Some landscapes simply move us in ways we cannot explain. At a conference on "the spirit of place" several years later, I learned those reactions might be a part of the landscape we too often ignore.

Most conference participants shared an interpretation of "the spirit of place" I had never explicitly considered. To most, "spirit" was used more in the sense of a spirit or spiritualism than in a science-based way of describing genius loci. To most, "the spirit of place" referred to the intuitive, the emotional, the mystical attributes of a landscape, not the rational, quantitative attributes with which I was most comfortable.

A product of middle-class, middle America, clothed by Dockers and Eddie Bauer, schooled in science and Christianity, I felt somewhat out of place in a congregation of those converted to another landscape view that saw greater importance in nonscientific ways of knowing than I had ever considered: a view that relied more on non-Judeo-Christian landscape values than those in which I had been schooled. The conference opened my eyes to new ways of seeing the landscape that challenged, constructively, many of my previous landscape values and assumptions.

The *spirit* of place they came to discuss was that inexplicable reaction we have each experienced in landscape. Awe, anxiety, humility, tranquillity, what triggers such reactions? At Gettysburg, it might be our knowledge of events that happened there. It could also be more, since other historical landscapes arouse little response, and many ordinary places with no known historical significance trigger profound reactions. What are we reacting to? Why do some people react to a landscape while others do not—why does our sense of place differ? Does a landscape possess a sense of place in and of itself, or is sense of place solely a subjective reaction to the landscape's physical attributes and sociocultural connotations? Before the conference I felt I knew the answers to those questions.

Scores of books have been written about sense of place. Most consider it from scientific-like bases and apply empirical analysis of existing landscapes and human behavior to explain it. They typically presume that sense of place, although in part subjective, emotional, and aesthetic-based, is effectively described and understood using the physical and social sciences.

Conference attendees proposed that there are other ways of knowing the landscape that are open to forces and influences beyond or excluded by such empiricism. Some rejected outright the science-based paradigm that dominates our current Western view of the world in favor of an Eastern or Native American philosophy. Others simply embraced alternate ways of knowing the land such as folklore, geomancy, mysticism, and superstition.

The conference led me to look more closely at my previous reliance on rationality, and the way it shaped my landscape point of view. It called into question my perspective, highlighting the strengths and limitations of that view. It suggested that our prevailing paradigm that emerged during the Industrial Revolution contributes to our sense of disconnection from the land. It proposed that the current magnitude of landscape disturbance contributes to that disconnection by separating us from nature. And it suggested that our traditional landscape values, and the way we use technology to intervene between our desires and the land's inherent characteristics, shield us from certain fundamental landscape forces and sensations our ancestors easily detected, responded to, and benefited from. Might our inexplicable sensations be small, inadvertent tastes of those forces, like the faint aroma of some tantalizingly scent carried on the wind? Might those forgotten forces elicit the inexplicable landscape reactions we sometimes feel, and contribute to the differences we detect in sense of place?

The conference suggested the landscape might be more than the sum of its collective physical and cultural parts, in the same way our personality, our character, our soul transcend our biochemical being. Is it too far-fetched to think the landscape might possess some of that same gestalt? Perhaps only part of the landscape story is written in the language of our rational worldview. Perhaps some small part is also written in an ancient, obscure language we've lost knowledge of. If so, our traditional landscape values and stewardship attitudes reflect that loss. Jonathan Alder, John Muir, and Aldo Leopold told us as much.

Today, at one point along the Big Darby Creek in Madison County, remnants of an Indian mound stand on a low bluff, sentinel over the river below. A long floodplain terrace forms the opposite bank, just the type where Indians might have stopped while traveling the Big Darby waterway. This was the landscape of Jonathan Alder and James Kilbourne. It was also a landscape familiar to Scoouwa (James Smith). He once wintered here, and on several occasions canoed down the Big Darby to its confluence with the Scioto, then up the Scioto to its headwaters shared with the Sandusky. Like Alder and Kilbourne, he too was impressed with the abundance of this paradise.

Scoouwa and his adopted Indian companions spent the winter of

1757–58 camped on the Darby Plains at the headwaters of the Sandusky, Scioto, and Big Darby Rivers. In April, as buds became blossoms and succulent spring growth replaced the withered and skeletal
grayness, they made a bark canoe and set off down the Big Darby.
Water level was very low, exposing the glacial erratics that litter the
shallow stream. To avoid splitting the canoe on the rocks, one companion, Tecaughretanego, concluded they would encamp on shore
and pray for rain.

Before addressing the Supreme Being, Tecaughretanego purified
himself in a sweat-house he constructed on the riverbank. He built
it by sticking long, supple branches in the ground to form a dome-
shaped lattice. This he draped with blankets and skins. Stones were
heated in a fire and rolled into the hut. Tecaughretanego followed
with a little kettle of water mixed with a variety of cured herbs. For
fifteen minutes he sang aloud as he sprinkled the water over the
stones, producing clouds of scented steam. He emerged, cleansed,
to burn tobacco and pray.

Several days later, the rain came and raised the creek. Scoouwa and
his companions paddled safely down to the Scioto, then up to the carrying (portage) place to reach the headwaters of their home. Perhaps
Scoouwa's small party stopped on the terrace opposite that remnant
mound visible today; perhaps there Tecaughretanego made his sweat-
house and prayed for spring rains to raise the water level. I've sat
on that bluff, looking down at the terrace and river, and imagined
their canoes passing by, and wondered about the paradise wilderness
through which they traveled so recently. Canoeists still paddle past
this point. They certainly see the landscape from a much different
perspective.

CHAPTER 15

Remaking Urban America

The clock, not the steam-engine, is the key-machine of the
modern industrial age.

Lewis Mumford

Today, most Americans live in one of four settings. Some live, as Jefferson envisioned, in relative isolation on a farm or homestead where the nearest neighbor might be beyond shouting distance. While nearly 90 percent of the population once lived in this setting, the percentage in that dispersed group shrank as society shifted from an agrarian to a diversified economy, as envisioned by Hamilton. Today few Americans live on a farm or homestead.

Others live in the galaxy of small towns scattered across the landscape. These rural communities range in size from a mere handful of houses, communities without traffic lights, to small cities with a distinct "downtown," albeit one only several blocks long. They all share several common characteristics: they are compact and small enough to walk across (though some are linear in form, not circular or rectangular), rural land surrounds them, and most residents work nearby.

Another setting forms the opposite end of the urban spectrum: the nation's largest, densest cities, such as New York. There, multifamily buildings such as high-rise apartments or condominiums, attached row houses, and walk-ups, house scores in the city core. Densities can be greater than twenty thousand people per square mile in New York, a remnant of urban growth that occurred before socioeconomic factors transformed American cities and gave birth to suburbs in the latter half of the nineteenth century.

Although we often consider ourselves an urban nation, most Americans today live in communities like Worthington, not in the high-rises and row houses of our dense central cities. We are actually a suburban nation—more suburban than any other nation. Physically and socially, suburbia is the essence of the American landscape. Suburbia is the heart and soul of America.

Contemporary suburbs are preferred communities of low densities composed mostly of middle- and upper-income, owner-occupied, single-family homes located near an urban core. They are usually surrounded by other suburbs, although rural land might make up part of their border, or they might edge the core. Suburbs are not self-sufficient cities, even those with small commercial and business districts. Many suburbanites commute to work outside the community, and much of their shopping, entertainment, and search for services occurs outside, too.

Most American cities are simply aggregations of suburbs clustered around a core that, until recently, served as the regional business, commercial, civic, and cultural center. Since industrialization, cities have grown primarily by suburban development on annexed rural land, or by the accretion of outlying suburbs and small satellite cities. This low-density sprawl offset a significant rise in density in the late 1800s and early 1900s in many original interior neighborhoods. Cities today are vastly larger in area and of much lower density that they were when industrialization began. Only a small portion of the overall metropolitan population now lives in the original core, even in cities such as New York, Chicago, Philadelphia, Boston, and San Francisco, where thousands live in downtown high-rises and row houses. Far more people live in the surrounding suburbs.

Suburbia is as distinctive a landscape as the skyscrapers of Manhattan, and as ubiquitous as the agrarian grid. Nationwide, suburbs share striking similarities in physical appearance, despite striking differences in their physical landscape and cultural heritage. The contemporary suburb is quintessentially American, even though its precursors date to ancient times. How did this happen? Why do we have such an apparent preference for the suburban landscape? How has it become so synonymous with the American Dream? The answer is found in our response to many factors—social, economic, philosophical, political, technological. A compromise between our attitudes toward wilderness, cities, and farms lies at its heart.

The rural society Jefferson sought to create spread people across the landscape in a vast checkerboard of small farms. Jefferson typified a deep disdain for cities, an antiurbanism in American thought: "I view large cities as pestilential to the morals, the health, and the liberties of man. True, they nourish some of the elegant arts, but the useful ones thrive elsewhere, and less perfection in the others, with more health, virtue, and freedom, would be my choice."[1] Like the wilderness, crowded cities such as New York were thought to be morally and physically threatening, despite their many potential cultural or economic benefits. Alexander Hamilton's more positive view of cities was not as widely held in the nineteenth century as the negative view, which may be true today.

Within the Jeffersonian reverence of rural life, an attitude emerged during the 1800s that perceived the working farm as aesthetically vulgar and offensive even though the act of farming was regarded as virtuous. To be a farmer was noble, but daily farm work was dirty, smelly, and backbreaking, as the generations of pioneers who cleared the eastern forests or turned the prairie sod on the Plains experienced. Disdain for these three landscape extremes, for wilderness, city, and farm, enabled the emergence in the mid-1800s (1825–75) of a new landscape set in the middle—the romantic suburban landscape.

The softening of traditional antipathies toward nature was a critical prerequisite for that emergence. As those values gradually softened, people adopted an aesthetic that identified contact with "nature" (meaning a romantic form of nature, sanitized and safely controlled by people) as physically reinvigorating and morally uplifting. Although people began to appreciate wilderness scenery, for landscapes where one lived and worked the aesthetic favored places where humans could control nature. Preferred landscapes were those where nature served people, landscapes where the inherent beauty of wilderness was polished and revealed by human industry; places where civilization replaced wilderness; lush landscapes where raw nature was refined by settlement; peaceful, pastoral places where people and nature coexisted harmoniously; places unlike the menacing wilderness, the congested city, or the vulgar farm. This aesthetic saw sylvan landscapes of shaded groves and grassy glades gently washed with a fresh breeze and cooled by a clear, tranquil stream as more desirable than the other landscapes. That was the message often interpreted in the writings of Thoreau and the transcendentalists, and at times misinterpreted in Muir's. That was the message conveyed by Thomas Cole and other popular Hudson River school painters. This aesthetic underpinned our preference for suburbs designed in the romantic style. Other factors were at work as well.

When James Kilbourne founded Worthington in the central Ohio wilderness of 1803, he did not establish a suburb; he established a new community. Suburbs as we know them did not exist then. Most Americans until the mid-1800s lived in one of the other three settings common today: on farms, in small communities like fledgling Worthington, or in large cities such as New York, Philadelphia, Boston, and Baltimore.

Little has changed since then in the remaining landscape of rural America. The basic fabric of open field, fencerow, woodlot, and farmstead has changed little from our landscape perspective. But these are not static landscapes. Change has occurred in field size, farming practices, and many other factors. Certainly physical change, such as reforestation, has profoundly affected some rural landscapes. Suburban growth outward from a city has transformed other rural landscapes. Still others have been transformed by irrigation or the location of a large industrial plant and accompanying suburban growth. Those rural landscapes that remain, however, closely resemble those that existed before mechanization.

Similarly, the landscape character—especially the distribution and density of land uses—of small town America has changed little since 1800. Surely some have grown to become cities and so have changed significantly. Others have been engulfed by the expansion of a nearby city and consequently had their character affected, though not necessarily overwhelmed. Nor are the small towns that remain static landscapes. Changes in lifestyle, including employment, shopping preferences, and the provision for the car, have triggered changes in the landscape. The vitality of many small town Main Streets has been sapped by outlying strip commercial development. Early residents would see change today, in individual properties and in provision for the car, but the basic town character would still be familiar to them. Worthington would feel familiar to Kilbourne today.

In contrast, the character of urban America has changed profoundly since the nineteenth century. Our big cities now radiate outward for miles from an inner core of tall buildings crowded tightly together. Away from the core, density drops quickly, making most of the metropolitan fabric a blanket of low-density suburban development draped over a vast region. The city has become a suburb; urbanity remains only in the core. Land uses within the sprawl are usually segregated by function, with discrete zones assigned for industrial, commercial, residential, and recreational uses. Zones are located in relation to one another based on their mutual compatibility and infrastructure needs, especially transportation. Most people with middle and upper incomes choose to live here, commuting miles to work, shop, and play. This suburban blanket eventually disintegrates almost imperceptibly at its outer fringe into the rural countryside.

The landscape of large American and European cities was much different before the full effect of industrialization. As Kenneth Jackson described in *Crabgrass Frontier,* the key to that difference was the dominant mode of transportation. Today, we rely on the car or mass transit to get around. They enable us to make the long commutes to work and play necessary to live in low-density neighborhoods far from the urban core. Before the auto and mass transit, most city dwellers lived within walking distance of the workplace. That made "walking cities" dense and congested which, in turn, made the use of the horse and carriage for daily commutes impractical for all but the affluent.

For example, in the early 1800s, London was likely the world's largest city, with a population of about 800,000 people; yet a person could easily walk the three miles from the outer edge of the city to the center. Densities in nineteenth-century European cities such as London, Vienna, Berlin, and Amsterdam normally exceeded 75,000 people per square mile in parts and were rarely less than 35,000. Densities in the largest American cities reached that range after the population explosion midcentury. Today, most American cities have densities well below 10,000 per square mile as a result of their preponderance of growth after the advent of mass transit and the proliferation of the automobile. (New York City is the principal exception.) By comparison, many foreign cities still possess characteristics of the walking city. Density in Shanghai's populated area nears 275,000 people per square mile.

The landscape the pedestrian crossed in the walking city was also much different from the one we now drive across in cars. Land uses in the walking city were randomly mixed with little rhyme or reason. Beyond districts necessitated by environmental conditions, such as waterfront warehousing, or some created by social standards, such as redlight districts, the city was more a jumble of land uses. Cities had no real core of greater density, nor did they have centers defined by a concentration of business, civic, or governmental functions. The random mix made the high density uniform.

Urban chaos was unavoidable because most people worked in small shops scattered throughout the city. Large factories and large business or government office complexes were just forming. For many people, the workplace was an integral part of the living space, often occupying the ground floor of the dwelling. This mixing made the creation of distinct residential districts impractical. Consequently, walking cities like nineteenth-century New York and London were dense, crowded, congested jumbles of intermingled land uses.

The boundary between that chaotic urban fabric and the rural countryside was sharp and distinct. Although that sharp boundary was once a function of the walls and fortifications that long ago provided protection, most large European cities by the nineteenth century had either expanded well beyond their walls or torn them down. The boundary between city and countryside persisted because people had to live within walking distance of the workplace and all the other land uses they needed on a daily basis. The most practical response to that situation was to crowd and intermix the various land uses tightly together. For that reason major American cities, lacking ancient barriers, exhibited the same sharp edge as their European counterparts.

The sharp edge was also related in part to the preference of the wealthy to live either in the city or far from it, instead of at the fringe. Housing has been built by the wealthy at the periphery of cities for thousands of years. There they could enjoy the advantages of clean, country life and the economic and cultural advantages of the adjacent city. Recognition of that ideal balance was expressed in a letter, written in cuneiform on a clay tablet, to the king of Persia in 539 B.C., "Our property seems to me the most beautiful in the world. It is so close to Babylon that we enjoy all the advantages of the city, and yet when we come home we are away from all the noise and dust."[2] The term "suburb" has been applied to such housing since the Middle Ages: Geoffrey Chaucer used a form of the word in *The Canterbury Tales* (c. 1386).

Throughout Western history, though, "suburb" connoted a place of poverty, squalor, and social undesirability, a place far inferior to the larger city it surrounded. To live at the urban-rural fringe in a lower density residential district segregated from other land uses required the ability to commute into the city core, as well as a desire to live at the fringe. In the early 1800s, few of the wealthy who could afford to commute had such a desire. Life in the core was preferred.

The European aristocracy preferred city life because the city was the center of cultural and spiritual life and offered safety, despite its crowding and congestion. In contrast, the slums outside the city walls were populated with poor vagabonds and undesirables who were barred at the city gate for political, economic, or social reasons. Our vernacular language is filled with pejorative terms related to this: an outcast is literally one who has been cast out of or denied access to a group or place such as the home or community; similarly, to live at the fringe of society is to be only a marginal, suspect member. The undesirability of settlement outside the city proper was at times reinforced by the location of obnoxious land uses outside the walls. There, smelly nuisances such as soap-making businesses, slaughterhouses, and tanneries were often located. Despite the long-standing practice of the wealthy to build grand country estates and summer homes in the fresh air far outside the city, or the quaint charm of rural settlements safely surrounded by fields and forests, life for social outcasts in shanty towns and "suburban" slums at the periphery of major American and European cities before complete industrialization was anything but paradise. In 1799, a description of "suburban" Philadelphia reflected that condition: "Persons who are disposed to visit the environs of this city, and more particularly on a warm day after a rain, are saluted with a great variety of fetid and disgusting smells, which are exhaled from the dead carcases [sic] of animals, from stagnant waters, and from every species of filth that can be collected from the city, thrown in heaps as if designedly to promote the purposes of death."[3]

That soon changed. By the end of the nineteenth century, life at the city fringe was no longer limited to the outcasts of society. It was rapidly becoming a preferred lifestyle. Urban America also began to sort out its confusion. Land uses began to segregate by type. City centers—downtowns—began to emerge, densities began to differentiate, and the sharp demarcation between city and countryside began to lose its edge as the city boundary dissolved into the rural fringe. Those profound shifts were driven in part by the Industrial Revolution and its associated advances in transportation technology.

Industrialization incrementally reshaped nineteenth-century urban and rural America as many interrelated social, economic, political, and technical forces pushed and pulled at the urban fabric. Those forces are too many and too tangled to examine in detail here. A quick glance at some will remind us of their impact on the quality of American life and the American landscape.

In the early 1800s, more than 80 percent of the American labor force was employed on the farm, while only a few people were employed directly in factory production (the remainder was employed in trade, transportation, and the professions). America was an agrarian nation in Jefferson's, Alder's, and Kilbourne's day. Hamilton's image of a diversified national economy was yet to take hold much beyond New England.

Most of what little manufacturing occurred in the early 1800s took place in small workshops or mills. Factories where many supervised workers operated machines under one roof were rare, although there were some textile mills in New England. In 1820, two-thirds of all clothing worn by Americans was made entirely in the home—the Age of Homespun was in full sway. Some home-produced commodities were earmarked for market rather than household consumption. That "putting out" system for the production of simple goods was popular on farms during the slack (winter) season, particularly in New England, but it did not displace the preeminence of agriculture or disrupt rural life. More sophisticated goods were produced mostly by skilled artisans working in small shops located in towns and cities. James Kilbourne's experience in Granby reflected this. Remember, he got his start working as an apprentice in a small Connecticut clothing mill and worked on the nearby Griswold family farm during the mill's slack season.

Over the course of the century, part-time, unskilled workers putting out goods from small workshops, as well as master craftsmen producing finished goods with the assistance of several journeyman and apprentices (like Kilbourne), gradually gave way to entrepreneurs employing workers to mass-produce low-cost goods for widespread distribution. Again, Kilbourne's life is illustrative. In Worthington in 1811, his business interests shifted from land speculation and surveying back to the clothing business that had made him wealthy in Connecticut (Kilbourne claimed he reentered the clothing business in response to a personal plea from President James Madison to produce woolen uniforms for the U.S. Army in the War of 1812; however, the record is less certain about the circumstance). That year he formed the Worthington Manufacturing Company, a diversified conglomerate that initially produced woolen clothing. Soon, a wide variety of household products was added. In it, we see a company perhaps halfway between the traditional household form of production and the emerging market form of production:

> The company indicated its intention to produce hats, leather goods, and pot and pearl ash for soap as well as woolen goods. Further, it agreed to purchase and sell local handicrafts and to open a variety store where a wide range of foreign and domestic items could be purchased. The purchase of property or real estate was also anticipated; in a barter economy land and household goods were often substituted for cash in payment of debts. Company shares cost $100 each, payable over a five year period. Each shareholder had a single vote in the annual election of a president, secretary, and three member board of directors. Sheep, labor, or building materials were acceptable in lieu of cash.[4]

The earliest stages of industrialization had reached the Midwest by the time the Worthington Manufacturing Company was established. Lexington, Kentucky, had forty-two companies; in Ohio, the river towns led the way. Cincinnati produced furniture and plows; Marietta built ships for river and ocean use. Fledgling industries had started in Zanesville, Steubenville, Lancaster, Chillicothe, and Dayton. In 1810, the state produced some $2 million worth of manufactured goods, most shipped to eastern markets.

Kilbourne's company flourished during the war as it supplied goods for the quartermaster corps of the Northwestern Army headquartered in Franklinton, and in the wave of settlement in the Northwest Territory that followed. Kilbourne brought in skilled and unskilled laborers from the East Coast to work in his factories, whom he often paid with company products or stock. In 1819, one customer recorded the following purchases from Kilbourne's conglomerate: barrels and casks, hay rakes, boots and shoes, a pail and tub, beans and corn, ironware and nuts, nails, shingles, a saddle and harness, woolen goods, cabinetry, hats, and lumber for building. While the company did not manufacture all those products but imported some for retail sale, it did mass-produce items ranging from woolen clothing to nails. It even briefly offered its own specie in response to the chronic shortage of credit and legal tender at the time.

The national economic bubble that carried the company's meteoric rise burst in 1819, and a general depression ensued. Like the post–Revolutionary War depression that ruined the father, the new depression ruined the son—the Worthington Manufacturing Company, and its president, went bankrupt. Twice Kilbourne had gone from rags to riches, once in Granby, Connecticut, and once in the frontier wilderness of central Ohio. In 1820 he was forced to begin again. Over the remainder of his life he struggled to regain financial stability while continuing to shape the region's future through local, state, and national political office, civic affairs, and business enterprises.

By the mid-1800s, the national conversion from a household economy to a market economy was well under way. About half of the nation's labor force remained on the farm, while the percentage of people employed in factory production quadrupled. Though a sizable shift, full industrialization based on mechanized mass production had not occurred. Most production relied on the small workshops of traditional methods, and most farmers tilled the land by hand or with beast.

By the end of the nineteenth century, America was an industrial nation;

the conversion carried forward as several prerequisite conditions evolved. For widespread industrialization, a market economy based on specialized production and trade had to develop in place of the former self-sufficient, household economy of colonial days. In that former economic system, individual households and small villages produced much of what they needed, and traded for the specialized goods they could not produce; and they consumed virtually all they produced, so there was little left over to sell. Commerce was limited outside the few large cities, creating little incentive for mass production and industrialization. In the absence of mechanization, the labor-intensive form of farming occupied almost the entire workforce.

Technological change, together with other forces such as population growth, triggered the transformation. Agricultural productivity rose dramatically during the 1800s as a result of technological innovations like the cotton gin (1793), the cast-iron plow (1813), the reaper (1831), the steel plow (1837), and cultivators, threshers, and seed drills. Yields also improved as farming moved to the more fertile soils in the Midwest. Yields were further augmented by better seed and livestock and by improved farming techniques. Those techniques spread by way of farming organizations like the Grange, or the local lyceums and fairs where people like Thoreau and Marsh spoke and people like Muir displayed their inventions. And they spread with the development of land grant agricultural and mechanical colleges that emphasized the application of emerging sciences and technologies to everyday needs.[5]

As practical knowledge and technological innovations flooded the nation's farm fields, yields in excess of immediate needs enabled the surplus to be sold, making farming more profitable while emphasizing efficiency and more specialized farming responsive to local conditions. Commercial farming replaced the earlier subsistence system: sheep dotted the eroded hillsides of New England; wheat and corn proliferated in the Midwest; stock farming flourished in the Plains; tobacco dominated farming in the upper southern states, rice the coastal areas of South Carolina, and sugar cane southern Louisiana, but, overall, cotton dominated the South where a distinctive socioeconomic system and landscape developed, as did the short-lived cattle kingdom of the Plains and prairies.

With changes in agriculture, and the conversion from a household economy to a market economy, came corresponding changes in general commerce and banking. Large-scale trade of agricultural commodities required more sophisticated banking and brokering than those necessitated by the simple horse trading of a local barter economy. The ability to obtain short-term credit also became critical. Despite Hamilton's grand plans for the nation's financial system, state banks battled the federal bank over monetary policy and the issuance of currency and credit well into the 1800s. Those battles were gradually resolved, enabling the new market economic system to emerge.

Industrialization also required the technology and raw materials for mass production. Complex, heavy machinery made of iron and steel was necessary, as were the precision machine tools to make the new machinery and the lubricants and fossil fuels to keep them running. Simple mechanical or animal power sources had to be replaced by steam engines that would run the factories, steamboats, and railroad locomotives.

Technological innovation was not limited only to mass production. The nineteenth century also saw the proliferation of innovations for daily household and business needs. Inventions like the telegraph, the lightbulb, the sewing machine, the telephone, and the cash register revolutionized home life and the workplace, making them more comfortable and productive, while also creating demand for even more inexpensive, mass-produced conveniences.

At the same time, improvements in transportation radically cut the cost of moving goods to market. The crossing of the National Road and the Ohio Canal in Zanesville symbolized Hamilton's and Albert Gallatin's dream of an integrated network of transportation corridors that bound the nation into one great diversified economy. Improvements in transportation technology, such as the development of the steamboat pioneered by James Kilbourne's father-in-law, John Fitch, hastened the conversion.

Railroads were the most important transportation improvement. Over the latter half of the 1800s, they reshaped the landscape and reoriented American thought.[6] Spurred in part by the Civil War and federal land grants, the railroads revolutionized the mass movement of material, natural resources, agricultural commodities, people, and information over long distances. More efficient than wagon roads, less constrained by weather than canals, and far faster than travel by either, the railroads linked the nation physically and psychologically, a linkage symbolized by the first transcontinental railroad that crossed the trestle spanning the Green River, below which Powell's flotilla once cast off.

In the midst of that economic and technological change, an early indication of a new urban form appeared on Boston's Beacon Hill. The development was begun by a land speculator around 1800 as a residential district for the wealthy near the fringe of the city (now in the heart of downtown). Beacon Hill helped establish three important trends that reshaped urban and suburban American. First, it signaled the segregation of housing districts from the random mix of urban land uses. Although done to accommodate the desires of the affluent, land use segregation by function spread to other categories as the Industrial Revolution unfolded and as urban mass transit developed.

Second, physical segregation also fueled socioeconomic segregation as the wealthy congregated in less congested residential enclaves located at the periphery of the city. As prosperity spread during the Industrial Revolution, their suburban flight was joined by others, including well-to-do bankers, businesspeople, and professionals. That gave rise to other suburban communities at the periphery of major cities, such as Cambridge and Somerville outside Boston.

Third, those exclusive enclaves began to shift the status of suburbs: instead of being socially and physically inferior to the city, they became superior and preferred. It was a slow shift, and places like Beacon Hill were the exception through the 1800s. Life in most of the suburban precursors remained wanting in comparison to urban life. William Dean Howells (1837–1920), a leading man of letters after the Civil War, documented the shortcomings.

Howells had experienced firsthand the relative advantages and disadvantages of urban, suburban, and small town America. Raised in two small Ohio towns, he knew what life was like in the rural communities that dotted the hinterlands. As a young man he moved to Boston, where he quickly rose to become editor of *Atlantic Monthly*. In Boston he moved back and forth between the new suburbs of Cambridge, Back Bay, Belmont, and Beacon Hill. In the 1890s, he finally moved to New York City, where he remained. Suburbs, he noted, lacked the cultural amenities of the major cities; in addition, their municipal services were far inferior. In 1871, he wrote: "I had not before this thought it a grave disadvantage that our street was unlighted. Our street was not drained nor graded; no municipal cart ever came to carry away our ashes; there was not a water-butt within half a mile to save us from fire, nor more than the one thousandth part of a policeman to protect us from theft."[7]

Services such as public waterworks (Philadelphia, 1799), free public education (Boston, 1818), and mass transit (New York, 1829) began in the cities and only later spread to the suburbs. Most of the early suburban precursors, and small satellite cities about an urban core, saw themselves as juvenile cities-in-the-making and aspired to become major cities in their own right. Until they did, most lacked the amenities and services of their large urban neighbors and so were commonly scorned. Still, communities like Beacon Hill, Cambridge, and Somerville, although not truly suburbs in the contemporary context, nevertheless exhibited several suburban characteristics: preferential status, peripheral location, mostly residential composition, and commuter orientation.

Perhaps the first true suburb in the contemporary sense was created in 1814 when regular steam ferry service began crossing the East River, linking Brooklyn with Manhattan. At the time, Brooklyn was agrarian with a population of less than five thousand people. Its development, foreshadowing that of suburban America in general, depended on low-cost, convenient mass transit that linked the suburban home with the place of employment. That link provided the wealthy with the opportunity to escape the endemic urban problems that plagued what Washington Irving called "Gotham," in favor of the respite found in rural life just across the river. Walt Whitman, as a young writer, watched all that from his office at the *Brooklyn Eagle,* overlooking the Fulton Ferry slip. In describing "Brooklyn the Beautiful" he wrote, "There men of moderate means may find homes at a moderate rent, whereas in New York City there is no median between a palatial mansion and a dilapidated hovel." He provided perhaps the first description of rush hour for the new suburban commuters: "In the morning there is one incessant stream of people—employed in New York on business—tending toward the ferry. This rush commences soon after six o'clock. . . . It is highly edifying to see the phrenzy exhibited by certain portions of the younger gentlemen, a few rods from the landing, when the bell strikes[:] . . . they rush forward as if for dear life, and woe to the fat woman or unwieldy person of any kind, who stands in their way."[8]

While land speculation and development boomed where "clean sea breezes and a glorious view of greater New York and its harbor" beckoned the upper and middle class, the resulting exodus from the city of New York to suburban Brooklyn was not uniformly hailed as a step forward for urban America.[9] Even then, some expressed concerns over the competition posed by the residential progeny as the flight of the wealthy from the mother city sapped its economic and social stability. Ironically, by 1890 Brooklyn grew into the nation's fourth largest city, a city of eight hundred thousand people, from which the next generation of suburbanites fled.

Before 1825, no American city had a mass transit system, other than ferries, that operated along a fixed route with an established schedule and a single fare, although there were carriages or hackneys for hire for short trips, and stagecoaches provided for long-distance travel. In 1829 the introduction of horse-drawn coaches, called omnibuses, that followed prescribed routes and schedules on Broadway in New York began the transformation of the walking city into the "streetcar city." Similar transformations began in Philadelphia (1831), Boston (1835), and Baltimore (1844). By the 1850s New York had nearly nine hundred licensed coaches operated by twenty-two separate companies, and the average waiting time between coaches on some street corners was only two minutes (at one corner on Broadway, the wait was fifteen seconds). The omnibus marked the origins of urban mass transit and quickly became an integral part of urban life.

The pivotal improvement that made the urban transformation and corresponding growth of suburbs possible arose from the same technology that was reshaping the national landscape physically and psychologically—the railroad. In the mid-1830s steam railroads began to be used for interurban transit around the New York area, foreshadowing the commuter trains common today. Intraurban rail transit soon began when the horse-drawn omnibus was attached to rails. Faster, smoother, and more efficient than the omnibus, the new horsecar had largely replaced its predecessor by 1860. By then, 142 miles of track had been laid in New York, and ridership reached one hundred thousand per day. The horsecar systems revolutionized urban America as they formed the first integrated, large-scale urban transit networks. The location of their tracks often affected adjacent land prices and the direction of development, just as freeways do today. The horsecar systems also promoted the segregation of urban land uses by providing most workers their first means of commuting to work.

The horsecar and the subsequent streetcar, or electric trolley, were the technological keys that unlocked the door to suburban America. Industrialization created a large group of well-to-do people who could afford to flee the failings of urban America for the fresh air and tranquillity of life outside the city gates. Improvements in urban mass transit created the means for those first suburbanites to commute from their suburban paradise to their urban workplace. These changes fragmented the interwoven fabric of urban America as land uses separated into residential enclaves, large industrial complexes, and business and civic centers.

The surge in the nation's population, which grew from 5.3 million in 1800 to 76 million by 1900, hastened those changes. The American population was 14 times greater at the end of the century than at the beginning (in contrast, our population will be only about 3.5 times larger, 265 million, at the end of the century than at its beginning). The rate of nineteenth-century population growth created an insatiable demand for products and workers. Those, in turn, triggered an atmosphere of technological innovation and the movement to mass production and industrialization. Population growth created the demand for goods that drove industrialization and supplied much of the labor to populate the factories. The urban labor force was supplemented by workers who left the farm as their jobs were replaced by mechanization and as greater economic opportunities appeared in the factories. By the end of the 1800s, the mass migration from the farm to the city was well under way.

Much of the nation's population growth was attributable to a human tide of immigrants that flooded the eastern cities. The flood began modestly at first. While the overall population grew from 9.6 million to 17 million from 1820 to 1840, only about seven hundred thousand immigrants made the crossing, mostly from the British Isles and German-speaking regions of Europe. Thus immigration constituted about 10 percent of the nation's population growth. The great wave arrived from 1840 to 1860, when 4.2 million immigrants disembarked, mostly on the Atlantic shore, in the New World. During the same period, the total population grew to 31.4 million; so immigration accounted for roughly 29 percent of that growth—the peak percentage contribution it would ever make. Three million arrived between 1845 to 1855 alone, including about 1.5 million Irish who fled the potato famine, and about a million Germans. Smaller contingents came from Switzerland, Norway, Sweden, and the Netherlands.

Many of the immigrants, especially the Irish, remained in the crowded slums of the large eastern cities, where they subsisted on low-paying, menial jobs and suffered religious and social discrimination. Other immigrants with Protestant, peasant backgrounds, like many Germans, moved to the Midwest and became successful farmers; those with a skill found opportunities to practice their trade in midwestern cities like Cincinnati, Columbus, St. Louis, and Milwaukee. The Homestead Act was seen in part as a release valve for this human tide.

The flood of immigration presented problems for the traditional walking city. By the mid-1800s, general prosperity and improvements in transportation made it possible for the affluent to move out of the city, which was growing increasingly crowded. Out-migration left the city core occupied more and more by those with lower incomes who often worked

at low-skilled jobs in large factories. That fundamental change in the mix
of people living in urban America continued well into the next century
and left a legacy of poverty and discrimination we still struggle to resolve.

Yet before the mass migration to the suburbs and the large-scale reshaping of urban American could begin, Americans needed another motivation to flee, a sociocultural reason to abandon the urban core for
life in suburbia. That reason emerged simultaneously with the softening of the traditional attitudes toward nature, and other socioeconomic
changes. In combination, they laid the foundation for the suburbs.

CHAPTER 16

Dream Weavers

'Mid pleasures and palaces though we may roam,
Be it ever so humble, there's no place like home.
                              John Howard Payne,
                              "Home, Sweet Home"

The physical and psychological character of the American home changed dramatically during the 1800s. At the start of the century, the house for many Americans living on the farm or working in the walking city provided simple shelter: a place to eat, sleep, and work, a place often shared with people outside the immediate family, a place often shared even with animals. What little leisure time one had was often spent at another location. Schooling and religious training occurred elsewhere too. Child-rearing and general family life also differed in the 1800s, since domesticity was not as central to home life as it would become.

By the end of the century, house became a home to far more families. Homes became larger with new attention to functional layout and use of space. People paid more attention to decoration and architectural style. As J. B. Jackson noted, homes preserved the past with attics, basements, and barns, and provided for the future with more rooms and larger lots. They reflected a sense of permanence, of familial stability, continuity, and self-sufficiency. Yet homes were also seen as real estate that might be sold in the future, creating new economic and architectural concerns reflected in the use of inexpensive balloon-frame construction that provided greater flexibility than did the previous brick, stone, or mortise-and-tenon construction. And the best location for that proper home was in a spacious suburb, not a crowded city.

Industrialization helped trigger those profound changes by relocating the place of employment outside the dwelling. The mass exodus of men commuting to work created a crisis in American home life that contributed to new domestic values, exemplified by John Howard Payne's 1823 song, "Home, Sweet Home," the most popular song of the period. Spouses became more isolated from each other during the working day. The wife became responsible for the residence. The family became increasingly independent and feminized, and this "woman's sphere" assumed a superior social and moral status over nondomestic institutions such as the man's workplace. The proper path for young ladies, encouraged to nurture extravagant hopes for their personal environment, was to provide care for a husband and children in a comfortable suburban setting.[1] By the late 1800s, domesticity developed as fully here as anywhere in the world. It remains a fundamental force in American life today, despite profound sociocultural change this century.

The reasons for our devotion to domesticity are many. Our traditional preference for the private ownership of land is paramount. This contributes to an equally strong preference for individual home ownership afforded by our prosperity.

The home is also the focus of traditional Judeo-Christian values, values in which the nuclear family unit occupied an exulted station, values that declared the family to be God's chosen instrument for propagation, the nurturing of the young, and the perpetuation of moral principles. Those traditional values, often overshadowed by economic and environmental necessity early in the nation's history, were forcefully reaffirmed during the 1800s.

At the same time, the home became a safeguard against the dangers of moral decay that loomed ahead as urbanization and industrialization spread. Proper home life offset the sinfulness and greed sure to follow. The home became a private, protective sphere, a personal bastion or refuge free from the many threats and intrusions of a rapidly changing world. In the home, the embattled family found fulfillment, serenity, and satisfaction, and the nation found salvation. In 1853, the Reverend William G. Eliot Jr. captured the sentiment: "The foundation of our free institutions is in our love, as a people, for our homes. The strength of our country is found, not in the declaration that all men are free and equal, but in the quiet influence of the fireside, the bonds which unite together in the family circle. The corner-stone of our republic is the hearthstone."[2] That same sentiment resonates today.

The new values promoted by the cult of domesticity were defined by many voices. Among the most persuasive were those of Sarah Josepha Hale, Horace Bushnell, and Catharine Beecher. Hale was the editor of

Godey's Lady's Book, a Philadelphia-based periodical aimed at middle-class America. She was a leader in the crusade to institutionalize the role of the woman as homemaker and queen of the household, and her influential writing was widely published in *Godey's* and in other journals. Her vision, like that of her mostly male compatriots, presented the male as the coarser gender, while the female was softer, more sympathetic; the woman was more moral, more nurturing and pure. Unless a governess, she was to be a wife and mother. Any other occupation was morally and socially unacceptable. A woman was to be dependent on a man. That was the natural and proper order of things.

Much of Hale's advice centered on the proper care and management of the woman's domain—the home. She joined a chorus of voices offering advice on good housekeeping, raising the activity to the equivalent of a moral and social obligation. From closet design to dusting, the proper procedure for every task was essential for an efficient, happy home in which the woman, the wife, the mother best fulfilled her new domestic responsibilities. Home care was now the housewife's job, and the home her place of employment.

Many of those voices, reacting to the revolution in science and technology, applied sciencelike study of every aspect of home management, room by room, day by day. Home economics was born in the words of authors like Hale. Her book *Manners: Or, Happy Homes or Good Society All the Year Round* (1868) joined advice columns like Laura Lyman's "Aunt Hunnibee" in *Hearth and Home* and those in Mary Virginia Terhune's popular magazines the *Homemaker* and the *Housekeeper* in foreshadowing popular household advice columns of today.

Instruction in home economics was also available by the mid-1800s in scores of small, private women's schools scattered mostly in the major eastern cities. By the 1870s, it had also spread to some of the new land grant colleges. By the end of the century, thirty universities had home economics departments, and many more offered courses on the subject. National conferences were also held annually by 1900, and organizations such as the National Household Economics Association, sponsored by the American Federation of Women's Clubs, had formed. But women were by no means the only people promoting new domestic values.

Horace Bushnell, like many of his day, brought a religious fervor to the topic of domesticity. According to David Handlin in *The American Home,* Bushnell believed the enormous changes in American life necessitated a new form of religious instruction centered in the home. Born into a Congregationalist family in rural Connecticut, Bushnell entered Yale University in 1823 uncertain about the Congregationalist doctrine taught there. Throughout his college years, he questioned the belief that religious truth was capable of intellectual demonstration using logic and reason. Instead, like the romantic authors he read, such as Samuel Taylor Coleridge, he thought Christianity was better grasped with intuition; in effect, it appealed more to the heart than to the mind.

On graduation, Bushnell's doubts were unresolved. He began work in New York, first as a teacher, then a journalist. Neither career appealed to him, and he returned to Yale in 1829 to study law. Two years later, a religious revival swept the campus and engulfed him. Reborn in the revival, and his doubts finally resolved, he transferred to the Yale Divinity School to begin a life of Christian service. By 1833, he was the minister of the North Congregationalist Church in Hartford.

From his pulpit he preached a doctrine that differed dramatically on the nature of children, the home, and religious instruction from that originally offered at Yale. His doctrine of "Christian Nurture" influenced people far beyond his congregation as other ministers all across the country echoed his powerful message from their pulpits. That message was best articulated in a single speech titled "The Age of Homespun," given in 1851 at the Litchfield County centennial celebration. Like Frederick Jackson Turner's Frontier Thesis, the Age of Homespun struck a chord with American thinking. It encapsulated many fears about the effects of industrialization on the individual and family. The speech also offered a remedy—a means within the grasp of every family to find domestic bliss.

According to Handlin, Bushnell called his speech "The Age of Homespun" because he thought the making of clothes at home exemplified the spirit of an age all but over in which the house had been largely self-sufficient—although not independent, as it was supported by neighborly cooperation and local trade. By 1851, the household economy and agrarian lifestyle from which the country formed were rapidly disappearing, and in their place an industrial, market economy was booming. Bushnell, with great insight, called the new system "The Day of the Roads." A new system of roads, railroads, steamships, and newspapers, he explained in another speech, was transforming the nation's isolation and backwardness. He believed the giant new network of communication, a "Road for Thought," would become "a vast sensorium springing out its nerves of cognition and feeling" that would uplift and enlighten all it touched.[3] He was not sentimental about the transformation. His generation had experienced the drudgery and difficulty of life during the Age of Homespun. But he was cautious.

He lamented the possible loss of the positive values fostered under the earlier difficult lifestyle, the "old simplicity" of life characterized by "severe values" that arose from "rough necessities." He feared that progress often brought a barbarism, a backsliding in civil and spiritual values, if the economy slumped in established areas after the new system took hold. Then "religion droops, good morals decline, hope, which is the nurse of character, yields to desperation, low and sordid passions grow rank in the mold of decay, one blames another, society rots into fragments, and every good interest is blasted."[4] He saw telltale signs of this in New England cities like Hartford.

Barbarism could also arise in new communities, such as those he visited in California. Emigration and transplantation, he believed, sapped the vital values of a people. Perhaps not in the first generation, or the second, but sooner or later a sense of social disconnection and disease set in and resulted in cultural rot. Thankfully, the remedies to those pitfalls of progress, and the best hopes of prosperity, rested in part with good education.

Another potent remedy rested in the strength of the family and the home. Here his beliefs diverged from traditional Congregationalist theology. Unlike the prevailing doctrine, Bushnell's concept of Christian Nurture believed children's minds and souls were malleable and should be shown the righteous path from the beginning. The child should not be left alone to find his or her spiritual way until the time he or she was old enough to make a conscious conversion to Christianity. He believed a child's Christian education best began in the home at an early age. Parental guidance was critical, and it formed the basis of the child's character and values.

The physical character of the house was important as well. Bushnell believed childhood should be a time of free play and pleasure within a pleasant setting. Parents, he advised, should make the house an attractive place to foster a spirit of domestic grace and childhood joy.

Catharine Beecher shared Horace Bushnell's religious fervor but was Sarah Josepha Hale's philosophical sister-in-arms. Born in 1800, Catharine was the daughter of Lyman Beecher, a famous Puritan Congregationalist preacher from Yale. Her seven brothers all became preachers, including Henry Ward Beecher, one of the nation's leading Protestant clergy during the century's third quarter. Her younger sister, Harriet Beecher Stowe, wrote *Uncle Tom's Cabin* (1852), and another sister, Isabella, was one of the nation's leading feminists. Catharine's childhood home was full of fire and brimstone and a missionary passion that infected the children.

When her mother died in 1816, Catharine assumed primary responsibility for the large Beecher household; Harriet was five, and Henry an infant. Later, Catharine's fiancé, a professor at Yale, died in a 1823 shipwreck. The next year, she began work as the head of a girl's school in Hartford. In 1832, the famous family left for Cincinnati, where she founded the Western Female Institute. Although the school closed after four years, Catharine remained in the Queen City until her death in 1878.

Catharine Beecher's influence on American family life stemmed from the twenty-five books she wrote. Her *Treatise on Domestic Economy, for the Use of Young Ladies at Home and at School* (1841) was perhaps her most important work. Immensely popular, *Treatise* was frequently used as a textbook and was reprinted dozens of times over the following decades. At the core of her philosophy was the belief that women were morally superior to men. She was not a feminist like her sister Isabella, nor did she agree with those such as Melusina Fay Peirce or Charlotte Perkins Gilman, who favored cooperative living that would free women from domestic enslavement. Instead, Catharine believed women should depend on and be subservient to men. She wrote, "Heaven has appointed to one sex the superior and to the other the subordinate station."[5] Her contradictory perception of women as morally superior yet subordinate to men was resolved by her belief that women best achieved their goals by being unassuming and gentle rather than by seeking immediate equality and self-realization as advocated by the feminists. In that way, she believed, a man would yield to a woman's wishes.

Treatise and her later books offered advice and instruction on everything from child care to the proper procedure for nearly every imaginable household activity, since she linked the home with personal piety and purity. Like Sarah Josepha Hale, Catharine was a champion of good housekeeping and home economics. Her book *The American Woman's Home* (1869), coauthored with her sister Harriet, told how proper family values were perpetuated by proper design of the home. The sisters offered plans for proper room layout, even the design of a storage cupboard on rollers.

Catharine also offered plans for the proper dwelling and the surrounding landscape setting. "Implanted in the heart of every true man," she wrote, "is the desire for a home of his own."[6] Homes, she advised, were to be substantial, with amenities such as parlors, dining rooms, separate sleeping areas, and indoor privies. The grounds, she asserted, should have a yard and extensive garden (she devoted five chapters in *Treatise* to the yard and garden).

She argued that the physical and social worlds were better separated into the female-dominated sphere of home life, presumably in the suburbs, and a male-dominated sphere of business in the city. *Treatise,* as Kenneth Jackson summarized, "provided a vision of a healthy, happy, well-fed, and pious family living harmoniously in a well-built, well-furnished, well-kept house." In later publications, Catharine acknowledged that her ideas were "chiefly applicable to the wants and habits of those living either in the country or in suburban vicinities as give space or ground for healthful outdoor occupation in the family service."[7] Her message that family life thrived best in a semirural setting was clear. There the rustic cottage and quiet, bucolic landscape best promoted proper family life.

Harriet, too, became an influential national spokeswoman on domesticity. Her books *House and Home Papers* (1865) and *The Chimney Corner* (1868) used the same matriarchal style used by Hale and others. In them, advice full of proper values and practical knowledge was passed during a dialogue between a wise matriarchal figure, like a grandmother, to an eager, well-intentioned, but bumbling young housewife.

As the nineteenth century progressed, the cult of domesticity redefined the nature of the proper American family and home. Ministers and missionary-like social advocates preached the new domestic gospel from pulpits and periodicals like *Mother's Magazine and Family Circle* and the *Happy Home and Parlor Magazine.* Popular books, such as Catharine Sedgwick's *Home* (1835), carried the crusade as well. In it, Sedgwick, the most widely read female author of the first half of the 1800s, told the fictional story of William Barclay. The story traced Barclay's heroic life from his sweet childhood in a "picturesque district" of New England called Greenbrook to his struggle to survive in the social, moral, and physical hostility of New York City. William and family eventually returned to his childhood home, leaving the evils of urban America for the pleasures of life in the rural countryside. *Home,* like much of the literature of the day, was a thinly veiled morality play that portrayed the emerging suburban lifestyle as far superior to city life.

While domestic tastemakers like Hale, Bushnell, Beecher, and Sedgwick redefined the family and home, they offered little advice on the physical appearance of the new suburban setting they sought to promote.

Another group of tastemakers, responding to and reinforcing the emerging domestic values, gave physical form to that setting. They created the physical and aesthetic archetype for the suburban landscape. Those tastemakers included, most notably, Andrew Jackson Downing and his disciples, and Frederick Law Olmsted.

In 1841, the same year Catharine Beecher's *Treatise* appeared, Andrew Jackson Downing published the first American book on landscape gardening. And like Beecher's treatise, Downing's *A Treatise on the Theory and Practice of Landscape Gardening* significantly shaped public attitudes toward the home. Both authors advocated the sanctity of the home set in a rural, suburban environment. In 1848 Downing wrote, "In the United States, nature and domestic life are better than society and the manners of towns. Hence all sensible men gladly escape, earlier or later, and partially or wholly, from the turmoil of cities." In 1850 he said, "[I] believe above all things under heaven, in the power and virtue of the individual home."[8]

But Downing emphasized the architecture and the aesthetic appearance of the home and grounds more than the moral rationale for the home or the good housekeeping and home economics of its proper management. His *Treatise,* written when he was only twenty-six years old, went through numerous editions and printings. It was followed, before his tragic death in 1852, by two other best-sellers, *Cottage Residences* (1842) and *The Architecture of Country Homes* (1850), and a popular book on fruit trees. He also wrote influential editorials for the *Horticulturist,* a monthly journal on rural art and taste that was popular with the middle class. Despite his premature death in a steamboat explosion on the Hudson River (he drowned trying to save other passengers), Downing became the single most influential individual in defining the pastoral aesthetic of suburban America.

Downing was born in 1815 in Newburgh, New York, a small Hudson River valley community about fifty miles north of New York City. There he lived his entire thirty-seven-year life (coincidentally, James Kilbourne had successfully marketed his personal property in central Ohio to Newburgh residents about ten years before Downing's birth). Downing was the youngest son of a self-made wheelwright and nurseryman, who died when the boy was seven. He and his elder brother inherited the family nursery, where young Downing received from his brother a thorough training in plant propagation and the ornamental use of plants. As a young man, Downing acquired a reputation as a horticultural expert; and the nursery he comanaged with his brother became one of the finest in the eastern United States. Downing was described as "tall and slight of figure, with large dark eyes and a quantity of long black hair, the fashionable dress of the [eighteen] forties helped to give him the appearance of a foreign gentleman. . . . His friendships were largely women—he was stiff and formal in the presence of men. He delighted in giving those little attentions which women so enjoy—a favorite flower by the breakfast plate, a charming rhyme upon some anniversary. . . . He was a fitting type for a romantic-loving period."[9]

Downing was apparently an elitist who spent his life trying to deny his humble roots. As a youngster he was adopted by the wealthy living in large estates along the steep slopes of the Hudson River valley. As a teenager he attended their parties and offered them free advice on the plants they should use to beautify their extensive properties. Lacking any formal training, he obtained his design ideas from reading English authors like Humphry Repton and John Claudius Loudon. Their ideas on how to create picturesque scenery appealed to his own preference for the valley's beauty.

His design ideas also grew from his snobbish social values. Handlin described Downing as having acquired from his elite friends "an admiration for the gentleman and a consequent disdain for anything that could be labeled 'common'. . . . Characterized as stiff, cold, proud, haughty, and reserved, Downing was so enamored of the life of the gentleman that he felt that any evidence of work was unpleasant. He even instructed that his lawns be mowed 'by invisible hands' at night so that his family and guests would never have to see any trace of this distasteful activity."[10] Yet it was an activity he, perhaps as much as anyone, made into a way of life for many Americans. His elitism changed little after the nursery failed in 1847, and he was saved from bankruptcy only by loans from friends.

Fortunately, horticulture was not his true calling: the philosophy and practice of landscape gardening were. Those he found more profitable. Recognizing the rapid growth in the number of people wishing to landscape their property, he offered his wealthy friends and the new middle-class suburban homeowner the advice they lacked on the architecture of the home and the landscape. He, more than anyone, defined the physical appearance of the romantic landscape style, the model of which he derived from the design of English estates and the English countryside. In *Treatise*, Downing wrote,

The development of the Beautiful is the end and aim of Landscape Gardening, as it is of all other fine arts. . . . Landscape Gardening differs from gardening in its common sense, in embracing the whole scene immediately about
a country house, which it softens and refines, or renders more spirited and
striking by the aid of art. In it we seek to embody our *ideal* of a rural home;
not through plots of fruit trees, and beds of choice flowers, though these have
their place, but by collecting and combining beautiful forms in trees, sur-
faces of ground, buildings, and walks, in the landscape surrounding us. It is,
in short, the Beautiful, embodied in a home scene. And we attain it by the
removal or concealment of everything uncouth and discordant, and by the
introduction and preservation of forms pleasing in their expression, their outlines, and their fitness for the abode of man. In the orchard, we hope to grat-
ify the palate; in the flower garden, the eye and the smell; but in the landscape garden we appeal to that sense of the Beautiful and the Perfect, which
is one of the highest attributes of our nature.[11]

Downing was a strong proponent of the English cottage, a house form
well suited to the rapidly growing middle class. With his help, it soon
became a national type. The form arrived in the New World mostly as
the work of English author and landscape gardener John Claudius Loudon, via his 1839 *Encyclopaedia of Cottage, Farm, and Villa Architecture.* Downing then spread it in *Treatise* as he derived many of his ideas
from Loudon's *Encyclopaedia,* including thoughts on the character of
the proper residential landscape.

Loudon advocated a lyrical, romantic view of nature in which the
setting of the residence assumed noneconomic importance. He believed
the house should be surrounded by a large, open, decorative lawn, a naturalistic yard of smooth manicured turf full of flowers and free of weeds.
Echoing ideas Andrew Jackson Davis published in his 1833 book, *Rural Residences,* Loudon helped transform the property around a residence
from a place for generating economic return, or from leftover land to
be avoided as unhealthy, to the suburban lawn, the purpose of which
was primarily aesthetic and recreational. That notion of yard marked a
dramatic shift from historical concepts of residential property in Western Europe and America.

Until the early 1800s, most property surrounding a house had utilitarian functions. Private lawns and gardens in the crowded walking city
were extremely rare, as little land was left around the dwelling for any
purpose, let alone decorative or recreational. What little vacant land remained, mostly in the rear of the property, was usually rancid and overrun by rodents, since that was where the garbage was thrown and that
was where the outhouse was located. Yet deplorable as they were, many

rear yards were filled with a back-alley dwelling or stable house. Nor did the earliest residential enclaves have yards. Homes in Boston's Beacon Hill and Back Bay were placed right on the sidewalk or had postage stamp–size garden areas between the front door and street. Side yards often were nonexistent in attached row houses; and for free-standing dwellings, side yards were only large enough to allow a carriage to pass to the stable at the rear of the property. Industrialization and the erosion of the walking city did little to change the situation—yards and dense urban areas were simply incompatible.

Nor were the plots of turf sometimes found around homes in small towns or on farms of the same recreational use and decorative public image as our neatly manicured contemporary lawns. Those too served more practical purposes. Grand rural residences also had little land for such nonutilitarian functions. Historically, the typical estate surrounding a medieval English manor house was put to productive use as pasture, cropland, vegetable garden, woodland, wasteland, and hunting reserve. Little was left as open lawn for decorative or aesthetic purposes. Any turf landscape surrounding the manor house was likely to be pasture maintained by grazing flocks of sheep.

In England, elaborate pleasure gardens were often added to grounds as the classicism of the Enlightenment spread at the end of the Tudor era and the start of the Stuart era (c. 1600). The new gardens drew on the design of French Renaissance gardens and grounds, such as André le Nôtre's designs at Vaux-le-Vicomte and Versailles, and on earlier Italian Renaissance designs for villas such as Villa Medici, Villa d'Este, Villa Lante, and Isola Bella. The resulting seventeenth- and early-eighteenth-century English gardens, such as those at Hatfield House in Hertfordshire, circa 1611, Melbourne Hall in Derbyshire, circa 1704, Bramham Park in Yorkshire, circa 1710, and the royal palace at Hampton Court along the Thames just north of London, were often ordered in an axial arrangement along a formal allée, and usually consisted of many separate symmetrical spaces, each a different type of garden—flower garden, fruit garden, an intricate maze formed from clipped hedges, a knot garden, even a lush area of lawn used for a bowling green.

In the later 1700s and 1800s, though, landscape and garden design changed with the advent of romanticism. Romanticism, recall, originated as a reaction against the classicism and empiricism of the Enlightenment, affecting philosophy, literature, and art, including landscape painting and landscape design. Early in the movement, painters such as Claude Lorrain (1600–82) infused their landscapes with a Virgilian, Arcadian, allegorical flavor. As the genre evolved, others, such as John Constable (1776–1837), brought a keen eye for detail to their depictions of nature; yet the landscape remained romanticized. The general style was popularized in America during the mid-1800s by painters such as Thomas Cole and Frederic Church of the Hudson River school.[12] Constable's paintings perhaps most typified nineteenth-century American perceptions of and preferences for the English landscape, as they might well today.[13]

The design of English estates and parks changed in reaction to the spread of romanticism. Beginning in the 1700s, landscape design, particularly in England, expressed a more naturalistic, picturesque style similar in character to that depicted in romantic landscape painting, notably the work of Claude Lorrain. The English landscape gardeners sought to make real landscapes look like the painted landscapes. This was done in the context of the site's physical character and cultural legacy, especially the enclosure movement and its effect on the remnant manorial landscape of the medieval period.[14]

The landscapes designed in the new style, like the paintings they imitated, contained enframed views of gently rolling pastoral fields partitioned into small segments by low stone walls, fences, or hedgerows, and dappled with shade from scattered clumps of stately trees. Interspersed was often a quiet pond or softly meandering stream crossed by a bridge or bordered by a temple or pavilion of classical style, with rustic cottages or a quaint country village in the distance.

The desired effect of the painting and design was one of people and nature coexisting harmoniously in a lush, idyllic world in which the human-created landscape appeared to have arisen organically from the inherent character of the land. Critical elements of a scene's composition included the sequence of light and dark (sun, shade), the progressive movement of the eye from foreground to background, a harmony of color and form, and an asymmetrical balance ("nature abhors a straight line" was the oft-quoted dictum). To create such a composition, the painter and designer freely manipulated the characteristics of the land. Vegetation was often removed in favor of selective, sometimes exotic, species in carefully crafted groupings; and ponds or streams were often inserted where none existed before. Consequently, the notion that a scene evolved from the land's organic character meant the painter or designer limited the manipulation of the land to those changes that appeared to most observers to blend unnoticed with the untouched landscape. Nature was what looked natural to the unknowing mind and the unseeing eye.

Like the transcendentalist's concept of nature, the painting and landscape design styles did not represent real nature. Instead they represented
an idealized, civilized concept of nature, a romantic notion of nature
where the landscape was always pristine, plentiful, and peaceful. They
created perfect landscapes devoid of nature's messiness, places free of
want and work, picturesque scenes of serene, Arcadian beauty where the
viewer found respite from the real world in a tranquil, Eden-like garden.

The break in England from the prevailing classicism to that romantic
imagery was gradual and progressive. Sir John Vanbrugh's design for the
grounds at Castle Howard, circa 1701, reflected aspects of an Arcadian
image similar to a Claude Lorrain painting. Alexander Pope's (1688–
1744) design of the gardens and grounds at Twickenham, circa 1724,
furthered the break; in 1731, he admonished garden designers to

> Consult the Genius of the Place in all;
> That tells the Waters or to rise, or fall;
> Or helps th'ambitious Hill the heavens to scale,
> Or scoops in circling theatres the Vale;
> Calls in the Country, catches op'ning glades;
> Joins willing woods, and varies shades from shades;
> Now breaks, or now directs, th'intending Lines;
> Paints as you plant, and, as you work, designs.[15]

His charge to "consult the genius of the place" had multiple meanings, among them that the design should fit the character of the land
and reflect its genius loci, or spirit of place. The site, not some classical
theory and imposed geometry, Pope said, should serve as the source of
the designer's inspiration. That concept became a hallmark of the informal, naturalistic style of landscape design, albeit interpreted through the
tint of romanticism. Downing called it "fitness."

William Kent's design at Chiswick House, circa 1727, continued the
break from classicism. The emerging naturalistic style developed further
with the work of Lancelot ("Capability") Brown (1716–83). While the
earliest practitioners of the style were more the product of previous periods, people bridging design styles, Brown was perhaps the first true designer in the new style. He acquired much of his approach while working for Kent at the Stowe estate. Brown's full creative genius emerged
after Kent's death in 1749, in his reworking of Kent's Stowe plan and
his redesign of the grounds at Blenheim Palace. There his pastoral scenery
became icons of the style. Capability Brown was the heart of the English Landscape style. He was its inspiration and the father of its guiding principles.

The style perhaps peaked in the work of Sir Humphry Repton (1752–1818). Articulate, amiable, inventive, he was the first to call himself a professional landscape gardener, yet he worked at a time when the prevailing style was contentiously divided in theory and practice. By the early 1800s, Brown's legacy had spurred a generation of shallow imitators who blindly sought to copy his principles without understanding their basis. Brown, it was said, copied nature, while his followers copied him. His principles became dogmatic rules imposed in every situation, without regard for genius loci, for fitness. Function was overwhelmed by stylistic fad.

Repton understood the principles and reacted disdainfully to the host of charlatans who had bastardized them. In response to the fallacious claims of the pretenders, Repton wrote several books based on his extensive work.[16] They reasserted, even elaborated on, the proper legacy of Brown's ideas. Yet his most influential writings were produced for his hundreds of private clients, for each of whom he prepared a "Red Book." Bound in red, each volume stated his recommendations in great detail, supported by a bit of theoretical discussion and illustrated with an innovative overlay of "before" and "after" sketches.

Repton was not an ideologue who clung tightly to narrow preconceptions, nor was he an iconoclast incapable of learning from more formal styles, as were many of his contemporaries. His work drew from many sources and was more flexible and freely tailored to the practical needs of his clients. He quoted the following stanza as a warning at the beginning of his *Sketches*: "Insult not Nature with absurd expense, / Nor spoil her simple charms by vain pretence [*sic*]; / Weigh well the subject, be with caution bold, / Profuse of genius, not profuse of gold."[17]

He recognized the many practical differences between the type of scenery created by landscape painting and the scenery created by proper landscape gardening, differences often overlooked by his contemporaries. He focused his writing on those differences since he felt other practitioners paid too little attention to them. Whimsical changes in landscape design taste at the expense of sound aesthetic or functional principles bothered him. He attacked unfounded changes for what they were— simply the latest landscape fashion fad, not refinements in good taste or sound site engineering. Practical, informed, open-minded, his writing, more so than his executed work, exerted a powerful influence on Loudon, Downing, and others (Loudon assembled and published Repton's professional writings in 1840, although he modified some of their meaning in his own work).

Brown's principles of the English Landscape style, reasserted and refined by Repton, affected the emerging American notions of the suburban lawn and residence in the mid-1800s. At the same time, they found expression in the romantic, pastoral design of cemeteries and public parks. Those designs reinforced the public's growing preference for the suburban ideal.

Before the advent of urban parks, cemeteries located separate from the church and designed in the English Landscape style became extremely popular as pleasuring grounds for Sunday outings. There the family could picnic among the shaded groves and sunny glades adorned by an arboretum-like palette of plant material and stroll along the winding paths of the sort first found in Boston's Mt. Auburn Cemetery (1831). The popular appeal of such places quickly spread with the construction of "rural" cemeteries like Laurel Hill in Philadelphia (1836), Greenwood in Brooklyn (1838), and Spring Grove in Cincinnati (1845). By midcentury, rural cemeteries swept the country. Downing reported in 1848 that thirty thousand persons visited Laurel Hill from April to December and, by his estimate, the same number visited Mt. Auburn, while twice as many visited Greenwood.

The style was also applied to the nation's first public parks in the latter half of the 1800s. Before then, large, recreation-oriented public open space did not exist in urban America. By midcentury, though, growing antiurbanism and concerns for public health and social welfare led to the development of parks where people could find relief from the problems endemic to the early industrial city. Joseph Paxton's pioneering design for Birkenhead Park, across the Mersey River from Liverpool, served as the model, both in concept and design. New York's Central Park, begun in the 1850s, was the first such public open space in America. Others were soon built in the romantic style in most major urban areas and in many smaller communities too.

By the mid-1800s, then, the ideas of Downing, Loudon, and Davis were affecting the design of residences, cemeteries, and parks. Downing was the central figure. While the core of his ideas derived from the two premier English landscape gardeners, Capability Brown and Humphry Repton, he was inspired by the romantic ideas of English art critic and philosopher John Ruskin, by the English landscape painter Raphael Hoyle, and by a tour of England. Downing led American landscape preferences toward the suburban single-family, detached home set in the center of a manicured grass carpet, shaded by stately specimen trees, and scented by flowering trees and shrubs. That miniature rural world was the perfect place to escape the city. It was the ideal setting to uplift one's aesthetic, social, civic, and moral values. It was the proper place for domesticity and child rearing.

Downing was not a prolific designer. Few residential landscapes can be definitively attributed to him, although in 1851 he was commissioned by Congress to prepare a landscape plan for the grounds of the Capitol, including the Mall, the grounds of the Smithsonian Castle, and the White House. Those plans were never implemented. Instead, his influence arose from his prolific writing.

Villas were his preferred type of residence, though he tempered his recommendations in his articles and books to reflect more practical and functional dwellings designed to promote health and a fitness to the land and local conditions. Beauty, he felt, came from nature, not some abstract set of aesthetic and architectural norms. Suburbia, he said (though seldom using the term), consisted of cottages centered on irregular lots with street frontages of at least one hundred feet. Each home sat amid a romantic landscape and each home had a garden. Broad, curvilinear avenues laid in harmony with the topography meandered through the development and led to a large, commonly owned park in the village center. That suburban landscape, he felt, best blended individual ownership, civic-mindedness, and nature. There, the commuter could breathe free and be uplifted by nature and the aesthetic beauty of his residence.

Downing's ideas were a national phenomenon, and he was an immensely popular star. Thousands of young couples setting up housekeeping bought his books. No one, whether rich or poor, as the influential Swedish novelist Fredrika Bremer was told during a visit to America at midcentury, built a house or laid out a garden without consulting her good friend Downing.

Downing's sudden death shocked the nation's aesthetic soul. Condolences arrived by the thousand from mourners across the country, including the president, and hundreds from abroad. The *New York Daily Tribune* called him "a man of genius and high culture." The *New England Farmer* wrote, "The death of no man in the nation could be a greater loss."[18] The agricultural press, once critical of Downing's criticism of the poor aesthetic quality of farm architecture (Downing had written about the "tumbledown mansion of Farmer Slack, who had no feel for beauty"), joined the national outcry. The *Southern Cultivator* called his sudden death "a national calamity"; the *Michigan Farmer* said, "There seems to be a void that no man can fill"; and the *Prairie Farmer* added, "His is a country's loss, and a country mourns him."[19]

His ideas continued to flourish after his death, spread by disciples such as Frank J. Scott and Jacob Weidenmann. Each published an influential book in 1870 that furthered Downing's interpretation of the suburban home and landscape.[20] By that time the middle class had joined the wealthy's flight from city to suburb. While Downing's work was always slanted toward the well-to-do, Scott and Weidenmann refined his ideas for the general public.

Jacob Weidenmann, a Swiss-born architect and engineer, designed public parks, rural cemeteries, and residential properties. He served for seven years, beginning in 1861, as the superintendent of public parks in Hartford, Connecticut. During that time he worked closely with Horace Bushnell on the design of a park bearing Bushnell's name. As a leader in the rural cemetery movement, he designed Hartford's Cedar Hill Cemetery and promoted cemetery design in articles and a book (*Modern Cemeteries*, 1888). After leaving Hartford for private practice in New York, he published his most influential book, *Beautifying Country Homes* (1870) and, in 1874, entered into a working agreement with Frederick Law Olmsted, with whom he collaborated until Weidenmann's death in 1893.

Frank J. Scott's significance exceeded Jacob Weidenmann's, although the two shared similar design philosophies. Scott began *Home Grounds* by stating, "The aim of this work is to aid persons of moderate income, who know little of the arts of decorative gardening, to beautify their homes; to suggest and illustrate the simple means with which beautiful home-surroundings may be realized on small grounds, and with little cost; and thus to assist in giving an intelligent direction to the desires, and a satisfactory result for the labors of those who are engaged in embellishing homes, as well as those whose imaginations are warm with the hopes of homes that are yet to be."[21]

After paying homage to Downing, to whom his book was dedicated, and Loudon, Scott noted the influence of two other precursors to the emerging suburban landscape style—cemetery and park design, the latter as practiced by Frederick Law Olmsted and Calvert Vaux in the design of New York's Central Park. Following his mentors, Scott's approach to landscape design, what he called "decorative planting," sought to make the landscape a "picture," similar to the landscape pictures painted in the English Landscape style. He described decorative planting as "the art of picture making and picture framing, by means of the varied forms of vegetable growth."[22]

In Scott's and Weidenmann's work, the suburban lawn and landscape reached their fullest glorification. The front yard became a communal landscape and national icon. Scott even popularized lawn furniture. He wrote, "A smooth, closely shaven surface of grass is by far the most essential element of beauty on the grounds of a suburban home. . . . Whoever spends the early hours of one summer while the dew spangles in the grass, in pushing these grass cutters [the recently invented lawn mowers] over a velvety lawn, breathing the fresh sweetness of the morning air and the perfume of the new mown hay, will never rest contented in the city."[23] Little has changed today.

More than Downing's *Treatise,* Scott's *Home Grounds* offered detailed instruction on every aspect of landscape design and maintenance, even down to the best mix of grass seeds. Scott emphasized the layout of the property and the aesthetic treatment of the grounds, while Downing placed greater emphasis on the architecture. Both contained an extensive discussion of plant material. Except for the moralizing, the books remain remarkably relevant. Dozens of contemporary counterparts line the shelves of every bookstore.

As the nineteenth century concluded, suburbs designed in the romantic style were well ensconced in the American psyche. Many forces contributed to that phenomenon. Over the century, popular attitudes toward nature gradually softened, making contact with domesticated nature acceptable, even desirable. Profound economic and technological changes swept the country as the national economy converted from an agrarian-based household economy to an industrial-based market economy. Industrialization brought greater affluence to a growing middle class, and a surge of technological innovations revolutionized the workplace and the home. More and more people could afford to flee the filth, noise, and crowding of the city core for the tranquillity of the suburbs, their daily commutes made possible by weblike networks of mass transit that tethered the two worlds together.

Fundamental changes also occurred in the nature of the family. New domestic social values spread by tastemakers such as Horace Bushnell and the Beecher sisters redefined the proper place of the home and home life in American culture, further motivating the suburban flight. And physical tastemakers such as Andrew Jackson Downing and Frank J. Scott offered a landscape image, an aesthetic and architectural model for the new suburban home and community.

The single-family home became a source of pride and a status symbol for its owners, a reflection of their moral and civic values. It was real property, a shield from a rapidly changing world, a source of psychological and financial security. No longer was it acceptable to have a run-down home; people had the time and money to invest in its upkeep. The new notion of home was to be of one set in a pastoral, romantic landscape. That was the ideal landscape, the epitome of beauty, a peaceful, healthful place for the nurturing of children. The suburban home became a paragon of virtue for middle-class America and an aspiration of American people of all economic levels.

As those values arose, the fabric of the American urban and rural landscape changed. Contemporary suburbia was born in concept and, with it, a new American landscape. Yet few of the dream weavers designed real landscapes. They mostly worked their magic with words. The person who most translated those values into physical reality was Frederick Law Olmsted. For breadth of accomplishment and lasting significance on our landscape, few people surpass his extraordinary life.

CHAPTER 17

America's Landscape Architect

What artist, so noble . . . with far-reaching conception
of beauty and designing power, sketches the outline, writes
the colors, and directs the shadows of a picture so great that
Nature shall be employed upon it for generations, before the
work he has arranged for her shall realize his intentions.

> Frederick Law Olmsted, *Walks and Talks of an
> American Farmer in England in the Years 1850–51*

Perhaps no single person, other than Thomas Jefferson, has had as profound an influence on the American landscape as Frederick Law Olmsted. Recall the pivotal role Olmsted played in the protection of Niagara Falls, or Yosemite and the first national parks. Recall as well his work on landmark projects like the Biltmore estate, where he helped launch young Gifford Pinchot's career as America's most influential land manager during the Progressive era, or his influence on the Morrill Land Grant Act that began the nation's grand experiment in public higher education. Those accomplishments alone secure his place in American landscape history. But they barely scratch the surface of a most fascinating and influential nineteenth-century figure, a person who quietly reshaped the landscape and, in so doing, reshaped the lives of millions. Surprisingly, serendipity perhaps more than purpose shaped Olmsted's remarkable life as he encountered many of the century's central themes and most important people. Olmsted was a conduit into which flowed currents of change, there to blend and eventually emerge as physical expression on the land. Today, his imprint is clearer than ever.

Olmsted was born on April 26, 1822, in Hartford, Connecticut, a city his Puritan ancestors founded after they immigrated in 1632 from their medieval home in Essex, England. Fred, as he was apparently called, was perhaps the seventh generation of Olmsteds in New England. The

American Olmsteds were farmers, seamen, even pirates. Fred's father was
a stoic, civic-minded businessman and part owner of a large dry-goods
store. Hartford was then a thriving commercial center of seven thousand
people, its conversion from a household economy to a market economy
well under way. Much like the model Hamilton envisioned, the capital
of the state was home to an arms manufacturer, an infant insurance industry, and one of New England's most important ports and distribution
centers. Yet the town was still dominated by devout Congregationalists,
theological (and biological) descendants of its Puritan founders.

When Fred was a child, his family frequently took leisurely outings
to enjoy the abundant rural scenery in the Connecticut River valley and
the New England countryside beyond. His father's affluence afforded such
lengthy sojourns, the family's principal form of recreation. They traveled
by horseback, carriage, stagecoach, canal boat, and steamboat. Pleasant
images of West Point, Trenton Falls, Niagara Falls, Quebec, Lake George,
and Geneva were etched in Fred's childhood memory.

At home Fred and his younger brother, John, learned to hunt, fish,
sail, and ride horseback. Again, the family's affluence left the boys idle
time for outdoor activities, which their father encouraged in the hope that
the vigorous challenges would fortify their physical and psychological
strength. Outdoor activities fed Fred's growing fondness for wandering
the countryside. Even before his teens he was comfortable wandering
alone many miles from home. With relatives sprinkled about, he rambled freely around the Hartford countryside learning about the land
firsthand.

Fred also acquired a love of books early on. Like Muir and Powell,
he too found in the libraries of neighbors a hodgepodge of new treasures. He ransacked his father's small library and rummaged about the
extensive collection at the Hartford Young Men's Institute, where his
father was a life member. In dusty attics and darkened shelves he discovered books by Virgil, Izaak Walton, William Gilpin, and Sir Uvedale
Price that promoted pastoral scenery; and he read Catharine Sedgwick's
novels that glorified the home as the proper place for child rearing and
domestic bliss. He read eclectically—dictionaries, encyclopedias, travelogues, as well as literature, scholarly tomes, and more than a dozen newspapers and periodicals.

Fred's wandering mind was further infected when he attended lectures
with his father at the Institute. There he likely heard many leading scholars of the day, including Bushnell, Emerson, and Marsh. From those
sources, surely the first seeds of his later work were sown.

The Olmsted household also instilled in Fred a sense of social respon-sibility, modesty, and civility that reflected his parents' genteel, nurturing character and quiet devotion to family and community. In many ways, the parents' openness to ideas, their fondness of leisure and cultural ac-tivities, their urbanity and tolerance of the boys' search for direction was unusual given the Olmsted Puritan heritage and the prevailing Congre-gationalist climate.

Formal education for Fred began at age four in a small private school of the sort called "dame schools," run by women. By age seven his schooling came under the tutelage of local clergymen, with whom Fred lived except during vacations between terms. Interspersed among the tutors were several short, unsuccessful stints at secular schools, some private, some public. As biographer Laura Wood Roper noted in *FLO,* his years spent under the moral and academic guidance of those tutors were not happy ones. At age fourteen he left their control, carrying to his grave a smoldering resentment of the fire and brimstone of their harsh treatment.

As friends went on to Yale, Fred was left behind, his eyesight weak-ened sufficiently by a severe illness when he was sixteen to make inten-sive study impossible. His education was continued instead in the tu-telage of one final minister. Fortunately the burning fear of damnation and hell's raging fires, a consuming fear that lit the lives of his former tutors, burned more dimly in the new clergyman. He saw life in a dif-ferent, more peaceful light, allowing Fred several years of learning at a more leisurely pace. The minister was also a civil engineer, so Fred stud-ied engineering and rhetoric. Mostly, though, Fred lived "a decently re-strained vagabond life, generally pursued under the guise of an angler, a fowler or a dabbler on the shallowest shores of the deep sea of the nat-ural sciences."[1]

As the lazy summer of 1840 came to an end, Fred was convinced fur-ther study in engineering was fruitless. That was not his calling. Adrift about his future, he felt his years with parochial tutors had instilled lit-tle of value, either academically or spiritually. However, his studies had afforded him ample opportunity to wander the countryside and satisfy his love of the land. As with Muir and Powell, Olmsted's real passion was rambling about the landscape. Yet those rambles had not produced a rugged physical body to match his hearty disposition. Like Muir and Powell, Fred remained lean, at times gaunt, throughout much of his life. As an adult, he suffered from insomnia and other maladies, as well as the residual effects of the common diseases and rudimentary medical treatment of his day. Fred emerged into manhood as a self-confident, straightforward gentleman of wiry frame, quick mind, and good nature. When he reached age eighteen, his adolescent education ended and his adult education began.

His father insisted he begin an apprenticeship with a French dry-goods importer in New York City, a privileged position arranged by his father in the slight hope the son would follow in the father's foot-steps. Eighteen months later Fred was home again. The merchant life was not for him either. The lessons learned in New York were not for-gotten, though. The young man from Hartford learned much from the urban transformation under way in Gotham. His work had taken him to the wharves and aboard sleek clipper ships from around the world to check consignments. And living in Brooklyn Heights, he had joined the first suburban commuters who formed what Walt Whitman called "the incessant stream of people" lining up to catch the Fulton Ferry to Manhattan.

That summer was spent sailing and generally idling with his brother; on conclusion John left for New Haven to begin his freshman year at Yale. Fred remained behind to idle the winter away. In October, he vis-ited John with important news: Fred announced he would go "before the mast" the coming spring as a seaman on a clipper ship bound for China. His decision was not surprising, given the family seafaring her-itage and his taste of international trade in New York. His father was supportive, too, thinking the experience might toughen the aimless young man. In the meantime, Fred made several more visits to New Haven and heard John James Audubon speak in Hartford at a dinner hosted by a family friend and member of the local Natural History Society.

On April 24, 1843, the nineteen-year-old set sail on the 330-ton *Ron-aldson,* an average-size American ship in the China trade. After brief stops at Java and Hong Kong, the *Ronaldson* anchored at Canton on September 9. On December 30, it set sail for home with a cargo of tea, cinnamon, and silk. By then the crew was happy to get under way, as most were sick from the foul conditions sitting in port. Fred was emaci-ated by frequent fevers and intestinal disorders and suffered from rheu-matic pains aggravated by the dampness that permeated everything. Those were the least of his worries once the return voyage began.

The captain's slightest displeasure resulted in violent disciplinary ac-tion. Floggings and other forms of legal and illegal maritime punishment were commonplace, unrestrained by the few passengers the ship carried to Canton. The defenseless crew also suffered from insufficient food, as much of the ship's stores had rotted. Brutalized and malnourished, they were savagely overworked as they set sail shorthanded. When the *Ronaldson* finally made port in New York in April 1844, Fred's father did not recognize the yellow, scurvy-affected skeleton standing before him on the pier.

The experience profoundly affected Fred's later sense of social justice and welfare. That sense became a prime rationale for his eventual design philosophy, and shaped landscapes like Central Park. It also motivated his participation in the conservation movement. His social conscience soon deepened further by his witnessing of the English social system, southern slavery, and the Civil War, his reactions to which led to other significant contributions to the country.

In the interim, he recuperated and returned to an idle, leisurely life filled with outdoor activities and meetings of the Natural History Society. Over the summer he also sprouted interests in farming and women. By fall, both interests had ripened. Winter and spring were spent on the farms of several friends learning farmwork and rural life. He did not see farming either in Jeffersonian or transcendental terms; instead, he saw it in scientific and business terms. Unlike other forms of business, he felt farming was a peaceful profession free of the greed and malice that often marked other occupations. Those carefree days of hopeful anticipation were unfortunately darkened by the gathering of an ominous cloud that loomed for years to come, as the health of Fred's beloved brother, John, began a steady decline.

At the end of the summer of 1845, Fred and John were again at the family home in Hartford. Both attended Yale in the fall, where Fred audited courses useful in farming, including some chemistry, mineralogy, and geology taught by Benjamin Silliman, and natural philosophy taught by Dension Olmsted, a relative. There, too, he debated religion and politics, particularly slavery, with John's friends, and he read authors like Emerson, Lowell, and Ruskin. The start of the new year found him back home in Hartford, supposedly continuing his academic studies independently. Women were his real focus. He also underwent a religious revival that awoke his long-dormant Congregationalist heritage. Like Horace Bushnell and John Muir, he would struggle for years to resolve the differences between his personal piety and the orthodox doctrine of his church; like them, his struggle would also affect his work and, with it, the American landscape. By April, though, he left home in the hope of securing work on a model farm in central New York State.

On April 16, he took a steamer up the Hudson River to Newburgh, then traveled overland to Albany. The next day he called on Luther Tucker, the editor of the *Cultivator*, a well-known agricultural journal. Their fortuitous meeting had unforeseen effects on Fred's life that would reach far into the future. According to Roper, when Fred arrived at the office he found Tucker talking to a slender man with dark hair and deep-set, melancholy eyes who had an air of mild hauteur—Andrew Jackson Downing. Fred knew of Downing, as did most of literate America, from Downing's phenomenally popular books that made him the leading ar-biter of rural taste.

Tucker, it turned out, was trying to convince Downing to edit a new monthly periodical called the *Horticulturist*. As a result of their intro-duction, Downing took friendly note of Olmsted and gave him a letter of introduction to George Geddes, one of the state's leading agricultur-alists, who ran a model farm in Onondaga County. In ten days, Fred had an apprenticeship on the five hundred–acre Geddes farm, "Fairmont."

Life at Fairmont was instructive, and the household filled with inter-esting young ladies. Fred found time to read widely, including Fredrika Bremer's *The Neighbours,* which impressed him greatly. By fall, he felt sufficiently confident in his farming skill to want a farm of his own. He considered one on Staten Island, and another near Sachem's Head, a tiny Connecticut community on a small peninsula jutting into Long Island Sound about fifteen miles east of New Haven. The Staten Island prop-erty was far superior to the smaller, exposed Sachem's Head property, but the family had stayed at a seaside inn in Sachem's Head on several occasions and the boys had sailed their sloop in the waters off its tip. In November, Fred's father bought him the bleak, run-down seventy-acre Sachem's Head farm. Fred took possession in February.

The following weeks of hard work and loneliness were punctuated during the summer by family visits, mainly from John, and by social ac-tivities at the Head House, a popular resort hotel nearby. There Fred played ninepins and discussed religion with Horace Bushnell. In Octo-ber 1847, with the crops safely stored and the farm prepared for the winter, Fred went to New York City to meet with the popular architect Andrew Jackson Davis to consult about house plans (Davis thought those Fred had drawn lacked character). On the way there, Fred stopped in Newburgh to buy apple and quince trees from Davis's friend and near namesake, Andrew Jackson Downing.

With the Head House hotel closed at the end of the summer season and the family busy in Hartford, Fred's mood grayed as the approach of winter darkened the skies and agitated the surf. He spent Thanksgiving with the family at Hartford, and there probably began to contemplate moving to the Staten Island farm he had originally considered. Its advantages remained its larger size and better condition. During that lonely winter, the farm's closer proximity to people became important too. Although still rural, the island was just across the river from New York City and home to many elite country estates. On New Year's Day of 1848, Fred's tolerant, perhaps indulgent, father bought him the 125-acre farm for thirteen thousand dollars. Once again, Fred began the tedious process of restoring the land.

Over the next two years, Fred worked diligently, gradually transforming the house and grounds functionally and aesthetically. The neighbors were sociable and helpful, enabling Fred to form friendships with some of New York's most prominent families, including the Vanderbilts. He formed his closest friendship with Dr. Cyrus Perkins and family. Cyrus was a retired New York surgeon and professor of anatomy at Dartmouth College. He lived with his wife and orphaned seventeen-year-old granddaughter, Mary Cleveland Bryant Perkins. Fred frequently dined with the Perkins and admired their beautiful home and its art collection, including a painting by the great pastoral landscape artist Salvator Rosa; the collection also contained a portrait of Daniel Webster, a longtime family friend (Mary recalled sitting on Webster's knee as a child while he recited Greek poetry to her). William Cullen Bryant was another close friend of the Perkins; his wife was Mary's godmother, and his daughter was her best friend. Fred found Mary engaging, but it was Fred's brother, John, whom Mary would first marry.

The fields and orchards on Fred's farm flourished under his careful supervision; and, unlike Sachem's Head, the house on Staten Island was large, comfortable, and continually filled with family and friends. Hands were hired, and the operation organized into a model of efficient production. Meanwhile, John, a frequent guest, and Mary Perkins fell in love as they read Bushnell's "Unconscious Influence" together. Fred, although enchanted with Perkins as well, remained friends with her while he flirted with others. Over the winter he made plans to erect a new cottage to house six Irish workers onsite, just the type Downing recommended in the *Horticulturist*. As 1850 began, Fred was also active in community affairs.

When the winter ended, John and Mary announced their engagement.

Before the wedding, John planned one final bachelor fling. He and his closest friend from Yale, Charles Loring Brace, announced they would go on a walking tour through England that spring and summer (Brace, a more zealous Congregationalist than John or Fred, later distinguished himself as a pioneering social worker in the New York slums). Fred was dumbfounded, not because of the trip's imposition on the pending nuptials, but because he had quietly harbored a passionate desire to walk through England after he read Silliman's *Journal of Travels in England, Holland, and Scotland* years earlier. The opportunity to go along proved too tempting, especially since he and Brace were good friends as well. So plans for the farm were postponed, and planning for the trip went ahead full steam.

The three set sail aboard the packet *Henry Clay* at the end of April. The crossing began eventfully, the three having been misled about accommodations by an unscrupulous booking agent, and the crew at odds with the captain. There was little to be done about the accommodations, but Fred's experience at sea helped him avert a mutiny before the handsome ship left harbor. As order was restored, the captain's violent outrage disgusted Fred, as had the captain's behavior on the *Ronaldson*. Thankfully the crossing, once under way, was calm, although slowed by hostile winds. On May 27, the *Henry Clay* made port in Liverpool and the threesome rambled off on foot to explore the English countryside.

They covered more than three hundred miles in twenty-three days, arriving in London on June 21, their sightseeing route having passed through Birkenhead, Chester, Wales, Shrewsbury, Hereford, Monmouth, Tintern Abbey, Bristol, Bath, Salisbury, Winchester, Portsmouth, and the Isle of Wight. They found the people warm and welcoming, the landscape delightful. In his journal from the trip, later published as *Walks and Talks of an American Farmer in England* (1852), Fred wrote of one sunrise in Chester, "Such a scene I had never looked upon before, and yet it was in all parts as familiar to me as my native valley. Land of our poets! Home of our fathers! Dear old mother England! It would be strange if I were not affected at meeting thee at last face to face."[2] He found the scenery of rural England a constant delight:

> The great beauty and the peculiarity of the English landscape is to be found in the frequent long, graceful lines of deep green hedges and hedge-row timber, crossing hill, valley, and plain, in every direction; and in the occasional large trees, dotting the broad fields, either singly or in small groups, left to their natural open growth. . . . The less frequent brilliancy of broad streams

or ponds of water, also distinguishes the prospect from those we are accustomed to. . . . In the foreground you will notice the quaint buildings, generally pleasing objects in themselves, often supporting what is most agreeable
of all, and what you can never fail to admire, never see any thing ugly or
homely under, a curtain of ivy or other creepers; the ditches and the banks
by their side, on which the hedges are planted; the clean and careful cultivation, and the general tidiness of the agriculture; and the deep, narrow, crooked,
gulch-like lane, or the smooth, clean, matchless, broad highway.[3]

Fred was fascinated with what he called "the commonplace scenery"
of the rural Midlands. The subtle, charming pastoral landscape he later
remembered as unsurpassed either in England or anywhere else. In *Walks
and Talks* he wrote:

> Certain striking, prominent points, that the power of language has been most
> directed to the painting of, almost invariably disappoint, and seem little and
> commonplace, after the exaggerated forms which have been brought before
> the mind's eye. Beauty, grandeur, impressiveness in any way, from scenery,
> is not often to be found in a few prominent, distinguishable features, but in
> the manner and the unobserved materials with which these are connected
> and combined. Clouds, lights, states of atmosphere, and circumstances that
> we cannot always detect, affect all landscapes, and especially landscapes in
> which the vicinity of a water is an element, much more than we are often
> aware. . . . Dame Nature is a gentlewoman. No guide's fee will obtain you
> her favour, no abrupt demand; hardly will she bear questioning, or direct,
> curious gazing at her beauty; least of all, will she reveal it truly to the hur-
> ried glance of the passing traveller, while he waits for his dinner, or fresh
> horses, or fuel and water; always we must quietly and unimpatiently wait
> upon it. Gradually and silently the charm comes over us; the beauty has en-
> tered our souls; we know not exactly when or how, but going away we re-
> member it with a tender, subdued, filial-like joy.[4]

While the pastoral scenery of rural England struck a deep emotional
accord, the disparity between rich and poor, between the aristocracy and
peasant, stuck an equally deep discordance. Never had he seen such deplorable conditions as those that some of the farm laborers in the western counties suffered, conditions so brutish they made "men whose tastes
were such mere instincts or whose purpose in life and mode of life was
so low, so like that of *domestic animals* altogether."[5] His revulsion focused not on the peasants themselves; he felt sympathy for them. Instead,
he was revolted by the social inequity and injustice the peasants suffered
with so little apparent concern from the aristocracy. The glorious splendor of the manorial estates and private parks left him wondering, "Is it
right and best that this should be for the few, the very few of us, when

for many of the rest of us there must be but bare walls, tile floors, and every thing besides harshly screaming, scrabble for life?"[6]

At age twenty-eight, Fred began to combine two fundamental feelings. The insight from that combination laid the foundation for his life's work. He combined his love of landscape scenery with his deepening sense of social responsibility. He was beginning to think the landscape should create both beauty and social justice. A landscape's form and use are often inseparably linked to larger social conditions. That he saw firsthand in the remnants of the manorial English countryside. That, too, he witnessed in the wretched slums of Liverpool and London; and that he soon saw in the southern slave states of America. But in the filth and crowded conditions of industrial England he also saw how a public park in an urban setting could create positive social effects. As Roper noted, the park at Birkenhead triggered the revelation.

The 120-acre open space constructed just several years earlier across the Mersey River from Liverpool was the first urban park he had ever seen. Designer Joseph Paxton artfully sculpted the flat, clay farmland into a gently rolling park of broad meadows and shady glens centered on a pleasant lake. Here Birkenhead citizens of all classes strolled the meandering walks and avenues of trees to rock gardens, cricket and archery grounds, and ornamental buildings. Here tax dollars had been used to construct a public park open to all.

Like Ruskin, beauty for Fred could not be separated from its social setting. In *Walks and Talks* Fred wrote of Birkenhead, "I was ready to admit that in democratic America there was nothing to be thought of as comparable with this People's Garden. . . . And all this magnificent pleasure ground is entirely, unreservedly, and for ever, the people's own. The poorest British peasant is as free to enjoy it in all its parts as the British queen. More than that, the baker of Birkenhead has the pride of an OWNER in it. . . . Is it not a grand good thing?"[7]

The development of such parks in England began only years before as part of a broader set of social reforms to improve the deplorable conditions in the country's industrial cities. The designs drew heavily on the romantic, pastoral style of Brown and Repton, and the "gardenesque" style of Loudon. The three wanderers saw other urban parks as well, including the Chester promenade and several parks in London. But no park was so successful in Fred's mind as Paxton's Birkenhead. There a public landscape was used masterfully to create the physical and psychological benefits of beautiful surroundings for the pleasure and refreshment of all people, not just the wealthy. Despite the deeply disturbing

English social inequities he observed, he admired the parks as a form of social responsibility ready to emerge in America.

The wanderers remained in London about a month, renting a furnished room where they slept. Days were spent exploring the city. Armed with letters of introduction from influential family and friends, they gained entry into the elite circle of London society. In the finest London parlors and salons they met influential politicians, socialites, scholars, and businessmen from England and America. In early July, though, they left for the continent, Fred intent on visiting more gardens and nurseries. For another month they traveled through France, Holland, Belgium, and Germany, spending most of the time in Germany. During those travels, Fred likely first learned of the one work in the romantic, pastoral style on the continent that matched, or perhaps even surpassed, the work of Brown and Repton—the estate of Prince Hermann Ludwig Heinrich, Fürst von Pückler-Muskau (1785–1871).

Back in England by early August, the three travelers ended their summer touring other parts of England, Ireland, and Scotland. On October 6, Fred and John sailed from Glasgow. Brace stayed behind for further travel. Seventeen days later, the young Olmsteds were home. Fred returned with all the seeds of his later work set.

In 1850 the nation to which he returned extended from coast to coast; yet while industrialization was well under way in the eastern cities, settlement remained stymied by the prairies of the Great Plains. Still, Thomas Hart Benton and William Gilpin preached the gospel of western expansion. In 1850, eighty-year-old James Kilbourne died in Worthington, Ohio, about a year after Jonathan Alder's death; twelve-year-old John Muir and his family had just immigrated from Scotland to the frontier forests of Wisconsin; sixteen-year-old John Wesley Powell and his family were just about to move from their frontier farm in Walworth County, Wisconsin, to a prairie farm in Illinois; forty-nine-year-old George Perkins Marsh was the U.S. minister to Turkey; Herman Melville was at work on *Moby-Dick*, while nearby Henry David Thoreau was fervently preaching the gospel of transcendentalism; and Gifford Pinchot and Frederick Jackson Turner were yet to be born. Andrew Jackson Downing was about to die in a tragic steamboat explosion (1852), the same year Harriet Beecher Stowe's *Uncle Tom's Cabin* helped raise the philosophical winds that would blow the nation apart.

By the time Olmsted returned to Staten Island, farming had lost its appeal. Although his work on the farm continued for years, a new interest drew him farther and farther afield. Fired by his revelations in En-

gland, his interest turned to America's pressing social and political issues, which he sought idealistically to affect through journalism. With
encouragement from a Staten Island neighbor, liberal publisher George
Palmer Putnam, he compiled his notes and letters from the European
tour into a two-volume book (*Walks and Talks*). Encouragement also
came from his friend Downing, who published pieces written by Olmsted on Birkenhead and English apple orchards in the *Horticulturist*.
And in October 1851, John Olmsted and Mary Perkins married.

Critics reviewed *Walks and Talks* warmly. Sales were lackluster. The
book, though, together with several magazine articles and newspaper
columns published by 1852, helped launch Olmsted into the New York
literary world, into which, with Putnam's and Downing's help, he was
quickly drawn deeper and deeper. There he found familiar faces among
the luminaries in his new world: William Cullen Bryant, Asa Gray, and
Benjamin Silliman.

The mild social slant of *Walks and Talks* sparked Olmsted's growing
moral and social activism. Images of maritime cruelty and English social injustice, combined with his own piety fueled by years of religious
debate, fired his passion. In his next project, the fire reached ferocity.
By the fall of 1852, just as *Uncle Tom's Cabin* appeared, Olmsted was
commissioned by the year-old *New York Daily Times* (now known simply as the *New York Times*) to travel through the South to write an unsentimental series of articles on the way slavery affected southern agriculture and the southern economy. The original idea for the series was
the brainchild of the newspaper's editor, who had been referred to Olmsted by a mutual friend, Charles Brace.

His next ramble, then, started from Washington, D.C., on December
11, 1852. Over the next four months his stops included Richmond, Petersburg, Norfolk, Raleigh, Fayetteville, Wilmington, Charleston, Savannah, Macon, Columbus, Montgomery, Mobile, New Orleans, Vicksburg,
and Memphis. During his return he looped about the base of the Appalachians, crossing northern Mississippi, Alabama, and Georgia; then he
followed the eastern face of the mountain range up the Atlantic Coast
states, returning home on April 6, 1853. Along the way he rambled freely,
traveling by rail, stage, and boat between cities, remaining a day here
and two days there. On longer layovers, he took day trips on horseback
into the surrounding countryside. His method of observation and data
gathering was much the same as that practiced during his previous ramble through Europe—he met a broad cross section of people busy at
their daily lives in the natural settings where those routine activities

occurred. Olmsted, according to Roper, was an unobtrusive, yet meticulous observer.

On return to the Staten Island farm that spring, he eagerly set about compiling his notes and preparing the articles for the *Times,* but his work was constantly interrupted by the needs of his land, which he found increasingly bothersome. A series of fifty articles in epistolary form resulted. Written under the pseudonym of "Yeoman" they appeared in the *Times* weekly for a year, beginning in February 1853 under the running title "The South." In his introductory article, Olmsted told his readers, "I wish to see for myself, and shall endeavor to report with candor and fidelity, to you, the ordinary condition of the laborers of the South, with respect to material comfort and moral and intellectual happiness."[8]

Public reaction to the series was mixed, but it garnered sufficient interest for the *Times* to commission another series to be based on a second southern tour. John Olmsted, who had recently returned from an extended honeymoon and convalescence in Europe with Mary and their newborn son, joined his older brother on the new tour. John was still sick, though, probably suffering with tuberculosis.

The tour began in mid-December when the two reunited travelers bought horses in Natchitoches, Louisiana, and rode west. From Texas they planned to ride on to California, hoping the time on horseback and convalescence in the warm, dry conditions of the Southwest would heal John's respiratory ailments. Texas they found fascinating, especially the German community between San Antonio and Austin in the west Texas district of Neu Braunfels. By May they were back in New Orleans, having covered two thousand miles. Plans to proceed to California were abandoned due to an outbreak of hostilities with Indians along their path.

In New Orleans, the brothers parted to follow separate return paths. John, whose health was little improved from the ride, headed directly home while Fred made his extended return into a third tour of the South. From May, until reaching home in August, he rode across the back country of central Mississippi, then through the Appalachian highlands of Alabama, Georgia, Tennessee, North Carolina, and Virginia, where he caught a steamer for New York.

The second series of articles, signed "Yeoman" as before, began to appear in the *Times* in March 1854 under the heading, "A Tour of the Southwest." The fifteen articles, based on correspondence Olmsted wrote to his editor from Texas, appeared irregularly over the succeeding three months. Olmsted's total pay for the two series was $720. He also wrote ten more articles on his back country return trip. These he sold to the

New York Tribune, which published the series, called "The Southerners at Home," over the summer of 1857.

Before his three southern tours, Olmsted's feelings on slavery were tempered by his puritanical practicality and the socioeconomic and political realities of the day. While he considered slavery morally wrong and economically wasteful, he felt the realities made sudden abolition impractical. Those views gradually gave way to a more fervent abolitionist stance in the aftermath of the trips.

On return from the South, Olmsted found the farm in disrepair and urgent need of attention, but the literary world had tightened its hold on him. For the next few years, writing and publishing occupied his heart and mind, while interest in the farm faded further. During that hectic time he had ready access to the *Times* and other outlets, which he used to voice his opinions on a number of important social issues. He also began work on expanding his articles on the southern trips into a trilogy of full-length books.

By 1855, Olmsted joined the fledgling publishing house Dix and Edwards as a partner (he virtually abandoned his work on the farm, which he eventually sold in the 1860s). Dix was a twenty-four-year-old friend of Brace and an acquaintance of Olmsted. Edwards was a businessman who handled the firm's financial affairs. Their primary publication was *Putnam's Monthly Magazine,* which the firm had just acquired. Olmsted served as editor. In 1856, Dix and Edwards published the first of the trilogy, by which time work was well under way on the second. John assisted with the supplemental research and editing, for Fred's time was split among too many projects. The second book appeared the next year, the pivotal year in Fred's life.

The year began with Olmsted's recent return from a lengthy tour of Europe to develop markets for the firm's publications (and to do some sightseeing on the side). There he had hobnobbed with many of the politically powerful and socially influential, including the American minister to England and next president James Buchanan, as well as the London literati, including Richard Henry Dana Jr. and William Thackeray, whose dinner parties were the rage. During his travels the news from home was not good. Dix and Edwards was in financial difficulty due to Edwards's highly suspect business practices. The firm briefly reorganized without Edwards that spring. Olmsted finally withdrew in June, closing the door on his publishing career, although his literary career was not yet complete, for work on the third volume remained.

As often happens, when one door closes, another opens. With the dissolution of his publishing career, and his continued disinterest in the farm, Olmsted was open to change. The opportunity came in August when the thirty-five-year-old began his third career—landscape architect. Unknowingly, he began a career that helped reshape urban America, a career that gave physical form to suburban America and reserved much of the remaining American wilderness.

On November 24, the turning point in Olmsted's life occurred just as his new career began. After years of inexorable decline, his beloved brother died. John's death took more than Fred's closest friend, his brother in the fullest sense, it also took much of Fred's boyish spirit and lighthearted caddishness. In his profound loss, he found something of great value. John's last words to Fred were "Don't let Mary suffer while you are alive."[9] As a human heart can be transplanted from a dead person to a dying person to renew life, John, with all his love, transferred his heart—his wife, Mary, and their three young children—to his brother. With all his love, Fred accepted the lasting gift. Fred and Mary were quietly married in Central Park on June 13, 1859, by New York's Mayor Daniel E. Tiemann.

The final book of the trilogy followed in 1860, its completion delayed by Olmsted's entry into his third profession. In addition, a compendium of the three was published in 1861. With its release, Olmsted's career as a journalist-publisher-author, as a member of the literati, concluded.

Olmsted's books were of no minor consequence in shaping American and English opinion on southern slavery. John Stuart Mill, to whom Olmsted dedicated one volume, wrote of it in the *Westminster Review*, "A work more needed, or one better adapted to the need, could scarcely have been produced at the present time. It contains more than enough to give a new turn to English feeling on the subject, if those who guide and sway public opinion were ever likely to reconsider a question on which they have so deeply committed themselves." Charles Darwin was a close student of Olmsted and much influenced by the books, as were American and English notables like John Eliot Cairnes, Charles Dickens, Edwin L. Godkin, and James Russell Lowell. Harriet Beecher Stowe said of another in the *Independent*, "[It is] the most thorough *exposé* of the economical view of this subject which had ever appeared."[10] Charles Eliot Norton asserted Olmsted's books were the "most important contributions to an exact acquaintance with the conditions and results of slavery in this country that have ever been published."[11] Even a skeptical John Greenleaf Whittier wrote warmly of them.

But the 1850s and 1860s were an inflamed time when passions ran hot on both sides of the slavery issue. Reason was a frequent casualty. Book sales were good, not great. Olmsted's overall unsentimental, unemotional, analytical tone perhaps did not appeal to the more sensational tastes of the popular reading audience, although in each successive volume the reticent author became a bit more impassioned. His book royalties alone provided insufficient income to support him in the aftermath of the Dix and Edwards failure. Nor was the farm profit sufficient motivation to hold his attention. Although handsome at times, receipts from the farm were undependable as a result of his half-hearted management as well as fluctuations in weather and commodity prices. Consequently, when a gentleman approached him in August with a tantalizing opportunity, he seized it.

Olmsted was taking tea at a seaside inn in Morris Cove, Connecticut, where he had retreated to work on the final volume, when Charles Wyllis Elliott approached. Elliott was the secretary of the first board of commissioners for Central Park. Construction of the park had just begun and the board was searching for a superintendent. Elliott, who happened to be taking tea at a nearby table, recognized Olmsted and thought he might make a perfect candidate. The two discussed the politics of the situation (everything in New York at that time was dictated by Tammany Hall politics) and the practical requirements. Elliott encouraged Olmsted to go to New York and apply, and he advised him to see the commissioners and obtain the support of his influential friends. Olmsted replied he would consider the position overnight on the boat trip back to New York. If he had no serious objections by morning, he would apply.

The next morning he had none and so set about the application process. Obtaining the support of powerful people was easy, considering his extraordinary circle of friends and acquaintances. Asa Gray wrote a letter on his behalf; James Hamilton circulated a petition in his support signed by local notables including Peter Cooper, Dudley Field, and Washington Irving. Horace Greeley, Russell Sturgis, and a half dozen other luminaries signed another petition. In addition, Olmsted's application was signed by nearly two hundred of the city's elite, including August Belmont, Albert Bierstadt, William Cullen Bryant, Willard Parker, and Whitelaw Reid.

On September 11, the commission met to consider the applicants: a son of John James Audubon, a city surveyor, a builder of Fifth Avenue houses, a son of the president of Union College, and Olmsted. Olmsted got the job with an annual salary of fifteen hundred dollars despite his having no formal training or experience in park design, construction, or management; but then, no American had these credentials for creating such a large park. Central Park would be the nation's prototype. Fears that Olmsted was only a literary figure lacking the practical skills necessary to oversee the project were assuaged by Washington Irving and others who knew him to be, of all things, a practical, down-to-earth man. His bits of engineering training, his farming and nursery experience, and his travels in Europe where he studied Birkenhead and other English parks supported their arguments. His years before the mast and rambles through the South also suggested he was a person of the sort they sought.

The origin of the foul, dank place for which Olmsted assumed responsibility was twisted in the political and social upheavals that gripped Gotham in the mid-1800s. Prior to 1857, no American city had a public urban open space devoted to recreation; while squares, plazas, small parks, and remnants of commons existed, none was of the extent, function, or character of Liverpool's Birkenhead Park. If urbanites wanted to recreate and refresh in a pastoral setting of grass and trees, they visited one of the popular new "rural" cemeteries, or they hired a carriage and traveled to the rural countryside. Otherwise they had no relief from the dirty, congested conditions of the American city as it metamorphosed from a preindustrial, walking form to an industrial, mass transit form. The need for such pleasant, spacious places became apparent to some who had recently traveled in England and seen its public parks.

The first to make the call for a public park in New York City was William Cullen Bryant. In July 1844 he published "A New Park" in his *New York Evening Post;* and he followed it with many more editorials championing the cause in New York. In 1845 he wrote:

> The population of your city, increasing with such prodigious rapidity; your sultry summers, and the corrupt atmosphere generated in hot and crowded streets, make it a cause of regret that in laying out New York, no preparation was made, while it was yet practicable, for a range of parks and public gardens. . . . There are yet unoccupied lands on the island which might, I suppose, be procured for the purpose, and which, on account of their rocky and uneven surfaces, might be laid out into surprisingly beautiful pleasure-grounds; but while we are discussing the subject the advancing population of the city is sweeping over them and covering them from our reach.[12]

Others followed suit, most notably Andrew Jackson Downing, who took up the call in the *Horticulturist.* In an 1848 essay titled "A Talk about Public Parks and Gardens," he advanced the idea that such parks would surely rival the rural cemeteries in popularity and should be maintained at public expense: "Get some country town of the first class to set the example by making a public park or garden of this kind. Let our people once see for themselves the influence for good which it would effect, no less than the healthful enjoyment it will afford, and I feel confident that the taste for public pleasure-grounds, in the United States, will spread as rapidly as that for cemeteries has done. . . . In short, I am in earnest about the matter, and must therefore talk, write, preach, do all I can about it, and beg the assistance of all those who have public influence, till some good experiment of the kind is fairly tried in this country."[13]

Their arguments took two slants, one mostly practical and the other mostly social. From a practical slant they argued for the preservation of open space for public parks in the heart of the city before all the vacant land was developed. Soon, they suggested, the opportunity would be lost, or it would be prohibitively expensive. From a social slant they stressed the benefits of parks for the health and well-being of the working people, in effect making the provision of parkland a basic civic responsibility. The novel idea found growing support with social reformers when Gotham's population nearly doubled in a decade and living conditions deteriorated as the city swelled with the flood of immigrants fleeing various European upheavals, especially the Irish potato famine in 1848. By the 1850 New York mayoral election, the issue was politicized as a part of both candidates' campaign platform.

As a result, in April 1851, the new mayor urged the city council to secure land for a park while open space was still available at a reasonable price. The council concurred and quickly identified a fine, 150-acre wooded tract, called Jones's Wood, abutting the East River between Sixty-fourth and Seventy-fifth Streets. By July, the state legislature had authorized the acquisition. But by then opposition had arisen to the plan from the very people who originally championed it—Downing had by then concluded the park should be at least five hundred acres in size; and Bryant, who originally identified Jones's Wood as a potential public park in 1844, argued it should be established in conjunction with a second park located around Croton Reservoir in the center of the island. Development of Jones's Wood stalled while lobbying for the additional park land intensified.

Two years later the state legislature authorized the city to acquire a central park bounded by 5th Avenue, 8th Avenue (now Central Park West), 59th Street and 106th Street. For many opponents who questioned the basic notion that public money should be used to fund public parks, two parks were—at least—one too many. Under heavy political pressure, the city withdrew plans for Jones's Wood the following year and proceeded only with the central park.

The next three years were ones of preliminary preparations, including land acquisition. Politics permeated every step. Initial acquisition costs exceeded $5 million. Project opponents contested every step. By 1857, the parcel was in place, a chief engineer appointed, a plan prepared by the engineer approved, and a supposedly nonpartisan commission of eleven prominent citizens appointed to govern its development. Little was accomplished due to lack of money for implementation and the incessant use of the project for the preferential hiring of workers to repay political favors. To get the project moving, the commission hired Frederick Law Olmsted as superintendent on September 11, 1857. It also decided to scrap the engineer's plan and hold an open design competition to obtain a new one. The decision to begin anew resulted, in part, from intense pressure exerted by Calvert Vaux, one of the nation's leading architectural tastemakers.

Calvert Vaux had worked as Andrew Jackson Downing's assistant, then partner, then successor. Downing first met twenty-six-year-old Vaux while in London in 1850. Downing was so impressed with the young English architect that he returned to Newburgh with him in tow. Following Downing's death in 1852, Vaux continued his partner's small Newburgh practice and its design philosophy. In 1854, Vaux married Mary McEntee, sister of Jarvis McEntee, a painter in the Hudson River school. Three years later, he moved the practice to New York. By then Vaux was very influential in his own right, promoting in popular books such as *Villas and Cottages* (1857) the same set of aesthetic and social values advocated by Downing, Davis, and others. During his distinguished architectural career he designed many of the grand homes of the Gilded Age, including portions of Olana, Frederic Church's famous estate overlooking the Hudson River.

Like his American mentor, Vaux was intensely interested in the Central Park project, particularly after Downing's death. But he felt the first plan was marred by "manifest defects."[14] Since Vaux was well-known to several commissioners on the park board, it is likely the board decision to obtain a new design reflected his insistent lobbying.

The public announcement for the competition appeared on October

13. Eight criteria governed the designs: (1) response to the $1.5 million budget; (2) provision for four vehicular crossings from east to west between 59th and 106th Streets; (3) design of a 20-to-40-acre parade ground with proper consideration for spectators; (4) design of three playgrounds, each 3 to 10 acres; (5) designation of a site for a future exhibition hall; (6) location of a large fountain and prospect tower; (7) design of a 2-to-3-acre flower garden; and (8) location of a site for winter iceskating. The winning design would receive a two thousand dollar prize.

The challenge facing the entrants was formidable. Each design shared common characteristics with the others, since they all shared the same site conditions and the same specifications for the 770-acre park (when the park was later expanded to extend beyond 106th Street to 110th Street, its size grew to 843 acres). The differences between the designs arose from the manner in which each solved aesthetic, functional, and social issues.

Thirty-three designs were submitted by the closing date. The jury awarded first prize to entrant number thirty-three, the last submitted, known only as "Greensward." The winning design, it turned out, was submitted by Frederick Law Olmsted and Calvert Vaux. After a few minor changes in the plan, the board officially approved it. The board also abolished the posts of superintendent and chief engineer, and assigned to Olmsted the duties of both as architect-in-chief, at an annual salary of twenty-five hundred dollars. He was authorized to call on the services of Vaux and six field assistants as needed. Work proceeded at full pace. By autumn 1858, he commanded an army of twenty-five hundred men busy at work on the site, one-half mile wide by two and a half miles long. The next summer, the army grew to thirty-six hundred.

The Olmsted-Vaux partnership arose at Vaux's instigation. The two were likely acquaintances as a result of Olmsted's friendship with Downing, although they had not formed a close relationship. Vaux's invitation made great practical sense. No one knew the site better than Olmsted, and no one was better connected politically than Olmsted. In addition, his various engineering-related skills and his familiarity with Birkenhead and the romantic style of landscape design popular in Europe reinforced Vaux's architectural training and design philosophy. Olmsted hesitated, afraid his participation might offend the chief engineer. But with his boss's blessing, the new partnership set about at night creating its submission to the competition by mid-January, while during the day Olmsted led the work crew he had transformed into an efficient force on site preparations.

The detailed story of the plan's full implementation over nearly twenty years is an amazing story of politics, personnel management, and massive site engineering (they moved 5 million cubic yards of soil, and installed 114 miles of drainage pipe). Olmsted and Vaux participated off and on almost the entire time. As was the case throughout their entire professional collaboration, Olmsted concentrated on site design while Vaux concentrated on the architectural design of structures. The partnership grew rapidly, building on the fame and stability afforded by the project. Other landscape design projects for residences, institutional grounds, and parks cluttered the firm as the Central Park project progressed. But the Civil War did not care, not for the concerns of John Muir, John Wesley Powell, nor Frederick Law Olmsted. Appeals to Olmsted's social conscience again thrust him against a challenge where his administrative and organizational abilities, and his influential political and social connections, proved profoundly important. In June 1861, Olmsted took a leave of absence to head the U.S. Sanitary Commission as executive secretary.

The commission, then in its infancy, supervised medical care for the Union Army. During the politically charged war years, Olmsted shaped the commission into an effective organization that survives today as the American Red Cross. His political battles for the commission in Washington, and his travels to the battlefields on its behalf, put him on a first name basis with the nation's leaders: Lincoln, Grant, Garfield. Throughout he kept tabs on the slowed progress of his beloved Central Park.

The first portions of the park opened to rave reviews in 1863. Unfortunately, political problems stemming from the appointment of a comptroller over the firm of Olmsted and Vaux prompted them to resign from the project. Olmsted also resigned from the Sanitary Commission as a result of poor health and mental exhaustion from its constant battles.

To recuperate, Olmsted moved west for new challenges and a change in scenery. There he managed the seventy-square-mile gold-mining properties of John Charles Frémont's Mariposa Mining Company at the mouth of the Yosemite Valley in California. His exposure to the Mariposa Grove of Big Trees and Yosemite touched off his interest in the American conservation movement, in which he was to play a central role. He also produced plans for other projects, including the campus for the University of California at Berkeley and an adjacent residential community.

Vaux enticed a reinvigorated Olmsted back to New York after two years at Mariposa, with a new working agreement with the Central Park board and a new contract to design a large park in Brooklyn called

Prospect Park. The honeymoon did not last long. Political shenanigans of the Tweed Ring triggered the firm's second resignation from the Central Park project in 1870. A year later they were rehired.

In 1872, after more than a decade of landmark work in New York and the nation, work in which they reshaped urban America with their pastoral designs for grand public parks, Olmsted and Vaux quite amicably dissolved their partnership for mutual convenience. In the aftermath, Olmsted continued as head of the Central Park project; Vaux became a consultant. In 1873, out of frustration, Olmsted resigned a third time, then agreed to remain as a consultant. Finally, in 1878, he fully and permanently withdrew in the face of renewed Tweed Ring activities.

It's fitting that America's prototype urban park was constructed in the center of its major metropolis. Although the design of Prospect Park was perhaps the better reflection of Repton's and Pückler-Muskau's philosophies, Central Park set the stylistic standard for American park design and reinforced the emerging American preference for the romantic, pastoral landscape. In 1869, the Olmsted and Vaux plan for a new sixteen hundred–acre railroad suburb along the Des Plaines River nine miles from the center of Chicago produced another American prototype, the prototype for the modern American suburb.

Others, to be sure, had proposed the suburb's social relevance and physical character in books and articles, and some suburban communities exhibiting many of those characteristics had evolved by then. Most of the early suburbs, however, used the same gridiron layout for roads that was common in most cities. The grid was a national obsession, not only in the national land survey but in urban planning as well. From New York to Chicago to San Francisco, the grid dominated street layout in big and small cities as it was laid over the landscape with the same rigidity and disregard for topography and physical features as Jefferson's Land Ordinance. Nor was the overall aesthetic character of early suburbs fully reflective of the romantic style that eventually dominated the landscape. The physical evolution of the suburb was not yet mature. Riverside, Illinois, was perhaps the first large-scale project that went from inception to fruition specifically designed in the new suburban style.

Yet it drew on important precedents set by other projects. Olmsted likely drew on Andrew Jackson Davis's plan for Llewellyn Park, New Jersey, an exclusive four hundred–acre development in the Orange Mountains overlooking Manhattan Island, designed about a decade earlier. The Llewellyn plan was remarkable for its use of gently curving streets,

a centrally located open space, and covenants controlling the development's governance and land use. Olmsted knew Davis well and was familiar with the Llewellyn plan; when hired to design the Berkeley property, Olmsted even said he proposed to lay out the property "on the
Llewellyn plan."[15] However, Riverside's influence on the American landscape superseded the influence of earlier suburban prototypes because
Olmsted went on to design many more like it, and Olmsted individually
became the nation's leading landscape designer. Olmsted's work included
scores of influential projects such as the Biltmore estate, the design of
Boston's "Emerald Necklace" system of parks, a campus for Leland Stanford's new university in Palo Alto, even the site design for the Columbian
Exposition in Chicago.

The creation of Riverside marked perhaps the first time all the economic, technological, social, and philosophical changes of the nineteenth
century combined in land planning. The year that work formally began
on Riverside (1869), John Muir first walked into Yosemite, John Wesley
Powell first descended the Colorado River, and shipments of cattle had
just begun from Abilene, Kansas. Frank J. Scott was about to install the
lawn as a national icon. The population of greater New York, the nation's (and soon the world's) largest metropolitan area, approached 1.4
million persons. There, commuter ferries and horsecars each carried about
50 million passengers per year. In the nation's heartland, Chicago was
the nation's railroad hub, growing faster in proportion to its size than
any major city in the world. There the railroad was transforming the
city into an industrial center that Rudyard Kipling later called "a real
city," unlike the others he had visited in the West. In response to its
corruption and crowding, he wrote, "Having seen it, I earnestly desire
never to see it again."[16] In addition to the dirty industry and foul stockyards, the railroads also brought Chicago a host of remote suburbs strung
along the rail lines like beads on a string.

In August 1868, Olmsted first visited the Riverside site owned by
Emery E. Childs and a group of eastern investors who formed the Riverside Improvement Company. The site was surrounded by the sea of
open prairie that covered much of Illinois, a subtle landscape devoid of
the blatant physical features of eastern landscapes like those at Llewellyn
Park. The Des Plaines River lent the site topographic and visual relief
from the prevailing monotony, although trees were spartan and survived mostly in groves along the river's shallow floodplain. While desolate from an eastern viewpoint, the site contrasted with the surrounding
prairie sea; nor was it isolated in that vast expanse—Chicago loomed

just over the horizon. Riverside already possessed the first stop outside the city on the main line of the Chicago, Burlington, and Quincy Railroad. There the developers planned to open one of the nation's first large-scale, preplanned suburban real estate projects, a project in which they gave almost free hand to Olmsted, Vaux, and Company.

Olmsted's preliminary report was quickly completed by September 1. In it he outlined the basic elements of the intended plan. The final plan completed a bit later covered the development's every detail from the layout of roads and lots, to the location of schools, parks, and walks, to the detailed specifications for the infrastructure, including lighting, drainage, and water supply; it even specified raising the height of the river's mill dam to make an impoundment large enough for boating and skating. The plan called for a broad approach boulevard that connected the community with Chicago, but this was never constructed. It proposed a 160-acre park with winding walks, a public drive, boat landings, and rustic pavilions running three miles along the river, and a series of small, informal parks and community spaces that wound throughout the development. To Olmsted, the essential qualification of the suburb was domesticity and the idea of habitation: "There are two aspects of suburban habitation that need to be considered to ensure success; first, that of the domiciliation of men by families, each family being well provided for in regard to its domestic indoor and outdoor private life; second, that of the harmonious association and co-operation of men in a community, and the intimate relationship and constant intercourse, and inter-dependence between families. Each has its charm, and the charm of both should be aided and acknowledged by all means in the general plan of every suburb."[17]

Olmsted believed the primary requirements of a suburb were good roads and walks, designed with pleasant openings and outlooks to suggest refined domestic life, secluded, but not far removed from the community. He felt the grid too stiff and formal for those purposes, recommending that roads gracefully curve through generous corridors to suggest leisure, contemplation, and happy tranquillity. He sought to promote rural attractiveness by offering generous lots, most measuring 100 by 225 feet, and an environment that combined urban conveniences with the domestic advantages of a charming rural landscape. To convey the desired feeling of spontaneity, he proposed that trees be planted at irregular intervals. To give a sense of openness, he insisted that homes be set back 30 feet from the street; and to suggest prosperity and elegance, he required that homeowners maintain immaculate gardens.[18]

The plan sought to provide privacy for the indoor and outdoor domestic life of individual families while maintaining a sense of community. That difficult balance was achieved through the careful location of homes, walks, roads, and plantings. "The grand fact," he wrote, that residents "are Christians, loving one another, and not Pagans, fearing one another," should be reflected in "the completeness, and choiceness and beauty of the means they possess of coming together, of being together, and especially of recreating together on common ground."[19]

Beyond the idealistic hopes of neighborliness, the plan's site engineering was technically sophisticated and its financial plan carefully calculated. It was, after all, serious business in addition to an aesthetic statement and form of social philosophy. The firm of Olmsted and Vaux made a topographic survey of the site, provided all landscape architectural and architectural services, and supervised construction. In return, they were to be paid 7.5 percent of the cost of site improvements. Those they estimated at $1.5 million, so the firm stood to make $122,500, or its equivalent in lots. The Riverside Improvement Company estimated it would have 225,000 feet of salable frontage, for which it paid an average cost of $1.52 to acquire and $5.00 to improve. That frontage, they calculated, would then sell for $40 to $60 per foot, netting a profit of about $1 million as the property was sold over five years, even after setting aside $3 million for unexpected contingencies.

Riverside was a popular success. Unfortunately, the developers fared badly financially. Apparently the initial cash flow was not what they expected, and within the first year of construction, they defaulted on payments to Olmsted, Vaux, and Company. Sales then plummeted after the great Chicago fire in 1871 and the ensuing credit crunch. By 1873, the Riverside Improvement Company went bankrupt during a general financial panic; however, the extraordinary quality of the basic design and site work already completed carried the project forward successfully with new owners.

As streetcar suburbs proliferated around urban America, few developments copied in toto the landscape style of the Riverside and Llewellyn Park prototypes, especially the curvilinear road layout. Variations on the theme were commonplace. The variations lessened, though, with the spread of the personal automobile after World War II. The dominant suburban design style gradually came to center on that established by the prototypes, as most of suburban America today bears witness.

Like Jefferson and Hamilton, Olmsted was a social engineer who sought to shape the landscape to his vision. He was strikingly similar to contemporaries John Muir and John Wesley Powell: each possessed a love of landscape and a passion for rambles to learn about it firsthand; each struggled to resolve personal piety, religious training, social attitudes, and interest in science; each used political influence and literary publications to advance his beliefs; and each offered America a new vision of the landscape through innovative ideas on conservation and land use.

Unlike Muir and Powell, Olmsted acquired that vision from his love of scenery and his social conscience. Those same sources shaped his design ideas for the American landscape, ideas Muir and Powell probably did not share. While Olmsted's plan for suburban American was not innovative, it was profoundly influential. It translated prevailing social concerns and aesthetic values into physical form, giving expression to common desires to flee the city and return to nature for physical and psychological cleansing. His designs made contact with nature possible in city parks or suburbs, so that we could salve our wounded souls, refresh and reinvigorate the body and spirit. They provided refuge of the sort seen in Claude Lorrain's idyllic landscapes, not the wild scenes of Central and South America painted by Frederic Church as his career waned when he, like Muir, sought to follow in the footsteps of Alexander von Humboldt. Neither was Olmsted's version of nature the crude form found in the forest or on the farm. Olmstedian suburbia was a middle ground, nestled safely between city and wilderness. Suburbia sat securely situated between the extremes of too much civilization and too much nature. People like Beecher, Bushnell, and Downing, even Thoreau, impressed those values on the American mind. Places like Central Park and Riverside made their physical imprint on the American landscape.

After a debilitating physical and mental decline, Frederick Law Olmsted, the principal landscape architect of suburban America, died on August 28, 1903. His passing attracted little fanfare for a life of such impact. Probably by 1903 his effect on the landscape was so deeply engrained in our landscape perception and expectations that it simply appeared natural, as if having always been there, and so was no longer noticed. Perhaps his role in shaping attitudes toward slavery was overshadowed by other, more emotional, voices and events. Perhaps his role in the conservation movement was overshadowed in a similar manner.

CHAPTER 18

City as Suburb

Whatever its imperfections, Levittown is a good place to live.
Herbert Gans, *The Levittowners: Ways of Life
and Politics in a New Suburban Community*

By the early 1900s, the profound economic restructuring and the corresponding sociocultural changes begun in the 1800s had, after nearly one
hundred years, transformed urban America. The dense, chaotic concentration of dwellings, small shops, and offices that dominated the large
walking city was a short-lived phenomenon. Certainly remnants of the
previous urban form remained in many sectors, but expansive new
growth responsive to the streetcar became dominant. First the horsecar
and commuter train and, by the late 1880s, the streetcar gradually transformed the old urban landscape into a more star-shaped pattern, as new
suburban development located within walking distance of the transit
routes radiated outward from the city's central core. Land beyond walking distance often sat vacant.

The walking city evolved into the streetcar city: a new urban landscape of lesser density but much greater expanse than its predecessor; a
mosaic where land uses began to segregate into zones based on compatible function; a pattern stitched together by streetcar lines that func-
tioned as conduits that drew housing development out from the core and
returned commuters to the core to work and shop. By this time metropolitan American was unique not because of the size of its cities, but
beçause of the extent of their suburban sprawl, the number of its commuters, and the proportion of its homeowners.[1]

Streetcars cut the chains that limited a city dweller's mobility. Eco-

nomical, clean, and convenient, they allowed people access to employment and services beyond their immediate neighborhood. Since the lines
radiated outward from the city's center, it was there that streetcars focused most of that new mobility. There, giant department stores arose
to satisfy the shopper's every need; there most business, government offices, and cultural facilities congregated, so the core remained the heart
of the city despite suburban flight.

The intense competition for land in the core that ensued drove property prices higher, displacing land uses unable to generate the profits
needed to meet the higher costs. Most forms of housing were driven either to the cheaper land in outlying suburbs, or into taller tenements.
Innovations in building technology, including the elevator, steel curtain
wall construction, electric lights, and telephones, in combination with
rising property prices, pushed buildings skyward.

At the opposite end of the line, people spent their leisure time at country clubs, racetracks, beer gardens, amusement parks, and resorts. Play-
grounds like Coney Island, often developed by the streetcar and commuter railroad companies as a means to spur ridership, sought to satisfy
society's growing demand for leisure activities. There city dwellers could
escape after work and on weekends. There, contact with nature, often
in a suburban setting at the city fringe, refreshed and reinvigorated the
city dwellers. Urbanites unable to afford a suburban home could temporarily enjoy social and physical benefits similar to those offered by
suburban life. And there the social theories and design philosophies of
people like Catharine Beecher and Andrew Jackson Downing, philosophies previously expressed in physical form only in picturesque rural
cemeteries, pastoral urban parks, and spacious suburbs for the affluent
found further expression.

In the late 1800s the private country clubs of the wealthy perhaps
best personified the leisure, suburban lifestyle—the good life. Like the
velvety, well-manicured suburban lawn, private country clubs were associated with what Thorstein Veblen called "conspicuous consumption"
in the *Theory of the Leisure Class* (1899). At the turn of the century,
there were only a handful of clubs, most of them located in wealthy
suburbs, such as Brookline (Boston), Lake Forest (Chicago), and South-
ampton (New York); during the second decade of the twentieth century
the number approached five hundred, and by the start of the Great Depression in 1930, it reached nearly six thousand. Ninety percent of the
clubs were private. The illusion of the English landscape that soon became their most popular design style reinforced society's preference for

the romantic landscape.[2] The country club and suburbia became synonymous, both in social structure and physical image. For most Americans, contact with little-touched nature had long been lost. Natural was
redefined to mean suburban lawns, manicured greens, and all-weather
tennis courts.[3] The image of contemporary communities like Concord
Estates had been born.

The wealthy, together with a lower caste of menial workers they
took along, were the first to flee the walking city for the fresh air and
tranquil spaciousness of remote suburbs like Riverside and Llewellyn
Park. As the Industrial Revolution and corresponding revolutions in
technology and sociocultural values continued in the post–Civil War
era, and as prosperity spread, more of the rapidly expanding middle
class joined the exodus. Though most could not afford a home in the
remote and elite suburbs, they could afford a home in a middle-class
neighborhood built on vacant land near a village more centrally located. Or they found a home in one of the new suburban communities
at the urban fringe that sprang from scratch as naturally as crops that
grew in the fields several seasons before. Like the rings that denote annual tree growth, the growth of those middle-class streetcar suburbs en-
cased the core, forming concentric rings expanding outward along the
trolley lines. Commuter access and the provision of municipal services
were the critical nutrients for that growth.

The relation between property value, commuter access, and city services was well known to real estate developers, corporate officers, and
municipal leaders by the turn of the century. For those in control of it,
profits or political power were the principal yields. They knew that the
value of inexpensive land at the urban fringe would skyrocket if the
land were integrated into the urban fabric by extension of a trolley line
and municipal services. Yet most felt the new developments satisfied a
social need to provide affordable housing and to reform and improve
the living conditions of tenement dwellers. Those who promoted the suburban flight did so for many reasons—some economic, some political,
some social.

Typical lot sizes in the new suburbs were small because residents
needed to crowd close to the streetcar tracks; in contrast, spacious suburbs for the affluent remained rare. Although the prototypes promised
gently curving streets, many new suburban neighborhoods were laid out
in a gridlike pattern. Primary residential streets ran perpendicular to
the streetcar lines to serve as feeder lanes down which the commuters
walked to catch the trolley. The lines defined the city's main thorough-

fares and were bounded by the city's premium property, particularly the closer the property was located to the city core.

Construction costs for basic homes also dropped during the nineteenth century following the advent of balloon-frame construction in the 1830s, the same construction technique that helped open the Great Plains to settlement. No longer was the city dweller who wanted an individual residence limited to an expensive custom home built by a cadre of master craftsmen. The new form of construction was simple, straightforward, and economical, and it lent itself to the standardization necessary for tradesmen to follow. It transformed the construction of individual residences from a specialized craft affordable only by the affluent into a large industry, in much the same manner that Ford's Model T assembly line transformed auto production from a highly skilled custom craft conducted in small shops and garages to a giant industry of mass production. As the new technique of home construction spread, the demand for inexpensive, single-family homes was fueled by social and physical tastemakers. An avalanche of books and magazines, including Downing's *The Architecture of Country Homes* (1850), Calvert Vaux's *Villas and Cottages* (1857), and *Godey's Lady's Book,* published standard house plans as they and others promoted the new concepts of domesticity and the new suburban aesthetic. By the turn of the century, entire homes became available via mail-order catalogue.

The flight of the wealthy from the city core, followed by the middle class, furthered the socioeconomic segregation of urban America. Where in the walking city people of various ethnic heritages and economic stations lived in close proximity, the mixing of rich and poor, black and white, Jew and Gentile lessened in the streetcar city. The melting pot of urban America began to dissipate into homogenous neighborhoods. The central city, abandoned by the well-to-do and the middle class, became home to the poor and to disenfranchised minorities. That segregation was soon reinforced, and in some cases institutionalized, by government programs for the thinly disguised purpose of promoting suburbs and single-family home ownership. In some ways, the preferential treatment of suburbs, and urban segregation, continues little abated today.

The population of American cities continued to grow during the streetcar era, even though densities were declining and suburbs were flourishing. Densities skyrocketed in some interior sectors as land was recycled to house greater numbers in taller and more tightly packed tenements, high-rises, and row houses. At some point, however, that rise was offset by the rapid pace of annexation and low-density suburban sprawl. New York City grew from 22 square miles in 1850 to 299 square miles in 1910, having absorbed Brooklyn, Queens, Staten Island, and parts of Westchester County now known as the Bronx. During the same period Chicago grew from 10 to 185 square miles and Philadelphia from 2 to 130 square miles. By the turn of the century, much of metropolitan America became aggregations of suburbs strung along streetcar lines extending outward from the remnant core of the original walking city.

But the streetcar city was short-lived too. By the 1920s another type of transportation had begun to dominate developmental form and reconfigure aspects of the old. The automobile transformed metropolitan America, creating the contemporary auto-oriented landscape. Today the preponderance of city and suburban development has occurred under its influence.

While streetcars followed prescribed routes, enticing development along with it, cars went wherever the driver wished within the network of roadways; development followed. The car made vacant land in the interstices between transit routes developable if the land had highway access, particularly if municipal services were offered. "The automobile so vastly changed the equation," wrote Kenneth Jackson, "that cities began to 'come apart' economically and functionally. . . . Indeed the automobile had a greater spatial and social impact on cities than any technological innovation since the development of the wheel."[4]

The transformation progressed in fits and starts. The first horseless carriages were built before the first electric streetcars, but they remained novelties for many years. In 1898 there was only one car in America for every eighteen thousand people, mostly crude, unreliable curiosities for the rich. Streetcars remained the primary means of intraurban transit, while railroads were used for most interurban travel. Outside the central city few roads were paved. Most were little more than dirt paths cleared of debris, making travel by car nearly impossible much of the year. The noisy contraptions frightened horses, leading many states and cities to pass laws limiting their use. A common ordinance in many communities limited speed to four miles per hour and required cars to be preceded by a person on foot carrying a lantern or flag to warn of its approach. At the turn of the century the future did not bode well for the automobile. In 1899 the *Literary Digest* captured the prevailing public opinion: "The ordinary 'horseless carriage' is at present a luxury for the wealthy; and altho [*sic*] its price will probably fall in the future, it will never, of course, come into as common use as the bicycle."[5]

Within a decade the public's skepticism and hesitancy toward the

newfangled horseless carriage gave way to an infatuation that would reshape our culture, our economy, and our landscape. In 1905 the song,
"In My Merry Oldsmobile," captured the nation's new love affair:

Come away with me, Lucille,
In my merry Oldsmobile,
Over the road of life we'll fly,
Autobubbling you and I,
To the church we'll swiftly steal,
And our wedding bells will peal,
You can go as far as you like with me,
In our merry Oldsmobile.[6]

The automobile captured a prominent place in the American heart,
and American ingenuity made the expensive, flimsy contraptions more
affordable and more reliable. By 1908, twenty-four American companies were producing inexpensive vehicles that were paragons of quality
by previous standards, including Henry Ford's first Model T. By 1913,
the number of cars rose to one for every eight people, and by 1927, to
one for every five. That proliferation posed a multitude of problems for
cities.

How were the new horseless carriages to be accommodated on the
notoriously rutted and dilapidated city and suburban streets, streets intended for horses, carriages, or streetcars? Country roads were even
worse. How was the crush of traffic on congested streets teeming with
cars, trucks, trolleys, wagons, and carriages to be regulated? And who
would pay for the many modifications to the public right-of-ways necessitated by the new vehicles? Should public taxes be used to pay for improvements necessitated by private vehicles, or should user fees be assessed? As the number of registered vehicles grew from fewer than ten
thousand in 1905 to 2.5 million in 1915, to 20 million in 1925, how was
the free-for-all in personal transportation to be organized and regulated?

The course chosen was for government intervention and the use of
public funds. Removal of the messy horse was for many the salvation of
the city and a proper use of municipal monies. The most urgent need
was for more and better paved roads that accommodated the higher
speeds and limited flexibility of mechanized traffic. A horse could deftly
step over an obstacle while a car could not. By the 1920s, most urbanites and planners considered the road to be the primary purview of the
automobile.

The thrust for street improvements was based on special interest as
much as the idealism of urban or social reform. Members of a broad

coalition of private-pressure groups, including the auto manufacturers, suppliers, and service industries, road builders, land developers, and merchants, knew highway construction fueled their economic interests. Publicly financed street improvements, they argued, would pay for themselves by increasing adjacent property tax revenues. City planners joined the chorus of commercial and civic supporters.

The private car and the newfound freedom of mobility it offered were the ultimate expression of independence and individualism. That unprecedented freedom disconnected the individual from much of the communal urban experience. Changes in urban form as the city metamorphosed to accommodate the car magnified the effect. The very notion of city as a cosmopolitan place, an urbane environment where a sense of communal rights and responsibilities tempered those of the individual, was altered by the car. Perhaps that cosmopolitan sense was never very strong in America. Perhaps it merely masked a more powerful, Jeffersonian aspect of our national psyche. Whatever the psychology, as our infatuation with the automobile spread, public attention turned away from policies and programs to improve cities in favor of policies and programs that promoted suburbs. The problem this created for urban America was not just that new programs aided suburbs; in addition, there were few corresponding programs to address urban issues such as housing and mass transit.

The Federal Road Act of 1916 was one such program that preferentially benefited suburbia. It offered federal funds to states that organized highway departments, reflecting the government's historical funding for railroads, canals, and early turnpikes. While the road improvements it triggered benefited both city and suburb, economically and socially, the program began an abandonment of federal assistance to cities that grew more pervasive concurrent with the middle-class flight from city to suburb. Roads were seen as public necessities warranting public subsidy. The streetcar was seen as a private enterprise that should pay for itself.

The Federal Road Act of 1921 provided further assistance for highways by designating two hundred thousand miles of "primary" roadway eligible for federal matching funds. It also created the Bureau of Public Roads to plan a highway network to connect all cities having a population of fifty thousand or more. By 1930 every state had enacted a gasoline tax to fund highway construction. Yet no such sources of revenues were established to aid urban mass transit, and other urban infrastructure and housing needs were often left to languish.

As highways improved around urban areas, the cost to commute from suburb to city in a private car as measured in time, convenience, and comfort, not to mention money, grew more attractive. The new highways stitching city to suburb smoothed the suburban flight. New roads and more cars doomed the streetcar city to obsolescence. As Mel Scott noted in *American City Planning,* the car crippled the city core and promoted construction of outlying commercial development. It spread the city over the largest area in history by encouraging families to relocate to distant developments. It heightened tensions between old and new communities as it hastened the annexation of outlying communities and promoted the premature subdivision of rural land during periods of prosperity. This sprawl created huge assessments to finance the requisite infrastructure that eventually became burdens upon all the taxpayers, and it complicated municipal governance.

Other forces were also at work reshaping urban and suburban America into our contemporary auto-oriented landscape. By the turn of the century the characteristics of the streetcar city created a serious problem rarely addressed in the walking cities. As the random mix of land uses in the walking city segregated into the residential, commercial, and industrial zones of the streetcar city, the earlier problems of incompatible land uses hopelessly mixed in the urban soup became potentially solvable. Where before the juxtaposition of incompatible uses such as housing and industry were nearly inevitable due to the random mix resulting from the lack of mass transit, and the other socioeconomic limitations of the time, in the early 1900s those limitations were largely lifted and the problems became avoidable through proper planning. For the first time, people no longer had to live next to obnoxious factories or offensive businesses. The avoidance of such nuisances was an important part of the flight to the suburbs. There the homeowner's newfound sense of domesticity could be satisfied in the tranquillity of green, spacious surroundings, and there the family's investment could be protected in a stable setting of similar people and property, *if* proper controls were placed on development.

Such land use issues were not limited to suburbs. As the streetcar lines concentrated business and commercial activity in the core, and tall buildings reached skyward, new public health and safety problems and economic issues arose from the new urban form. The construction of more obnoxious factories during industrialization often exacerbated nuisance problems over increasingly large areas. As cities grew outward into surrounding rural land, they often annexed or absorbed noisy, unsightly land uses that before were benign but became bothersome when

engulfed by suburban development. In addition, social reformers sought
ways to ease the demeaning conditions in the crowded tenements through
the establishment of building standards. Others sought to improve urban
life through the beautification of the city with green, open spaces and
tree-lined boulevards. Traffic engineers sought solutions to the chaos on
city streets created by the car, and city officials looked for more efficient
ways to provide infrastructure and municipal services (particularly fire
protection and health care) throughout the sprawling metropolitan maze.
The key to all those efforts was proper land use planning that regulated
the location and character of development.

Yet neither the new urban nor suburban form was governed by comprehensive land use planning or regulation. In the walking city, development was mostly unplanned and unregulated. People built whatever
they wanted, wherever they wanted, in a land use free-for-all with remarkably few controls or constraints. Until the early 1900s the regulation of private property necessary to enable centralized planning seemed
to many a frontal attack on the American Constitution and one of our
most cherished freedoms. By the early twentieth century, suburban home
ownership created a collective economic and social self-interest that overcame the philosophical resistance to governmental restrictions on use of
private property. Pressure for greater control triggered responses. One
became popular with the public: zoning to protect neighborhoods from
invasion by factories, stores, and apartment houses for the general purpose of protecting property values and avoiding nuisances caused by the
juxtaposition of incompatible land uses. Land use in this context meant
the uses or activities associated with a property and its structures, and
the physical character of those structures.

New York City enacted the first modern zoning ordinance in 1916.
Within ten years, cities of all sizes all across the country adopted similar
ordinances. Most laws were based on a model developed by a special ad-
visory committee appointed in 1921 by the secretary of commerce, Herbert Hoover. Published in 1924, it remains the heart of land use regulation in American cities, although revised and updated by another model
several years later:

> Section 1. Grant of Power.—For the purpose of promoting health, safety,
> morals, or the general welfare of the community, the legislative body of cities
> and incorporated villages is hereby empowered to regulate and restrict the
> height, number of stories, and size of buildings and other structures, the per-
> centage of lot that may be occupied, the size of yards, courts, and other open
> spaces, the density of population, and the location and use of buildings, struc-
> tures, and land for trade, industry, residence, or other purposes.

Sec. 2. Districts.—For any or all of said purposes the local legislative body may divide the municipality into districts of such number, shape, and area as may be deemed best suited to carry out the purposes of this act; and within such districts it may regulate and restrict the erection, construction, reconstruction, alteration, repair, or use of buildings, structures, or land. All such regulations shall be uniform for each class or kind of buildings throughout each district, but the regulations in one district may differ from those in other districts.

Sec. 3. Purposes in View.—Such regulations shall be made in accordance with a comprehensive plan and designed to lessen congestion in the streets; to secure safety from fire, panic, and other dangers; to promote health and the general welfare; to provide adequate light and air; to prevent the overcrowding of land; to avoid undue concentration of population; to facilitate the adequate provision of transportation, water, sewerage, schools, parks, and other public requirements. Such regulations shall be made with reasonable consideration among other things, to the character of the district and its peculiar suitability for particular uses, and with a view to conserving the value of buildings and encouraging the most appropriate use of land throughout such municipality.[7]

The U.S. Supreme Court, and the lower courts, ruled incrementally on the legality of zoning throughout the early 1900s. The principal test case on the general validity of comprehensive zoning regulations was decided in 1926 in *Village of Euclid et al. v. Ambler Realty Co.*[8] When the case arose several years earlier, Euclid was a small Cleveland suburb of about seventy-five hundred people. Most of the land in the village's fourteen square miles was agricultural or undeveloped. At the time, Ambler Realty Company owned sixty-eight acres of vacant land at the western end of the village, land it was holding for later industrial development. Such development was advancing westward directly toward the property. In the interim, the Village Council adopted a comprehensive zoning ordinance that divided the community into zones that regulated and restricted the use of property, the size and height of buildings, and the lot size. The new ordinance restricted the Ambler property, then bounded on two sides by residential subdivisions, to nonindustrial use.

Ambler argued that the ordinance unfairly devalued its land, without compensation, from ten thousand dollars per acre for industrial use to twenty-five thousand dollars per acre for residential use. In effect, Ambler stated that the ordinance was an arbitrary and unreasonable attempt by the community to erect "a dam to hold back the flood of industrial development and thus preserve a rural character in portions of the Village which, under the operation of natural economic laws, would be devoted most profitably to industrial undertakings."[9] To what extent, and under what conditions, could zoning restrict one's constitutional guarantee to the right of property? *Euclid v. Ambler* was a critical test case.

The court ruled that zoning was constitutional, even though it might adversely affect property values. The key was whether the zoning was a legitimate exercise of the police power to protect public health, safety, morals, and welfare. The court recognized that police power was flexible and must adjust as society becomes more complicated; the pace of change in the city precluded fixed or rigid rules. So where was the line separating the legitimate versus the illegitimate use of this power to be drawn? The court said the line was fuzzy by necessity and had to reflect the specific situation. Regulations that might be appropriate in one city, like New York, might not be in another, like Euclid: "a nuisance may be merely a right thing in the wrong place,—like a pig in a parlor."[10]

Today the courts have established five bases for valid zoning: it must serve a valid public purpose related to the police powers; the means used by the zoning must be related to the purposes; it cannot constitute a "taking" that leaves no reasonable use of a property without compensation; it must be done in accordance with a comprehensive plan; and it cannot violate any other constitutional provisions.

The effect of zoning on urban and suburban America was immediate and profound. Some effects fell outside zoning's set of valid purposes. Today in communities like Concord Estates, zoning and land use–related regulations restrict not only the density of neighborhoods and the mix of land uses but also the appearance of a home, including aspects of its architectural design, its color, and the palette of exterior materials. Standards restrict the height and placement of fences around one's property or prohibit them outright. They prohibit the placement of a small storage shed in one's yard, or the use of clotheslines, or even the parking of personal vehicles, such as RVs, trailers, and boats, on one's driveway. Such standards can have direct or indirect implications related to income, ethnicity, education, age, and religion. Many communities now struggle to find a reasonable balance between individual property rights and legitimate community interests, a balance often tilted by fundamental socioeconomic and cultural concerns. Zoning is a local form of land use planning and control, instituted and implemented by residents of the community. It is the primary tool we now use to shape the suburban landscape to our liking.

Both zoning and highway subsidies promoted suburban development, and both reflected popular suburban preferences while further engraining the preferred pastoral landscape style. Another form of government involvement initiated during the Depression to address the housing crisis that gripped the nation had the same effect. And it too was colored by larger socioeconomic issues.

The catastrophic collapse of the national economy during the 1930s had far-reaching effects on America, including its landscape. Nowhere was that more evident than in suburbia, mostly because home mortgages before the Depression were much different from the ones today. Mortgages required large down payments, often one-third to one-half the loan amount. Mortgages also lasted a short period, often only five years. This required a large balloon payment at the end of the period to pay off the remainder. Homeowners typically refinanced to meet the payment. That was not a problem so long as interest rates were little changed and the general economy was strong. In contrast, mortgages today have low down payments, often less than 10 percent, and have long payoff periods, often thirty years, at the end of which the entire debt is eliminated. The major differences between home financing then and now arose in reaction to the Depression and the thousands of tragedies it triggered.

In 1926, a typical American home cost about five thousand dollars. By 1932, that same home sold for only thirty-three hundred dollars, wiping out the equity savings of millions of middle-class families. Mortgage foreclosures in 1926 numbered about 68,000. In the midst of the Depression, in 1932, the number jumped to 250,000. By 1933, fully half of all the home mortgages in the country were technically in default, and foreclosures ran at over a 1,000 per day. From 1928, just before the Depression began, to 1933 at its depth, new home construction dropped 95 percent, and expenditures on repairs fell 90 percent.

All this happened to a nation firmly committed to suburban home ownership. As President Herbert Hoover said at a national conference in 1931 to address the nation's housing crisis, "I am confident that the sentiment for home ownership is so embedded in the American heart that millions of people who dwell in tenements, apartments, and rented homes . . . have the aspiration for wider opportunity in ownership of their own homes."[11] By the early 1900s, the dream of a comfortable single-family home set in pleasant suburban surroundings was firmly ensconced in the American psyche. The 1920 national census reported that more than half the nation's population lived in urban areas—America was no longer an agrarian nation of farmers; it was an industrial nation of city and suburban dwellers.

In the early 1930s, the collapse of the nation's housing market was widely perceived to be a major contributor to the country's economic and social crisis, and its resurgence one of the keys to recovery. To address those concerns, President Hoover convened the National Conference on Home Building and Home Ownership in 1931. More than four hundred specialists participated. Four recommendations emerged, ones that suggested a radical shift in public policy regarding housing. The recommendations promoted a major boon to speculative builders, appliance manufacturers, automobile companies, and other industries related to suburban home construction.

The first recommendation called for banks and savings and loan associations (S & Ls) to use long-term, self-amortized mortgages instead of the typical five-year mortgages with large balloon payments at the end. Second, the government should encourage the banks and S & Ls to lower interest rates to make residential mortgages more affordable. Third, the government should aid private efforts to house low-income families. And last, ways should be found to reduce the cost of new home construction to make homes more affordable for middle-class Americans. While noble in intent, the recommendations called for unprecedented public intervention in what before was considered a private industry outside the legitimate sphere of government.

But desperate times spurred extraordinary measures. The government began innovative policies and programs to relieve the housing crisis and assist families facing the tragedy of foreclosure. Initial efforts, such as the Federal Home Loan Bank Act (1932), the Emergency Relief and Construction Act (1932), and the Greenbelt Town Program conducted during the mid-1930s by the Resettlement Administration, proved inconsequential. Another, the Federal Housing Administration (FHA) created in 1934, found the heart of the American Dream.

Before the FHA, first mortgages were limited to one-half or two-thirds of the appraised value of the property, so down payments of one-third to one-half were required. The lack of mortgage credit reflected a nineteenth-century attitude that disdained mortgages; the well-to-do were to purchase homes outright. By World War I that attitude softened to accept a limited amount of debt. Yet the new short-term mortgages and large balloon payment meant home ownership remained out of reach to many middle-class Americans. At the time, interest rates ranged between 6 and 8 percent which, though comparable to current rates, made the cost of credit prohibitively expensive to people of moderate means. Last, there were virtually no minimum construction standards so, with little standardization in the construction industry, homes were more expensive and varied in structural integrity. Home ownership remained mostly for the affluent.

In line with the Hoover Conference recommendations to remedy those problems, the FHA hoped to improve housing standards and construction, facilitate sound home financing on reasonable terms, and stabilize the mortgage market. Since the unemployment rate was about 25 percent, and much higher in the housing industry, the unstated goal was to jump-start the national economy and return as many Americans to work as possible. To accomplish those goals, the Roosevelt administration sought a program that worked within the private sector to stimulate housing construction without government spending, a formidable challenge, yet one for which a partial solution already existed. The FHA simply extended parts of that model.

The Home Owners Loan Corporation (HOLC) was created in 1933, about a year before the FHA, for similar reasons and with similar goals: to declare the protection of home ownership as national policy; to protect the individual homeowner from foreclosure; and to lessen the burden of high interest and principle payments. In the 1930s, the HOLC made available billions of dollars of government money to individuals to refinance tens of thousands of mortgages in danger of default or foreclosure, or to provide low-interest loans to enable owners to recover homes lost through forced sales. By 1935, the HOLC had financed one-tenth of all owner-occupied, nonfarm residences in the country. But the terms of HOLC loans differed dramatically from those typically offered by banks and S & Ls.

HOLC loans extended the life of the loan to about twenty years and eliminated the balloon payment at the end. The loans were also self-amortizing with uniform payments spread over the life of the debt. That significantly reduced the monthly payment and provided greater financial security for the homeowner. It also raised a problem.

If loans lasted decades, how might the value of the property, and the loan, be affected by changes in the surrounding neighborhood? Any prudent mortgager wanted reasonable assurance that the investment would not be adversely affected by undesirable changes. Consequently, the HOLC began the practice of appraising the neighborhood where a mortgage was considered on the basis of two factors: the area's socioeconomic composition, including the residents' occupation, income, and ethnicity; and the quality of the housing, such as the age, type of construction, price, sales demand, and state of repair. Such appraisals had

328 Visions of Paradise

been normal real estate practice since the advent of mortgage lending. What made the HOLC appraisals unique was their rigor and level of documentation.

An army of carefully trained appraisers mapped urban America, neighborhood by neighborhood, rating each into one of four categories based on its desirability for investment. The "First" category, also labeled "A" or "green," included new, homogeneous neighborhoods that would be in demand in good times and bad; homogeneous meant populated by "American business and professional men," and "American" euphemistically meant white and Christian. Blacks and Jews were not considered "American" regardless of income or education. The "Second" category ("B" or "blue") included older neighborhoods that had reached their peak but were expected to remain stable and desirable for many years. Neighborhoods in the "Third" category ("C" or "yellow") were "definitely declining." Those in the "Fourth" ("D" or "red") were areas "in which the things taking place in C areas have already happened."[12]

Underpinning the appraisal was a succession-like conceptual model that considered undesirable change in urban neighborhoods likely. The decline was based on three assumptions. The first assumed that property values dropped as the property aged and the structure became obsolete. The second assumed that a downward progression in the desirability of residents accompanied the decline in value. The third, with hints of Jeffersonian antiurbanism mixed with hints of Beecher's and Bushnell's social theories of suburban domesticity, assumed that innercity neighborhoods were less desirable and stable than suburban ones. While the logic was flawed, it reflected the suburban attitudes seen in many zoning ordinances, and the view of most middle- and upper-income Americans.

The neighborhood appraisal maps were used by the HOLC when considering loan applications, yet it insisted there was no implication that good loans did not exist for, or could not be made in, the Third and Fourth categories. Evidence indicates the HOLC did not discriminate between categories. Other lending institutions did, however, and the practice of "redlining" began: private banks and S & Ls who were privy to the government's Residential Security Maps used them, or other materials with the same assumptions, as the basis for discriminatory lending practices. Those prejudicial practices persisted until the Civil Rights Movement in the 1960s, and probably still persist in places today. Redlining became a self-fulfilling prophecy that drove innercity decline as mortgage money for home purchase or repair was diverted from older urban neighborhoods populated with low-income families or ethnic minorities to new home construction or purchase in the suburbs. The FHA magnified that tragedy.

The FHA differed from the HOLC in one critical way: while the HOLC made mortgage loans to individuals using government money, the FHA did not. As Roosevelt wished, the FHA worked through the private sector to accomplish its goals. The FHA only guaranteed a loan made by a private lender to an individual, eliminating the lender's risk while ensuring the loan's profitability. To receive that guarantee, the loan had to conform to conditions that expanded on those used by the HOLC, including the use of the HOLC's appraisal techniques and often its Residential Security Maps.

The FHA required the mortgage to be long-term, usually twenty-five to thirty years, and fully self-amortizing. It specified that loans needed down payments of only 10 percent or less. It also established minimum construction standards for the property that it enforced by onsite inspection to protect its long-term investment beyond the neighborhood appraisal. The phenomenal results of the revolutionary program were nearly immediate.

FHA construction standards spread through much of the home construction industry, even to non-FHA housing. Those standards helped to raise the minimum quality and reduce construction costs. Mortgage interest rates fell nationwide an average of several percentage points as lenders lowered rates to reflect their diminished risk. Housing construction and employment accelerated rapidly. Where in 1933 new housing starts numbered 93,000, in 1937 they numbered 332,000, and by 1941 they had reached nearly 620,000. The advent of World War II, however, temporarily interrupted the nation's rush to suburbia and redirected the economic recovery.

The Depression-era success of the FHA led to its expansion after the war with the Servicemen's Readjustment Act of 1944. The "GI Bill" established the Veterans Administration (VA) and offered a FHA-like mortgage program to help 16 million soldiers purchase suburban homes. Today the FHA and VA programs are commonly considered as one, since they are so similar. Together they promoted a renewed rush to suburbia in the economic boom after the war. Housing starts jumped more than tenfold, from 114,000 in 1944 to 1,692,000 in 1950. In the 1950s and 1960s, almost half of all homes in suburbia used FHA-VA financing, and the percentage of families who were homeowners climbed from less than half in 1934 to nearly two-thirds by 1970. The programs remain at work today.

Had the FHA existed in 1930, millions of Americans would have been spared the tragedy of bankruptcy and foreclosure. The FHA made suburban home ownership a part of the American Dream for the middle class and many of lesser means. Perhaps no government program has had as direct an impact on so many Americans the past half-century. But the remarkable benefits came with insidious side effects.

Like government subsidies to highway construction, the FHA gave preferential treatment to suburbs over cities. The FHA directed capital toward construction of new single-family homes in the financially safe and secure suburb, leaving the city wanting. Often, FHA loans for new suburban homes were more attractive than small loans for the renovation of existing urban homes. Restrictions were also placed on the use of FHA funds for rental properties. These effects inadvertently hastened the decay of urban America and drove more of the middle class to the suburbs.

The program's most insidious side effect was its tolerance of redlining for racial segregation. Using a system to rate the borrower, the property, and the neighborhood, patterned after the HOLCs, the FHA sought to protect the long-term market value of its investment. That reasonable purpose was perverted by the popular means—which favored all-white subdivisions—used for its accomplishment. FHA institutionalized redlining within the public and private sectors. Although now illegal, the damage done by its use from the Depression to the time of the Civil Rights Movement left a lasting legacy in urban and suburban American that may require generations to undo.

While some government programs have promoted urban redevelopment since the Depression, their effectiveness has paled in comparison to those, such as highway subsidies and FHA funds, that promoted suburban development and home ownership. From direct benefits such as tax deductions for mortgage interest and real estate taxes, to indirect benefits such as federal subsidies for sewer and water-line construction, the government throughout the century has placed a higher priority on suburbanization than urbanization. Whether purposeful or not, the effect has been the same—suburbs have proliferated while inner cities have struggled to remain economically and socially viable.

After a hiatus of nearly twenty years during the Depression and World War II, the transformation of the streetcar city to the automobile-oriented city resumed at a quickened pace. As it did, government intervention in urban and suburban development joined other long-standing forces driving that transformation. The century-old popular preference for the suburban lifestyle was reinforced by government programs for its promotion, some local such as zoning, others federal such as FHA-VA and highway construction. The American love affair with the automobile was rekindled by the resurgence of prosperity in a peacetime economy. The car had evolved from a luxury and curiosity for the wealthy, to a preference for the well-to-do, to a convenience for the middle class, to a daily necessity for the millions of new suburban commuters. The combination of those forces gave rise to a new suburban landscape in the postwar period, the landscape that now clothes most of metropolitan America: the auto-oriented city.

An easy way to understand this fabric is to simply note where you went last week, why you went there, and how you got there. For most, a weekly travelogue documents the nature of the auto-oriented city. Most Americans use cars on a daily basis to commute to work, go shopping, and reach leisure activities widely scattered throughout the metropolitan area. The amount we move about the city during our daily routines, mostly in cars, far exceeds that of our parents and grandparents. That decentralized landscape and its related auto-dependent behavior are the most distinguishing features of contemporary American cities.

Land uses in our current auto-oriented cities are segregated by type and dispersed across the city far more than even in the streetcar city. Where, before, many land uses were still mixed in relatively close proximity, zoning has now separated most into discreet districts. The new landscape is a mosaic of residential, commercial, industrial, and institutional districts sprinkled about based mostly on transportation access. Business and commercial uses tend to line the primary highway corridors that link the segments together; warehousing, industrial, and distribution uses tend to cluster around transportation hubs located at the fringe. But single-family, detached housing forms the landscape's basic pattern, with multifamily housing, schools, local parks and small neighborhood-oriented businesses inset. Housing density tends to decline away from the city center and toward the interior of the interstices between highway corridors.

The making of this mosaic dissolved many of the centers where some land uses once concentrated. The auto-oriented core is far less the business and commercial heart of the city than it was when all the streetcar lines converged there. More and more businesses of all types and sizes have dispersed throughout the landscape as the individual car, coupled with a weblike highway system, has offset the traditional advantages of the urbane central city. Now customers and employees often reach a business located in the periphery more easily than one located in the congested core. And such suburban locations often offer the business a variety of economic advantages, such as less expensive land and fewer constraints on expansion, as well as a variety of other locational amenities, including a more pleasant "natural" setting closer to employee homes. The advent of the "information superhighway" will likely have a similar dispersal effect, further decreasing the need for many businesses to crowd tightly together, or for clients and customers to physically meet. Mail order, FAX machines, cellular phones, and the Internet are transforming business practices and may subtly reshape the landscape.

The degree to which a city exhibits these characteristics varies based on many factors. The primary one is the timing of the city's growth relative to the predominant type of transportation. Those that were once large walking cities, such as New York, often retain remnants of that former character in the city core. There districts of more intermixed land uses and much higher densities still persist as old tenements and townhouses mix among new high-rise housing, offices, and businesses. Overall density in such cities is greater than in those that developed mostly after the advent of mass transit and so assumed a more diffused form initially. Similarly, remnants of streetcar characteristics can still be seen in the interior suburban rings and along original arterials of most American cities.

For example, the current population density in New York City is approximately twenty-three thousand people per square mile, while the next highest densities in American cities range from eleven thousand to fifteen thousand in Boston, Chicago, Philadelphia, and San Francisco. After that, densities in many old industrial cities, such as Baltimore, Cleveland, Detroit, and Pittsburgh, range from six thousand to nine thousand. Densities are lower still in cities whose real growth occurred following World War II as the car became commonplace in the postwar baby boom. Sunbelt cities such as Dallas, Houston, Phoenix, San Antonio, and San Diego all have densities of only two thousand to four thousand, as do midwestern cities such as Columbus and Indianapolis.

Such low-density suburban sprawl is both the cause for and result of our reliance on the car, and it has produced a fascinating auto-oriented landscape and car culture. What do suburbs say about American society? What do the seemingly endless streets crowded with convenient but somewhat gaudy strip shopping centers tell about our culture? What meaning is to be found in RVs, mobile homes, trailer parks, drive-in movies, drive-thru banks, fast-food restaurants, convenience stores, interstate highways, and motel chains? What does the suburban shopping mall mean versus the giant downtown department store or the small family-run store that once served the neighborhood? And how should we consider a city with virtually no center, or a look-alike landscape tailored to the car? A last look at our landscape offers some thoughts to these and other questions regarding its meaning.

· Hearing the Echoes ·

Trains are magical, especially to a youngster: the deep, pulsating reverberations of the powerful engines, strong enough to vibrate buildings and ground; the hissing sounds as steam escapes from dozens of tiny places about the beast; and the high-pitched screech of metal wheels pinching against metal rails. Trains are at once enchanting and mysterious, frightening and enticing. They tempt us with unknown adventures in faraway places in a much different way than the car or plane does.

The 1897 Union Depot in Columbus is among my most vivid memories. Its cavernous concourse, strangely devoid of life, dwarfed my senses. Sadly my memories are only of the station's melancholy

twilight. It must have had a much different atmosphere in the early
1900s when anxious crowds pressed against the long bank of ticket
windows and filled the waiting area with a nervous sense of anticipation for the 150 passenger trains that stopped each day. I only remember one window opened, and a scant few arrival-departure times listed
on the giant board. Imagine the rush of excitement that must have
raced through the throng when an anonymous voice announced the
awaited train over the loudspeaker. Might the din from all the commotion have muffled the sound, adding to the anxiety as people strained
to hear the call? My memory hears only a voice that droned hypnoti-
cally in the background, a voice seemingly bored with the slow, inexorable pace of decline, a dulled voice resigned to the inescapable fate
of obsolescence. Perhaps the voice was only a phantom of my fantasy.

The station had many of the qualities of an empty cathedral. Rows
of heavy wooden benches ran continuously the length of the concourse. A dim, filtered light cascaded down from above. And a vast
emptiness filled the space made more noticeable by the haunting echo
of solitary footsteps on the terrazzo floor. The floor had a special quality, as if all its characteristics were exaggerated. Somehow it seemed
far flatter, harder, and colder than other floors. Expansive marble
floors in deserted public places prompt the same response.

The layout of the station had people above, trains below; an arm
of the concourse extended over the tracks. For those who, like me,
waited above, this made wonderful theater when a train arrived. First
we felt the mechanical monster's distant approach, and our senses
sharpened, teased by its slow pace and unsettled by the subtle quak-
ing. Then we heard it: a low-pitched rumble accompanied by a stac-
cato of shrill calls. Finally, through the dirty windows and the jets of
steam fleeing the bowels of the diesel and the attached cars, we saw
tantalizing glimpses of it slithering below like a long snake. A conduc-
tor soon opened the concourse doorway, and the handful of passengers milling about descended the long flight of stairs leading down to
trackside. There, as we stood amid the heavy wooden luggage wagons
on the thick concrete platform, the serpent seemed less menacing and
mysterious. It was just a machine of moving parts we could see and
touch and enter. Our mutual exposure to the elements accentuated
this mechanical reality.

The grand station sat on the site of its two predecessors, originally
located in 1850 along the primary north-south road at the northern
city limit to avoid disruption of the city proper (although it bisected

a neighborhood of working-class Irish immigrants). As train traffic increased dramatically in the 1850s and 1860s, the tangle of tracks and trains crossing the street became obstacles to the city's northward expansion. Even then, the city's growth responded directly to the location of tracks and their associated businesses. Those century-old patterns continue to shape the city today, just as they do much of America.

To solve the conflict, a giant viaduct was built in the late 1800s to carry street traffic over the tracks. When the 1897 station was constructed, the prominent Chicago architect Daniel Burnham faced the station's front over the viaduct with a row of low business buildings connected by an arcade. People passed through a grand arch to reach the actual terminal. The ornate facade was designed in the Beaux Arts–style Burnham popularized in famous projects like the 1893 Columbian Exposition.

By the 1960s, the station was largely abandoned, passenger traffic usurped by interstates and airlines, and freight service moved to a new yard at the city's fringe. The derelict station was razed in 1976 to make room for a convention center, its valuable site recycled to a higher use in the succession of urban land.

The new convention center and the attached hotel are typical of most modern buildings. They offer no sense of place or history. No attempt was made in their design to respond to the site's crucial role in the city's history. Although the convention center plan called for the preservation of a portion of the High Street arcade, the boarded-up buildings were demolished, controversially, late one night. Hopes for sensitive restoration and adaptive reuse were destroyed. Community activists salvaged one arch. It now stands out of the way in Arch Park, a tiny memorial isolated like a traffic island several blocks away. The park memorializes many things beyond the legacy of an era gone by, including the needless breaking of potentially meaningful links with our landscape.

J. B. Jackson calls structures and spaces, like the convention center and hotel that stand in self-imposed isolation from their surroundings, "autonomous" spaces. Their internal orientation ignores the physical and social fabric and proudly proclaims a form of independence from the community, as if security guards and an address like "One Corporate Plaza," instead of a street address, absolves the structure of communal responsibility. Our cities are filling with fortresslike monoliths whose architecture turns, at times figuratively and at times literally, a blank wall to the street and the urban community. "Signature" buildings intended to make an aesthetic statement are "hung" in our cities like Picasso paintings. Instead of uplifting and enriching, the statement is often one of self-importance at the expense of community. Although the intent may differ, the psychological statement made by the physical form of these buildings too often amounts to: public not welcome; community of little concern. Such buildings have little sense of place. They could be located just as well anywhere in the world.

Autonomous places aren't limited to city cores. Exclusive business, commercial, residential, and recreational developments that seek to subtly separate themselves from their surroundings are becoming more and more common in the suburbs. Their elaborate entry features and gatehouses, and their internal road layouts that only link with the outside world at one or two points, have the same isolating and disconnecting effect as the corporate cathedrals in the central city. These autonomous places tear apart rather than stitch together a positive sense of community, and they offer little sense of physical or cultural continuity.

A second convention center that did try to reflect the legacy of the site in a contemporary context was recently built next to the first. Its design sought to express the transportation and telecommunication conduits that flow through the site's past, present, and future. The huge structure was segmented into a series of flowing, linear forms that mimic the tangle of tracks and freeways nearby, and the tangle of telecommunication conduits that now shape our cities. The forms front on the street with separate facades that match the scale of the pedestrian character of Burnham's arcade. It's a fascinating concept. Public reaction to the product has been mixed.

Until construction of the interstate highway system a generation ago, trains were the skeleton of the nation, a template like Jefferson's grid that shaped America and the American landscape during their formative years. That physical effect continues today, even in the train's near obsolescence. For those who have ridden one, the romance remains too. Train travel has an intimacy lost in the sterility of modern air or interstate travel. Unlike cars and planes, trains still travel at a leisurely pace through the heart of the landscape, providing an intimate look at our day-to-day world. A few trains still pass beneath the High Street rise, now mostly an anachronism. The two convention centers occupy the station's former site. Parts of the former rail yard to the east lie empty; a grassy field grows where the tracks

were removed. A tangle of freeway ramps now fills much of the resonant void. The rest of the land awaits recycling into some higher use
when the many economic and political conflicts are resolved. Like the
rapid rise and fall of cow towns, many industrial zones that gave early
form to urban America arose and fell with fundamental shifts in the
national economy. Many now stand silent until new uses are found
in the next shift. But the echoes of those former uses will reverberate
far into the future. New development should hear those echoes, not
ignore them.

CHAPTER 19

A Last Look

We are the children of our landscape.
　　　　　　　　Lawrence Durrell, *Justine*

Freedom and opportunity little constrained by the bonds of history have been the hallmark of the American landscape since settlement of the continent by Europeans. Here freedom and opportunity inherent in the ownership of land, in the use of that land, and in untethered movement across the continent, shaped the landscape. Here new ideas and old desires altered a virgin landscape with few of the complications posed by centuries of habitation in the Old World. These characteristics remain true today. Ours is a landscape made by a society devoted to individual "life, liberty, and property." Ours is a landscape free from the interference of war with foreign forces on our soil and the imposition of alien settlement patterns by a foreign conqueror. These features lend our landscape its most distinguishing characteristics.

It is at once a landscape of order and disorder. From far above, America exhibits the striking order imposed by use of the grid in the design of city and countryside. Even areas originally parceled using an unsystematic survey method, such as metes and bounds, exhibit an order from thirty thousand feet as they respond to the land's inherent physical features. Seen from a plane, America is a beautiful mosaic of shapes within shapes: squares and rectangles, circles and slivers. At first glance, the scene is also ordered at ground level. Individual homes share common landscaping conventions that respond to a social code of conduct, whether enforced through legal regulation or local custom. Farms do too. Neatness and tidiness are paramount.

This order has profound ecological and psychological connotations. Its imposition removes much of nature's messiness, reducing diversity, stability, and efficiency. Order imposes sharp lines and boundaries on an ecological world of gradients. It simplifies and compartmentalizes the land into a patchwork of homogeneous forms that require human manipulation to maintain. Physical reality is abstracted into lines on two-dimensional maps, lines like those that delineated initial parcels as the public domain was distributed, or those on a community master plan today. These lines then shape the land's physical form as they govern subsequent human action. Our quest to control nature has replaced the Native American sense of cooperation. Domination, not accommodation, of the land and the landscape values of many minority cultures has shaped the American landscape.

Yet within this order is disorder that stems from inexorable natural processes. These processes, temporarily disrupted by our actions, reexert themselves wherever maintenance slacks on urban land or wherever rural land lies fallow. Messiness reappears between the cracks in pavement, in derelict lots and abandoned fields, and with it, complexity and richness. This landscape disorder has connotations similar to signs of social incivilities. Our aesthetic reaction is to see mostly wasteful idleness in both the land and its human caretakers. In some instances, the apparent loss of control triggers apprehension, even fear. Such signs remind us that the order we impose is fleeting and the control we seek is ephemeral.

Our freedom and mobility are also continual sources of landscape disorder. Idiosyncrasy routinely casts aside common conventions, as seen in auto-oriented commercial strips, especially those in small towns. Mobility contributes to this as ideas spread across the country, triggering land use and landscaping practices alien to our locale. Ultimately, convenience, economics, and individual taste often take precedent over codes of conduct.

Ours has always been a temporary landscape, a landscape focused on the future not the past. Newness still dominates. While we struggle to preserve remnants of our past and create a sense of history, frequent movement and the rapid pace of change too often hide rather than highlight their value. This creates an adolescent-like landscape, a fluid, energetic place of dynamic change not like older, more stable, more "mature" landscapes. This transience, though, almost always occurs within a narrow range of "intensive" land uses associated with an industrial society, rather than within the type of uses associated with most Native

American cultures. Urbanization, recreation, agriculture, and other forms of resource extraction permanently replace wilderness and little-touched land; only the specific type of development and its character change.

Ours is an egalitarian, democratic landscape. Historically, one piece of land was considered the same as the next, not in size or physical character (although those have been common perceptions in principle), rather in the fact that all land was property, a commodity to which a common bundle of attributes and expectations were attached. Land was to be used to promote economic growth and prosperity. Land was to be distributed to and owned by citizens. In principle, America belonged to the people. Today, most of it does.

And ours is a rational landscape shaped by the values and perceptions of the Enlightenment. Land is known intellectually. Economic, functional, and practical concerns overwhelm appeals to aesthetics, emotion, and intuition. Science and reason serve as the basis for our actions, not folklore, myth, or superstition. Engineering technology fills the gap between the inherent landscape and our wants and wishes of it.

Freedom and opportunity, order and disorder, equality and democracy, permanence and transience, and rationality, each of these characteristics stems from our traditional Judeo-Christian, Euro-American landscape values, values little changed since James Kilbourne's day. Land remains a commodity for our use and benefit, property to be bought and sold. We remain separate from and superior to the land. And we continue to see the land mostly in nonemotional, nonhistorical terms. Some values have changed. Our overt moral and practical hostility toward wilderness has lessened. So has our blindness to action and outcome. We no longer believe in limitless abundance and resiliency. Our landscape behavior, though, has yet to fully reflect these changes.

Those traditional values foster a frontier, cowboylike landscape mentality in which consumptive use dominates, albeit from a Pinchotesque appreciation of nature. Extraction and manipulation take precedent over conservation or preservation on private property. Land and its uses remain disposable and are abandoned and routinely recycled when socioeconomic forces warrant. Planned obsolescence often prevails over permanence. Economics dictates use as we manipulate the land to fit our purposes rather than adjust our purposes to fit the land. Leopold and Powell would be appalled, though for different reasons. Yet attitudes toward the public domain have shifted profoundly. The government has changed from real estate agent to landlord. Our historical compulsion to dispose of the public domain to drive economic growth and westward

expansion, or to use it to benefit private business, albeit for public purposes, is being replaced by a desire to preserve large tracts for recreational or ecological reasons. Muir would be pleased.

The struggle to balance individual property rights and responsibilities with those of others—a struggle created by the interplay of those
values—has become increasingly complex. Our free use of private property has been tempered by a remarkable level of public rule and regulation, restraints that simultaneously limit that freedom and protect it.
How do we weigh idiosyncratic preferences regarding the use of one's
personal property against the rights of other individuals or the public?
On what basis do we establish the use of public property? These problems lie at the heart of local zoning and land use planning, and national
land policy and programs.

Our traditional values have also promoted a contradictory and in-
consistent set of landscape perceptions that complicates our behavior.
Like our attitudes toward Yellowstone, our attitudes toward the landscape in general mix fact and fiction. Consequently our landscape behavior remains schizophrenic. Intellectually, we are more informed about
the land than ever before. Science and the media have opened our eyes
to the world around us. Some of the resulting insight and actions are
excellent, many are trivial or contradictory. We reduce the production
of some greenhouse gases and recycle some household consumer products yet continue a consumption-based lifestyle. We set aside wildlife
refuges and seek to protect endangered species yet continue to devour
land for suburban sprawl. Our head may suggest one type of landscape
behavior, but it is our heart we usually hear. Personal experience, emotion, and economic self-interest are the more persuasive bases for forming opinion and affecting behavior than intellectual, secondhand knowledge derived from science and the media.

What is the nature of this experience? It's most often the landscape
around our home, workplace, or where we shop and recreate; or the local landscapes we see out the window of a building or car, bus, or train.
Less frequently, we see the landscape during longer travels for business
or pleasure, again mostly in fleeting glimpses out the window of a car
or plane, or during relatively short exposures outside. We rarely experience the landscape intimately beyond our immediate neighborhood.

What is the nature of the landscape we experience, whether at home
or away? It's mostly the standard romantic suburban landscape, or the
standard business landscape of city cores, or the standard rural landscape of commercial agriculture or resource extraction. Landscapes truly

responsive to local forces, physical and historical, are very rare. So too are landscapes little touched by humans. They lay beyond the regular experience of most Americans. The preservation of "wilderness" in our national parks and forests, and state lands too, is critical for our landscape values, although not for the exact reasons promoted by Frederick Jackson Turner and Teddy Roosevelt. Those public lands simply offer most people their only practical opportunity to experience little-touched nature, separate from the perpetuation of the positive American values in Turner's Frontier Thesis. While every landscape reflects its physical and sociocultural legacy somewhat, few do so significantly. In many ways, we are creating an American landscape sadly devoid of significant meaning, symbolism, and local identity. Diversity and richness are lost in the proliferation of autonomous places. The resulting homogenization diverts attention from the landscape's complete legacy, limiting a sense of physical and sociocultural continuity, disconnecting people from the land.

Most problematic, though, is the general disinterest in the land fostered by those traditional landscape values and our mobility. We still place little value on the land beyond its utilitarian and economic worth. Few of us draw delight and inspiration from it. Few study its legacy and meaning. It takes time and effort to learn about landscape, especially given the predominance of generic landscapes. Constant movement complicates the problem and makes setting deep landscape roots all the more difficult. Ours is a uniquely auto-oriented society and landscape. Provision for the car dominates development. Access is paramount. Yet mobility means more than daily commutes or long-distance moves. It includes the movement of ideas and information as well. This movement, both physical and intellectual, too often drives us toward conformity and homogenization rather than choice and individuality.

The prospects for the future of our landscape are embedded in that mobility and those values. Movement offers the opportunity to know new landscapes by relocation or by learning about remote places. Fascination with the foreign and unfamiliar can prompt us out of our landscape lethargy, enabling greater appreciation and literacy. This requires the desire to learn about the landscape and the opportunity to experience a distinct local one rather than the generic versions. Our landscape freedom ensures that ample remnants remain. People will always personalize their property, lending the landscape a vitality despite the pressure toward conformity and the prevalence of common types. The vernacular landscape responsive to the individual and the locality is more compelling and rewarding than the mass-produced landscape. The former foretells the future and reflects the past, the latter only repeats the present.

Powerful forces will continue the century-old reconsideration of our traditional landscape values and behavior: our devotion to property and consumption, to science, to separation and superiority, to disturbance and dominance supported by technology. Control and competition have served us well. However, the economics of resource depletion, the politics of population growth, and the recognition of our ecological impacts will push us farther along the path charted by Muir, Leopold, and others. Ironically, our rationality, a hallmark of those industrial values, will move us toward landscape values and behaviors that emphasize a cooperative caring for the land. It will gradually break down the common barrier erected between people and landscape, as it did for Muir and Leopold. This will eventually lead to a new stewardship ethic and a new relationship with the land, just as it did for them.

Perhaps then our stewardship beliefs and behaviors will more closely align. Land will be much more than property, and people will be much less separate from it. Rachel Carson called this relationship the web of life. That's really what Elzéard Bouffier sought to restore when he planted trees on the barren Provence mountainsides—the deep connection of people to the land. Such changes in our fundamental landscape values are prerequisites for meaningful change in our environmental behavior. Justice Holmes was right: morality cannot be legislated with laws and regulations. Reliance on rules to modify that behavior by requiring recycling or establishing quality standards for air and water, while important, will remain limited in effectiveness until our landscape values change. Then those rules become formalities. Nature preserves and bioreserves, preservation and conservation, even enlightened urban and suburban land use planning, treat the symptoms not the cause. Only with changed values will we be able to focus on enhancing landscape health rather than on repairing landscape injury. Only then will we become more fully connected to the land and better able to realize the wonder of our world.

Notes

PRELUDE

1. Consideration of the landscape as our autobiography and as a book we can "read" is not new; see, for example, Meinig, *Interpretation of Ordinary Landscapes,* and Watts, *Reading the Landscape.* The epigraph beginning this section is Meinig paraphrasing Lewis in *Interpretation of Ordinary Landscapes,* 2. The succeeding reference to Lewis also refers to this source.

2. Stilgoe, *Common Landscape of America,* 3. See also Meinig, *Interpretation of Ordinary Landscapes,* 1. The meaning of the term "landscape" has been the subject of much attention; see, for example, J. B. Jackson, *Discovering the Vernacular Landscape,* 3–8; Stilgoe, *Common Landscape of America,* 3–29; Conzen, *Making of the American Landscape,* 1–8; and, in general, Meinig, *Interpretation of Ordinary Landscapes.*

3. My distinction between space and place is similar to that proposed by Yi-Fu Tuan in *Space and Place.*

4. Jean Giono, "The Man Who Planted Hope and Grew Happiness," *Vogue* 88, no. 123 (March-April 1954): 108, 157–59. My characterization of the story is based on the version contained in Giono, *Man Who Planted Trees.*

5. I use Concord Estates, rather than the actual name of a specific community, to avoid confusion that might stem from the difference between that community's original boundary versus the common perception of its current boundary. While the core of the area I call Concord Estates is Muirfield Village, that original core development is now ringed by other subdivisions that comprise an extensive suburb.

6. John Stilgoe, as quoted in Lambert, "Safari on a City Street," 36.

7. Meisner, *Landscape Architecture in Ohio,* 4.

8. Thoreau, *Walden*, 79. This viewpoint is derived from that described by Mitchell in *Ceremonial Time*.

9. Meinig, *Interpretation of Ordinary Landscapes*, 12–13.

10. Ibid., 6.

11. J. B. Jackson made the point in *Discovering the Vernacular Landscape*, x; see also *Figure in a Landscape*.

CHAPTER 1. PARADISE LOST, PARADISE FOUND

My primary source for the discussion of Jonathan Alder was Johnda T. Davis's edition of Alder's journal (Alder, *Journal of Jonathan Alder*); Howe, *Historical Collections of Ohio*, and Foster, *Ohio Frontier*, also contain extensive excerpts from Alder's journal. I obtained Kilbourne's life story primarily from Berquist and Bowers, *New Eden*; and Lee, *History of Columbus, Ohio*. Sources for material on the general landscape values of Native Americans and Euro-Americans include Boorstin, *Americans: The Colonial Experience*; Cronon, *Changes in the Land*; Conzen, *Making of the American Landscape*; Hurt, *Ohio Frontier*; Jakle, *Images of the Ohio Valley*; Meinig, *Shaping of America*, vol. 1; Nash, *Wilderness and the American Mind*; Nobles, *American Frontiers*; and Oelschlaeger, *Idea of Wilderness*. See also Boorstin, *Americans: The National Experience*; de Crèvecœur, *Letters from an American Farmer*; Flint, *History and Geography of the Mississippi Valley*; Foster, *Ohio Frontier*; Opie, *Law of the Land*; and Worster, *Wealth of Nature*.

1. The term "frontier" is subject to many meanings. I am cognizant that its vernacular use to mean a sharp geographical line separating "wilderness" from "civilization" oversimplifies important characteristics, social, cultural, environmental, as Gregory Nobles detailed in *American Frontiers*. He ably showed how it is more accurately considered a zone of complex interaction and mixing among cultures. Please apply that meaning where appropriate.

2. Alder, *Journal of Jonathan Alder*, 5.

3. Ibid., 105, 116.

4. Ibid., 107.

5. Ibid., 108.

6. Ibid., 61.

7. Cronon, *Changes in the Land*, 61.

8. Ibid., 65.

9. Little, Personal Papers, 8–11.

10. The European landscape values and settlement practices transferred to the New World were as diverse as the many groups who made the crossing. The subsequent acculturation into the American values described herein was similarly complex yet dominated by those of northern and western Europe, especially Britain. See Zelinsky, *Cultural Geography of the United States*; see also Countryman, *Americans*; and Meinig, *Shaping of America*, vol. 1.

11. de Crèvecœur, *Letters from an American Farmer*, 24–25.

12. However, as Cronon was careful to qualify the point, there were impor-

tant variations within the way early colonists, and Indians, considered natural resources as commodities, as well as the closely related concepts of land sovereignty, and the ownership of communal and private property. These complex philosophical and legal concepts have many varied historical antecedents in both cultures. Those of European origin often reach back to the manorial customs of medieval England. Property rights evolved rapidly, both in the Old World and New, during the seventeenth and eighteenth centuries as a result of emerging legal and economic philosophies related to capitalism and democracy. John Locke's influence on both is a case in point. Yet despite some significant variance during early colonization, the differences in Euro-American concepts of property and ownership by Kilbourne's day lessened, and the concepts centered on those commonly held today. These are my focus. See also Cantor, *Imagining the Law;* Opie, *Law of the Land,* chapter 2; Platt, *Land Use and Society,* part 2; and Worster, *Wealth of Nature,* chapter 8.

13. Cronon, *Changes in the Land,* 57.

14. Ibid., 56.

15. Ibid., 56–57.

16. Flint, *History and Geography of the Mississippi Valley,* 113.

17. Cronon, *Changes in the Land,* 37–53.

18. Venable, *Footprints of the Pioneers,* 117–18. For further discussion of the pioneers' attitudes toward, and clearing of, the forest, see Conzen, *Making of the American Landscape,* chapter 8; Cronon, *Changes in the Land,* chapter 6; Jones, *History of Agriculture,* chapter 2; as well as de Crèvecœur's *Letters from an American Farmer.*

19. Although these historically were the predominate interpretations in Judeo-Christian theology, alternatives were offered. Most notably, St. Francis of Assisi (1181–1226) preached a more sympathetic perspective of nature similar to many Native American and Eastern philosophies. But his view was suppressed by the European medieval Christian church. For an alternate interpretation of the Bible's, and Judeo-Christianity's, stewardship ethic, see Denig, " 'On Values' Revisited." See also Hook, "Ecological Christian Stewardship."

20. Nash, *Wilderness and the American Mind,* 8, 24–25.

21. Annette Kolodny proposed another closely related motivation for and perception of New World colonization, as well as settlement westward across the continent, in her intriguing books *The Lay of the Land* and *The Land Before Her.* In *The Lay of the Land* she wrote:

> Along with their explicit hopes for commercial, religious, and political gains, the earliest explorers and settlers in the New World can be said to have carried with them a "yearning for paradise[," a yearning] . . . to experience the New World landscape, not merely as an object of domination and exploitation, but as a maternal "garden," receiving and nurturing human children. . . . At the deepest psychological level, the move to America was experienced as the daily reality of what has become its single dominating metaphor: regression from the cares of adult life and a return to the primal warmth of womb or breast in a feminine landscape. (*Lay of the Land,* 4–6)

Her thesis bears on my discussions of landscape illusions and myths, especially those related to the depiction of the landscape in art and literature and to the psychological origins of the popular American pastoral, romantic landscape.

Others share aspects of her thesis, especially related to perception of the New World and the West as utopian paradises; see, for example, Marx, *Machine in the Garden;* Smith, *Virgin Land;* and Worster, *Wealth of Nature.*

22. Berquist and Bowers, *New Eden,* 49.

23. Jakle, *Images of the Ohio Valley,* 119.

24. See, for example, *Louis Bromfield at Malabar;* for more on the landscape and sociocultural differences in Ohio resulting from the influence of settlers from the east and south, see Jakle, *Images of the Ohio Valley;* Knepper, *Ohio and Its People;* Meinig, *The Shaping of America;* and Peacefull, *Geography of Ohio.*

CHAPTER 2. ACTIONS AND OUTCOMES

Primary sources for this chapter include Atwater, *History of the State of Ohio;* Hildreth, *Pioneer History;* Howe, *Historical Collections of Ohio;* Jakle, *Images of the Ohio Valley;* Jones, *History of Agriculture;* Lafferty, *Ohio's Natural Heritage;* and Smith, *Scoouwa.* William Cronon's *Changes in the Land* provides an excellent overview of Native American and Euro-American modifications of (pre)colonial New England, and it parallels many parts of this chapter. See also Foster, *Ohio Frontier;* Hurt, *Ohio Frontier;* Nash, *Wilderness and the American Mind;* Viola and Margolis, *Seeds of Change;* and Worster, *Wealth of Nature.* For discussion of the generic nature of human landscape disturbance, see Goudie, *Human Impact on the Natural Environment.*

1. Fox, *The American Conservation Movement,* 151; and Nash, *Wilderness and the American Mind,* 96–97.

2. Atwater, *History of the State of Ohio,* 87.

3. Hildreth, *Pioneer History,* 485.

4. Smith, *Scoouwa,* 100.

5. Gist, *Christopher Gist's Journal,* 47; see also p. 55; and Jones, *History of Agriculture,* 6.

6. Cronon, *Changes in the Land,* 142–43.

7. Hildreth, *Pioneer History,* 497. In *Historical Collections of Ohio* (1857), Henry Howe described a typical "grand hunt" during the early 1800s:

> A large tract of wild land, the half or fourth of a township, was surrounded by lines of men [usually two to five hundred], with such intervals that each person could see or hear those next to him, right and left. The whole acted under the command of a captain, and at least four subordinates, who were generally mounted. At a signal of tin horns or trumpets, every man advanced in line towards the center, preserving an equal distance from those on either hand, and making as much noise as practicable. From the middle of each side of the exterior line, a blazed line of trees was previously marked to the center as a guide, and one of the sub-officers proceeded along each as the march progressed. About a half or three-fourths of a mile from the central point, a ring of blazed trees was made, and a similar one at the ground of the meeting, with a diameter at least equal to the greatest rifle range. On arriving at the first ring, the advancing lines halted till the commandant made a circuit, and saw the men equally distributed and all gaps closed. By this time, a herd of deer might be occasionally seen driving in

affright from one line to another. At the signal, the ranks move forward to the second ring, which is drawn around the foot of an eminence, or the margin of an open swamp or lake. Here, if the drive has been a successful one, great numbers of turkeys may be seen flying among the trees away from the spot. Deer in flocks, sweeping around the ring, under an incessant fire, panting and exhausted. When thus pressed, it is difficult to detain them long in the ring. They become desperate, and make for the line at full speed. If the men are too numerous and resolute to give way, they leap over their heads, and all the sticks, pitch-forks and guns raised to oppose them. By a concert of the regular hunters, gaps are sometimes made purposely to allow them to escape. The wolf is now seen skulking through the bushes, hoping to escape observation by concealment. If bears are driven in, they dash through the brush in a rage from one part of the field to another, regardless of the shower of bullets playing upon them. After the game appears to be mostly killed, a few good marksmen and dogs scour the ground within the circle, to stir up what may be concealed or wounded. This over, they advance again to the center with a shout, dragging along the carcasses which have fallen, for the purpose of making the count. It was at the hunt in Portage, that the bears were either exterminated or driven away from this vicinity. (346)

8. Lafferty, *Ohio's Natural Heritage*, 13. See also Lee, *History of the City of Columbus*, 13.

9. Hildreth, *Pioneer History*, 497.

10. Lafferty, *Ohio's Natural Heritage*, 11. See also Lee, *History of the City of Columbus*, chapter 17, for an extensive discussion of organized hunts and bounties in central Ohio; Cronon, *Changes in the Land*, chapter 5, provides a good overview for colonial New England.

CHAPTER 3. DESIGNS FOR A NATIONAL LANDSCAPE

General background material was drawn primarily from Divine, *America*. Primary sources on the Land Ordinance include Boorstin, *Americans: The National Experience*; Conzen, *Making of the American Landscape*; Hurt, *The Ohio Frontier*; J. B. Jackson, *Figure in a Landscape* and *Landscapes*; Johnson, *Order upon the Land*; Meinig, *Shaping of America*, vol. 1; Opie, *Law of the Land*; Reps, *Making of Urban America*; Stilgoe, *Common Landscape of America*; and Thrower, *Original Land Survey*. Major sources on land speculation and distribution, in addition to those above, include Berquist and Bowers, *New Eden*; and Tunnard, *City of Man*. Clawson, *Federal Lands Revisited* and *America's Land and Its Uses*; Worster, *Wealth of Nature*; and Zaslowsky and Watkins, *These American Lands*, provide additional insight on general American land use and land policy. Nobles, *American Frontiers*; Ellis, *American Sphinx*; and Billing, "Thomas Jefferson vs. Alexander Hamilton" provided background on Jefferson and Hamilton.

1. For an examination of (pre)colonial landscapes, see Boorstin, *Americans: The Colonial Experience* and *Americans: The National Experience*; Conzen, *Making of the American Landscape*; DeVoto, *Course of Empire*; Meinig, *Shaping of America*, vol. 1; and Stilgoe, *Common Landscape of America*.

2. Tunnard, *City of Man*, 105.

3. Dickens, *Martin Chuzzlewit*, 353–54.

4. Ibid., 378.

5. Ibid., xxxii.

6. The Native American occupancy of the Ohio territory was unbroken for millennia, until local groups were displaced during the Beaver Wars in the mid-1600s. Shortly afterward, groups from the surrounding territories recolonized the region. They were the direct ancestors of the Indians occupying Ohio when Euro-Americans first arrived.

7. Jefferson, *Notes on the State of Virginia*, 171–72.

8. Ibid., 172–73. The association of this set of positive values with an agrarian society—and the critical role it played in environmental stewardship and civic development—has roots reaching back to classical Greece. Jefferson's fear for the loss of our agrarian roots, and related values, is shared by contemporary authors who speculate their loss may trigger the same dire cultural and environmental consequences observed in ancient times. See Berry, *Unsettling of America*; Hanson, *Other Greeks*; Hughes, *Pan's Travail*; and Jackson, Berry, and Coleman, *Meeting the Expectations of the Land*.

9. Jefferson, *Notes on the State of Virginia*, 173.

CHAPTER 4. WESTWARD THE COURSE OF EMPIRE

General background material was drawn from Divine, *America*. I obtained material on the character, exploration, perception, and settlement of the Great Plains from Ambrose, *Undaunted Courage*; Boorstin, *Americans: The National Experience* and *Americans: The Democratic Experience*; Conzen, *Making of the American Landscape*; DeVoto, *Course of Empire*; Ellis, *American Sphinx*; Johnson, *Order upon the Land*; Meinig, *Shaping of America*, vol. 2; Nash, *Wilderness and the American Mind*; Nobles, *American Frontiers*; Opie, *Law of the Land*; Petulla, *American Environmental History*; Reisner, *Cadillac Desert*; Smith, *Virgin Land*; Stegner, *Beyond the Hundredth Meridian*; Truettner, *West as America*; Webb, *Great Plains*; West, *Way to the West*; and Worster, *Rivers of Empire, Wealth of Nature*, and *Unsettled Country*.

1. Meinig, *Shaping of America*, 10.

2. Ellis, *American Sphinx*, 205.

3. DeVoto, *Course of Empire*, 390, 392.

4. Ellis, *American Sphinx*, 206.

5. Divine, *America*, 232.

6. DeVoto, *Course of Empire*, 390.

7. Ambrose, *Undaunted Courage*, 101.

8. Smith, *Virgin Land*, 20–21.

9. Lewis and Clark, *Journals*, viii.

10. Ibid., 132, 137, 139.

11. Webb, *Great Plains*, 155–56.

12. Ibid., 153.

13. Smith, *Virgin Land*, 176.

14. Parkman, *California and Oregon Trail*, 55.

15. Webb, *Great Plains*, 159.

CHAPTER 5. LANDSCAPE: MYTH AND REALITY

Primary sources were those listed for the previous chapter.

1. Worster, *Wealth of Nature*, 4.
2. Worster, *Unsettled Country*, 65.
3. Divine, *America*, 348. O'Sullivan apparently used nearly the same language in justifying American claims to the Oregon territory; see Meinig, *Shaping of America*, 211.
4. Smith, *Virgin Land*, 21, 23.
5. Ibid., 25.
6. Ibid.; see also p. 37.
7. Ibid., 37; see also Truettner, *West as America*, 101.
8. Reisner, *Cadillac Desert*, 50.
9. Ibid., 37, 42.
10. Ibid., 37; see also, Opie, *Law of the Land*, 99–100.
11. Reisner, *Cadillac Desert*, 42.
12. Ibid., 39, 40.
13. Ibid., 41.
14. Stegner, *Beyond the Hundredth Meridian*, 2.

CHAPTER 6. PILGRIMS' PROGRESS

Sources for this chapter include, in addition to those listed for the previous two chapters, Day, "'Sooners' or 'Goners'"; Flint, *History and Geography of the Mississippi Valley;* Lapping, "Federal Rural Planning and Development Policy"; Reps, *Making of Urban America;* Steiner, "Evolution of Federal Agricultural Land Use Policy"; and Turner, *Frontier in American History.*

1. Webb, *Great Plains*, 220–21.
2. See Marx, *Machine in the Garden;* and Smith, *Virgin Land.*
3. Flint, *History and Geography of the Mississippi Valley,* 191–92.
4. Webb, *Great Plains*, 300.
5. Ibid., 239.
6. Day, "'Sooners' or 'Goners,'" 196–97.
7. Reps, *Making of Urban America*, 376.
8. Ibid.

CHAPTER 7. LOOKING AHEAD, LOOKING BACK

Sources for this chapter include, in addition to those from the previous three chapters, Powell, *Report on the Lands;* and Turner, *Frontier in American History.*

1. Powell, *Report on the Lands*, vii.
2. Ibid., viii–ix.
3. Ibid., ix–x.
4. Adams, *Education of Henry Adams*, 416. Although the statement was

apparently not made by Adams himself, he and many other prominent people of the day were in similar awe of King; see, in general, Adams, *Education of Henry Adams;* and Stegner, *Beyond the Hundredth Meridian.*

    5. Turner, *Frontier in American History,* 37.

    6. Ibid., 1.

    7. Ibid., 37–38.

    8. Cronon, Miles, and Gitlin, *Under an Open Sky,* provides an excellent re-interpretation of western settlement within the context of Turner's Frontier Thesis; see also Nobles, *American Frontiers.*

CHAPTER 8. THE VIEW FROM AFAR

Primary sources include Divine, *America;* Halpern, *On Nature;* Koster, *Transcendentalism in America;* P. Miller, *American Transcendentalists;* Nash, *Rights of Nature;* and *Wilderness and the American Mind;* Oelschlaeger, *Idea of Wilderness;* Palmer, *Looking at Philosophy;* Van Doren, *History of Knowledge;* and Worster, *Nature's Economy.*

    1. Thoreau, *Thoreau's Vision,* 144, 145.

    2. Hawthorne, *Scarlet Letter,* 132.

    3. Ibid., 134.

    4. Melville, *Moby-Dick,* 201.

    5. C. Miller, *American Transcendentalists,* 220.

    6. Thoreau, *Walden,* 74.

    7. Worster, *Nature's Economy,* 100.

    8. Thoreau, *Thoreau's Vision,* 145, 149.

CHAPTER 9. THE VIEW FROM WITHIN

Primary sources on Humboldt and Darwin, and the effect of American science and technology on the landscape, include Boorstin's trilogy, *Americans;* James and Martin, *All Possible Worlds;* Nash, *Rights of Nature;* Van Doren, *History of Knowledge;* and Worster, *Nature's Economy* and *Wealth of Nature.* Material for George Perkins Marsh was drawn primarily from Lowenthal, *George Perkins Marsh.* Material on Muir came mostly from Fox, *American Conservation Movement;* Halpern, *On Nature;* Nash, *Rights of Nature* and *Wilderness and the American Mind;* Oelschlaeger, *Idea of Wilderness;* and Worster, *Wealth of Nature.*

    1. Humboldt (1769–1859) was a German scientist and geographer whose remarkable work made contributions in fields ranging from botany to geology, astronomy to climatology. He traveled for five years in Central and South America, collecting sixty thousand species new to Europeans, generating vast amounts of climatological data, and producing many important observations and theories, including complete theories of taxonomy, isotherms, and terrestrial magnetism. He climbed many of the Central and South American mountains, mapped many of the rivers, and first described the effect of altitude and

physiography on the stratification of plants and animals. On his return trip he stopped in the United States, where he formed a lasting friendship with Thomas Jefferson. A prolific writer and speaker, Humboldt published thirty volumes on his explorations of Central and South America. Of that collected work, volumes 28–30, called *Personal Narrative of Travels,* were particularly influential world-wide and extremely influential on other emerging scientists. Darwin said this work changed his life. Humboldt soon became renowned around the world. In Europe, his fame was second only to that of Napoleon. Perhaps no other scientist since has achieved the same fame. His books quickly sold out and were reprinted again and again worldwide. The scholarly, scientific, and political nobility of the Western world, including Jefferson, sought his company, while common people lined the streets to applaud his passing.

Yet Humboldt's methods did not fully reflect the new "rules" of reductionist-based science that emerged during the Enlightenment. He was partly a self-taught amateur and generalist like those since Aristotle who studied nature. He was also partly a university trained, professional scientist. Humboldt can be seen as the culmination in the evolution of the old paradigm. He marked its pinnacle. He was also its endpoint, as Darwin's theories soon fundamentally changed the last key tenet of the old paradigm—the traditional, theology-based interpretation of the relationship between God, humankind, and nature. That change to a more secular, science-based interpretation, combined with other precursors, marked the beginning of our current paradigm.

Just three weeks after publishing his masterpiece, *Kosmos,* in 1859, Alexander von Humboldt died (the fifth and final volume was published posthumously by his brother, in 1862).

2. Marsh, *Man and Nature,* 3.

3. Ibid., 36.

4. Ibid., 9.

5. Environmental historians continue to document these historical effects today; for example, in *Pan's Travail,* a fascinating study of environmental problems in ancient Greece and Rome, J. Donald Hughes wrote, "Environmental changes as a result of human activities must be judged to be one of the causes in the decline of ancient Greek and Roman civilization, and in producing the stark conditions of the early Medieval centuries" (194). He also noted that such landscape change "has been clear to observers experienced in land use management since the versatile George Perkins Marsh" (1).

6. Marsh, *Man and Nature,* 3, 36–37.

7. Lowenthal, *George Perkins Marsh,* 246.

8. Ibid., 251.

9. See, for example, Brooks, *House of Life* and *Speaking for Nature,* and Downs, *Books That Changed America.*

10. Lowenthal, *George Perkins Marsh,* 246.

11. Marsh, *Man and Nature,* xxii.

12. Ibid., xxii (all quotes).

13. In *The Wealth of Nature,* Worster described the Campbellite sect and the potential importance of its doctrine on Muir's later life; see 191–96.

14. Fox, *American Conservation Movement,* 43.

15. Ibid., 44.
16. Ibid., 47.
17. Ibid., 52–53.
18. Muir, *Yosemite*, 1.
19. The Carrs had moved to Oakland in 1869, where Ezra began teaching at the University of California at Berkeley after he had been fired from the University of Wisconsin, so Muir and Jeanne Carr continued to correspond regularly.
20. Fox, *American Conservation Movement*, 5.
21. Ibid., 5.
22. Ibid., 56.

CHAPTER 10. A TALE OF TWO PARKS

This chapter was based on Simpson, "Tale of Two Parks." Primary sources therein included Chase, *Playing God in Yellowstone;* Chittenden, *Yellowstone National Park;* Haines, *Yellowstone Story;* and Nash, *Wilderness and the American Mind.* Additional insight was drawn from comments by John D. Varley, research administrator for the U.S. Department of the Interior, National Park Service, Yellowstone National Park.

1. Langford, "The Wonders of Yellowstone," 2.
2. Haines, *Yellowstone Story,* 98, 99.
3. Ibid., 101.
4. Ibid., 98.
5. The exact nature of that famous fireside chat and its role as the source of the national park concept may be as much legend as reality. Prior to that discussion, little of the public domain had been set aside to ensure public access. To the contrary, the government's primary goal since the Land Ordinance had been to dispose of the public domain as quickly as possible. Perhaps the first example of such protection, then, was the establishment of the Hot Springs Reservation in Arkansas. Established by the federal government in 1832, the 3,535-acre reserve maintained public access to the springs for therapeutic and recreational use. By the mid-1800s municipalities also began to establish public pleasuring grounds, such as New York City's Central Park, designed by Frederick Law Olmsted and Calvert Vaux in 1857. At the state and national level, the first example of setting public land aside for recreational use occurred when the Yosemite Valley and the Mariposa Big Tree Grove were protected in 1864 as part of a state trust.

Euro-American interest in Yosemite began concurrently with the California gold rush, although the valley's first sighting by a white person probably occurred when Joseph Rutherford Walker's party crossed from the east side of the Sierra in 1833. As white prospectors flooded the Sierra foothills following the discovery of gold at Sutter's Mill in 1848, conflicts with the local Indians who fought to protect their homeland became common. Consequently the state created the Mariposa Battalion to crush Indian resistance. The battalion entered the Yosemite Valley on March 27, 1851, in search of Indians.

Once the valley was cleared of its native people, word of the valley's spectacular scenery soon spread and attracted a growing number of tourists who arrived on horseback and even on foot. James Hutchins opened the valley's first businesses to cash in on the sprouting tourist trade; Muir later built and operated a sawmill for him. But the commercial development, both tourist-oriented (hotels) as well as resource-related (livestock grazing and logging), quickly began to damage the scenery. In response, a small group of visionary conservationists familiar with the area petitioned Congress through California Senator John Conness to intercede and grant the state the Yosemite Valley "for public use, resort and recreation . . . inalienable for all time" as a trustee of the federal government. Consequently, on June 30, 1864, Lincoln signed the bill protecting about 6,000 acres; the state formally accepted the grant about two years later.

During those two years, Olmsted was particularly instrumental in furthering the preserve's protection by acting as the influential chairman of the commission that oversaw it, and by preparing the primary report that detailed its resources and the philosophical basis for preservation. That report, "The Yosemite Valley and the Mariposa Big Tree Grove" (1865), reprinted in *Landscape Architecture*, in many ways first defined the national park concept. In her introduction to the reprint, Olmsted biographer Laura Wood Roper wrote, "In it he made explicit and systematic the political and moral ideas, already implicit in this novel act of the American people executed through their Congress, which not only justified their unexampled action but established it as sound precedent. In it also he made specific recommendations for turning the reservation to public use and enjoyment in accordance with the philosophy of the gift. *With this single report, in short, Olmsted formulated a philosophical base for the creation of state and national parks*" (13). See also Nash, *Wilderness*, 106.

So although some precedent for the concept of setting aside public land for recreational purposes existed prior to the Hedges's plan to protect Yellowstone, it was, nonetheless, a novel idea. See Chittenden, *Yellowstone National Park*, chapter 10; and Haines, *Yellowstone Story*, 130, for additional detail on the origins of the national park idea.

6. Langford, *Diary of the Washburn Expedition*, 117–18.

7. Haines, *Yellowstone Story*, 135.

8. Langford, "The Wonders of Yellowstone," 10.

9. Ibid., 12.

10. Ibid., 7.

11. U.S. House of Representatives Committee on the Public Lands, *Report: To Accompany Bill H.R. 764*, February 27, 1872, 1–2.

12. 17 Stat. 32 (1872). Management policies for Yellowstone are also guided by the Organic Act of 1916 (39 Stat. 535) and subsequent updates, which established the National Park Service (NPS). The Organic Act requires the NPS to "promote and regulate the use of the Federal areas known as national parks, monuments and reservations . . . by such means and measures . . . [as necessary] to conserve the scenery and natural and historic objects and the wild life therein and to provide for the enjoyment of the same in such manner and by such means as will leave them unimpaired for the enjoyment of future generations" (ch. 408, section 1). Many have interpreted this mandate in such a way as to create contradictory management goals similar to those in the Yellowstone Act; see Sax,

Mountains without Handrails, chapter 1. However, based on a detailed analysis of the legislative history, Robin Winks proposed that those who wrote the act had a clear set of priorities that promoted the preservation of scenery over other secondary considerations; see Winks, "Dispelling the Myth."

13. Nash, *Wilderness and the American Mind,* 150.

14. United States Senate, *The Report of Lieutenant Gustavus C. Doane,* 5–6.

15. Haines, *Yellowstone Story,* 179.

16. Chase, *Playing God in Yellowstone,* 17.

17. Ibid., 4.

18. For a discussion of many of the philosophical issues related to wilderness preservation in our parks, as well as public expectations and uses of our national parks, and the related politics, see Nash, *Rights of Nature;* Runte, *National Parks;* Sax, *Mountains without Handrails;* and Wirth, *Parks, Politics, and the People.*

CHAPTER 11. THE GREATER GOOD

Principal sources for this chapter include Fox, *American Conservation Movement;* C. Miller, "Greening of Gifford Pinchot"; Nash, *Rights of Nature* and *Wilderness and the American Mind;* and Pinchot, *Breaking New Ground.* See also Halpern, *On Nature;* Oelschlaeger, *Idea of Wilderness;* Petulla, *American Environmental History;* Rolston, *Environmental Ethics;* Worster, *Nature's Economy;* and Zaslowsky and Watkins, *These American Lands.*

1. Fox, *American Conservation Movement,* 21.

2. Humboldt's *Kosmos,* Darwin's *Origins,* Parkman's *The Oregon Trail,* as well as Hawthorne, Thoreau, and Whitman, Boswell, Carlyle, Shelley, Coleridge, and Wordsworth were favorites.

3. Nash, *Wilderness and the American Mind,* 132–33.

4. Fox, *American Conservation Movement,* 131.

5. Ibid., 98–99.

6. Nash, *Wilderness and the American Mind,* 131.

7. Fernow was actually the third head, succeeding Hough and another Marsh disciple, Dr. N. H. Egleston.

8. Fox, *American Conservation Movement,* 110.

9. Pinchot, *Breaking New Ground,* 31–32.

10. Pinchot recounted the timing of his decision a bit differently in his autobiography, *Breaking New Ground.* There he said he had already selected forestry as a career, on the advice of his father, when he matriculated at Yale in 1885. His father sought other means of penance as well. He became an early member of the American Forestry Association, and the Pinchot family later generously endowed the nation's first forestry school at Yale. The family also hosted the Yale Summer School of Forestry on its land in Pennsylvania, land that, ironically, still bore the scars of the type of timber speculation from which James and Cyril had profited.

11. Pinchot, *Breaking New Ground,* 22.

12. Although Gifford had briefly visited Vanderbilt's Biltmore estate outside Ashville, North Carolina, on his return trip from the South the previous year, James's longtime friendship with Olmsted was instrumental in arranging the meetings.

13. Pinchot, *Breaking New Ground,* 48.

14. Fox, *American Conservation Movement,* 113–14.

15. 30 Stat. 35 (1897).

16. Fox, *American Conservation Movement,* 139.

17. Nash, *Wilderness and the American Mind,* 163.

18. Ibid., 164.

19. Fox, *American Conservation Movement,* 140.

20. Nash, *Wilderness and the American Mind,* 168.

21. Muir, *Yosemite,* 196–97.

22. Nash, *Wilderness and the American Mind,* 161.

23. C. Miller, "Greening of Gifford Pinchot," 13.

24. Pinchot, *Breaking New Ground,* 509–10.

25. Nash, *Wilderness and the American Mind,* 180–81.

CHAPTER 12. WIDENING THE CIRCLE

Primary sources include Leopold, *Sand County Almanac;* Martin, "Voice for the Wilderness"; Nash, *Rights of Nature* and *Wilderness and the American Mind;* Oelschlaeger, *Idea of Wilderness;* and Worster, *Nature's Economy* and *Wealth of Nature.*

1. Leopold, *Sand County Almanac,* xvii, xviii–xix.

2. Pinchot, *Breaking New Ground,* 32.

3. The Taylor Grazing Act (1934) essentially closed the public domain to sale and settlement. It also placed 142 million acres into grazing districts managed by a new Grazing Service it created. In 1946, the Grazing Service was merged with the obsolete General Land Office to form the Bureau of Land Management (BLM) in order to manage what remained of the unassigned public domain after 150 years of sale or disposal to individuals under the scores of land laws, and after withdrawals for parks, forests, grasslands, wildlife refuges, veterans, defense installations, Indian reservations, land grants to states, colleges, canal companies, and railroads, and miscellaneous other purposes. The public domain is now around 650 million acres, roughly the size of India, with about half of it in Alaska; most of the remainder lies west of the Mississippi River in the lower forty-eight. The BLM administers about 240 million acres of national resource lands; the National Park Service 90 million acres of national parks, historic sites, and monuments; the Forest Service 191 million acres of national forests and grasslands; and the Fish and Wildlife Service 92 million acres of wildlife refuges. In fact the federal government owns on average about 50 percent of all land, excluding Indian reservations, in the western states; this ranges from a low of about 30 percent in Washington to a high of almost 90

percent in Nevada. See Worster, *Wealth of Nature,* chapter 8; and Zaslowsky and Watkins, *These American Lands.*

4. Leopold, *Sand County Almanac,* 138–39.
5. Martin, "Voice for the Wilderness," 73.
6. Nash, *Wilderness and the American Mind,* 186.
7. Nash, *Rights of Nature,* 60.
8. Rolston, *Environmental Ethics,* 65.
9. Nash, *Wilderness and the American Mind,* 189.
10. Leopold, *Sand County Almanac,* 189–90, 238–39, 240.

CHAPTER 13. LITIGATING THE LAND

Material for this chapter was drawn primarily from Anderson, *NEPA in the Courts;* Findley and Farber, *Environmental Law in a Nutshell;* Hagman, *Urban Planning;* Nash, *Rights of Nature;* and Stone, *Should Trees Have Standing?* I obtained additional insight from Professor Ken Pearlman, Department of City and Regional Planning, Ohio State University.

1. The Sierra Club actually recommended the remote Mineral King site as an alternative to a site nearer Los Angeles originally proposed by the U.S. Forest Service in the mid-1940s, so the club's later role in the controversy was many-sided, as was the controversy itself. See Sax, *Mountains without Handrails,* 67–70.
2. The case was 405 U.S. 727 (1972); Rogers C. B. Morton was then secretary of the interior. The quote is from Stone, *Should Trees Have Standing?* 61. (All quotations from the Supreme Court decision for *Sierra v. Morton* were taken from the reprint of that decision contained in Stone; page numbers refer to those in Stone.)
3. Ibid., 62.
4. 5 U.S.C. 551 *et seq.,* 701 *et seq.*
5. 348 U.S. 26 (1954).
6. *Berman v. Parker,* at 33.
7. *Data Processing v. Camp,* at 154. The three cases are *Barlow v. Collins,* 397 U.S. 159 (1970); *Association of Data Processing v. Camp,* 397 U.S. 150 (1970); and *Flast v. Cohen,* 392 U.S. 83 (1968).
8. 42 U.S.C. 4321 *et seq.*
9. See Anderson, *NEPA in the Courts,* 7.
10. Ibid., 7–9.
11. 42 U.S.C. 4331.
12. Ibid.
13. 42 U.S.C. 4342. Many of the CEQ's functions were later assumed by the Environmental Protection Agency (EPA). The EPA was created by executive order of the president shortly after NEPA was passed. Today the CEQ continues, but in a lesser capacity than originally specified.
14. 42 U.S.C. 4347.

15. After about a decade of angry confrontation in the courts, all parties to the adversarial struggle for the truth and the public good via the courts generally reached agreement on the basic requirements of the EIS—its content, timing, and use by the federal agencies, and its preparation and review process. Today most EISs are produced openly and completely, with relatively little controversy. Yet for all its significant effects, the extent to which NEPA resulted in improved environmental stewardship is not clear. Has our national environmental decision-making been improved, or has it become "rule by the minority" as some critics contend? Is the "public good" really best defined by or achieved through the adversarial process? After all, the process relies on both sides playing by the "rules" and participating actively. Do they? Has agreement on the substance and procedure of an EIS lessened the uncertainty initially created by NEPA and, therefore, reduced the law's effect on decision-making by rendering the process mechanical and predictable? Beyond the EIS, the government also committed itself in NEPA to a fundamental set of goals and policies. Has it been diligent in their pursuit regardless of whether they are enforceable by the courts? Has it been steadfast in their implementation? The record since NEPA's passage twenty-five years ago is spotty and subject to varying evaluations.

16. Stone, *Should Trees Have Standing?* 73, 76–81.

17. In determining whether there was injury sufficient for them to grant standing in environmental disputes, the courts later accepted as a type of harm that which resulted from a very tenuous chain of causation and, so, approximated the effect the Sierra Club sought. In a 1973 case known as the *United States v. SCRAP* (*United States v. Students Challenging Regulatory Agency Procedures,* 412 U.S. 669), the court made it clear that standing "is not to be denied simply because many people suffer the same injury." It also ruled the test for standing to be qualitative, not quantitative; thus the magnitude of the injury in fact was not important as long as *some* injury occurred, even if only indirectly related to the challenged action (see Findley and Farber, *Environmental Law in a Nutshell,* 4–5). The court reaffirmed its SCRAP decision in the 1978 Duke Power case (*Duke Power Co. v. Carolina Environmental Study Group, Inc.,* 438 U.S. 59), again accepting a very remote injury as sufficient for standing. In it, the court ruled that the plaintiffs need only demonstrate a "substantial likelihood" that the judicial relief they sought from the proposed governmental action would reduce their environmental injury (see ibid., 5–7).

However by the 1990s, the winds of popular, political, and judicial opinion shifted. In a nonenvironmental case in 1990, the court stated the SCRAP decision was "probably the most attenuated injury" ever accepted for standing, and that the decision "surely went to the very outer limit of the law" (*Whitmore v. Arkansas,* 495 U.S. 149; see ibid., 7). In another case later that year, the court seemingly asserted new limitations on the test for injury. In *Lujan v. National Wildlife Federation* (*Lujan v. National Wildlife Federation,* 498 U.S. 871), Justice Antonin Scalia's opinion for the court appeared to signal that the court would be less lenient in its acceptance of across-the-board type environmental litigation that seeks to protect the general public interest; instead, the court appeared to be assuming a more restrictive application of the standing law, requiring the plaintiff to establish specific injuries (whether economic or otherwise) to specific individuals from specific government actions on specific lands. The effect of the Lujan decision may well be to force a piecemeal approach to environmental litigation as opposed to unified litigation against broad government actions (see ibid., 7–10).

In the end, the greatest legal significance of *Sierra Club v. Morton* was its strong reaffirmation that noneconomic considerations were legitimate public concerns. In reiterating that concept, first addressed in *Berman v. Parker* more than fifteen years earlier, and subsequently reinforced by the three standing-related decisions around 1970, the court kept the legal doorway open while it compelled the government to give intangible considerations weight equal to that of traditional economic considerations in all its decisions. However, the final decision did not modify the test for standing as specifically proposed by the Sierra Club, or as Justice Douglas and Professor Stone advocated through the alternate interpretation, which called for extending rights to nature, as Aldo Leopold had proposed.

18. See Holmes, *Collected Legal Papers,* 170.

CHAPTER 14. THE EMOTIONAL LANDSCAPE

This chapter was based on Simpson, "The Emotional Landscape." Among the major sources therein were Caudill, *Night Comes to the Cumberlands;* and U.S. Department of the Interior, *Surface Mining and Our Environment.*

1. U.S. Department of the Interior, *Surface Mining,* 51–52.
2. 91 Stat. 445 (1977).
3. L. C. Dunlap, "An Analysis of the Legislative History of the Surface Mining Control and Reclamation Act of 1975," in *Rocky Mountain Mineral Law Institute Annual* (Albany, N.Y.: Matthew Bender, 1976), 11. See also M. E. Schecter, "The Surface Mining Control and Reclamation Act of 1977: Its Background and Its Effects," *New York Law School Law Review* 25 (1980): 953.
4. Dunlap, "An Analysis," 11.
5. 79 Stat. 5 (1965).
6. U.S. Department of the Interior, *Surface Mining,* 56–62.
7. Ibid., 105–6.
8. Parts of this discussion were based on personal conversations with former Congressmen Wayne L. Hays, Ken Hechler, and John L. Seiberling, and with James A. Curry, formerly with the U.S. Department of the Interior, Office of Surface Mining, Reclamation and Enforcement, and the TVA.
9. See Executive Office of the President, Council on Wage and Price Stability, *A Study of Coal Prices* (Washington, D.C.: Executive Office of the President, Council on Wage and Price Stability, 1976).
10. United States Senate Committee on Interior and Insular Affairs, Council of Environmental Quality, *Coal Surface Mining and Reclamation: An Environmental and Economic Assessment of Alternatives, S.R. 45,* 93d Cong., 1st sess., March 1973, serial no. 93–8.
11. United States House of Representatives, *Regulation of Strip Mining*

Hearings before the Subcommittee on Mines and Mining of the House Committee on Interior and Insular Affairs, 92d Cong., 1st sess., September 1971, 60.

CHAPTER 15. REMAKING URBAN AMERICA

Primary sources for this chapter include Berquist and Bowers, *New Eden;* Divine, *America;* K. Jackson, *Crabgrass Frontier;* and Reps, *Making of Urban America.* Secondary sources include Lapping, "Federal Rural Planning and Development Policy"; Steiner, "Evolution of Federal Agricultural Land Use"; and Van Doren, *History of Knowledge.*

  1. K . Jackson, *Crabgrass Frontier,* 68.
  2. Ibid., 12.
  3. Ibid., 16.
  4. Berquist and Bowers, *New Eden,* 117.
  5. Prior to that time, most colleges, such as those that offered Marsh and Powell so little of interest, were small, liberal arts oriented, and church affiliated. They were often among the first institutions established in new communities like Worthington, and they proliferated westward with the advance of settlement throughout the 1800s. A source of much community pride, they reflected a widespread sense of civic-mindedness and optimism, as well as the prevalent boosterism of the period. The local college symbolized the community's "arrival"; for many, it also denoted the town's moral and cultural stability (even superiority), and the town's physical permanence. Community boosters pointed proudly to it to promote further growth.
  Most colleges were modeled after those in Europe. Admission was often based on one's knowledge of the classics and one's social standing. Curriculums focused on subjects like rhetoric, grammar, mathematics, history, geography, philosophy, logic, and literature and emphasized the classics of ancient Greek and Latin culture. Such knowledge defined an educated person at the time, yet had little direct relevance to the typical person's daily needs. Today our landscape is dotted with the descendants of those colleges; Harvard, Yale, and Dartmouth are well-known examples.
  However, the typical person working on a farm or in a small business wanted and needed training in agriculture, home economics, or basic business skills. The curriculums of most colleges, including the few but growing number of state-charted public colleges, began to better address the day-to-day needs of the local community, but change was slow relative to the pace of change in society. By the mid-1800s, a whole new type of college that broke the old model was needed to better address the practical requirements of the growing nation.
  The Morrill Land Grant Act, signed into law by President Lincoln on July 2, 1862, reflected that need and created a fundamentally new, and distinctly American, form of higher education—the act established the nation's network of public land grant colleges. Championed throughout the 1850s by Vermont Senator Justin S. Morrill, the landmark legislation was finally passed for a variety of political reasons related to the Civil War, as well as in response to public pressure for agricultural colleges from influential people such as Thomas Clemson,

Horace Greeley, Andrew Jackson Downing, and Frederick Law Olmsted, and lobbying by state agricultural and horticultural societies. The resulting act provided each state thirty thousand acres of the public domain, per member of Congress, to sell or otherwise use "to the endowment, support and maintenance of at least one college where the leading object shall be, without excluding other scientific and classical studies, and including military tactics, to teach such branches of learning as are related to agriculture and the mechanic arts, in such manner as the legislatures of the States may respectively prescribe, in order to promote the liberal and practical education of the industrial classes in the several pursuits and professions of life." (See Steiner, "Evolution of Federal Agricultural Land Use," 351.)

Eastern states devoid of public land were given scrips to federal property in western states to be sold at the market rate of $1.25 per acre. In total the federal government gave the states more than 13 million acres, much to the dismay of the sparsely populated western states who consequently received less land than the more populated eastern states. Adding insult to injury, the federal largesse of land in the midst of the western states meant the states also lost the land as a potential source of revenue to finance internal improvements, although the land was usually sold for settlement by the recipient state to raise funds to support the land grant university. Western states often saw that land as their own and hoped the federal government might grant it to them as it had granted millions of prime acres to states and companies for other purposes.

The act's intent to promote practical, public education in agriculture and the mechanical trades for the working and middle classes, and for the sake of advancement of knowledge in those fields, reflected the nation's agricultural past while anticipating its industrial future. Although America was still predominately agriculture-based, industrialization was well under way. The new agricultural and mechanical colleges would therefore train the workforce to till the fields and run the factories more efficiently. And as the government had done so frequently in the past, it again used its most plentiful resource, land, to promote the nation's economic growth and development. Today, land grant colleges are a resonant symbol of our nation's past and our present persona; their physical and symbolic connections to the land are not coincidental.

6. See for example, Marx, *Machine in the Garden;* and Stilgoe, *Metropolitan Corridor.*

7. Howells, *Suburban Sketches,* 16. Although Howells criticized early suburbs' lack of municipal services, his writings also reflected emerging suburban ideals and indicated the effects of transportation on the urban and suburban experience, as well as the startling appearance of the steam locomotive in the pastoral suburban solitude. Of his suburban community he wrote, "Then, indeed, Charlesbridge appeared to us a kind of Paradise. The wind blew all day from the southwest, and all day in the grove across the way the orioles sang to their nestlings. The butcher's wagon rattled merrily up to our gate every morning; and if we kept no other reckoning, we should have known it was Thursday by the grocer. We were living in the country with the conveniences and luxuries of the city about us." (Ibid., 12.)

8. K. Jackson, *Crabgrass Frontier,* 28.

9. Ibid., 29.

CHAPTER 16. DREAM WEAVERS

This chapter was based mostly on Crandell, *Nature Pictorialized;* Downing, *Treatise on the Theory and Practice of Landscape Gardening;* Handlin, *American Home;* J. B. Jackson, *Figure in a Landscape* and *Landscapes;* K. Jackson, *Crabgrass Frontier;* Newton, *Design on the Land;* Reps, *Making of Urban America;* Repton, *Art of Landscape Gardening* (a combined edition that includes both *Sketches and Hints* and *Theory and Practice*); F. Scott, *Art of Beautifying Suburban Home Grounds;* M. Scott, *American City Planning since 1890;* and Tunnard, *City of Man.*

1. K. Jackson, *Crabgrass Frontier,* 48–49.
2. Ibid., 48.
3. Handlin, *American Home,* 7.
4. Ibid., 7, 8.
5. K. Jackson, *Crabgrass Frontier,* 62.
6. Ibid., 62.
7. Ibid., 62, 63.
8. Ibid., 63, 64.
9. Downing, *Treatise,* ivA.
10. Handlin, *American Home,* 42.
11. Downing, *Treatise.* 18–19. While Downing and others differentiated between the "beautiful" and the "picturesque" in landscape design, and placed great significance on this difference, the distinction is inconsequential for our purposes; see Downing, *Treatise,* 51.
12. The Hudson River school of painting was a loose confederacy of landscape painters in the mid-1800s whose work typically reflected a similar style. By style I mean their paintings' content, composition, color, and technique, as well as the way the artists manipulated these attributes to interpret their subjects. Many of this group—including the school's founder, Thomas Cole, and Frederic Church—lived along the Hudson River in upstate New York.

The style was typified by allegorical scenes like the proverbial pastoral clearing carved from the surrounding forest wilderness. The values the Hudson River school painters spread by way of their immensely popular paintings were derived in part from the English Landscape style used by artists like John Constable, as well as the picturesque landscapes of eighteenth- and nineteenth-century English gardens, parks, and estates designed by "Capability" Brown and Humphry Repton. Underlying those roots were the seventeenth-century pastoral paintings of Claude Lorrain and the appearance of the English manorial landscape (see note 14, below). The Hudson River school painting style is important to the American landscape because of its influence on nineteenth-century public aesthetic values, especially those related to the nonurban (wilderness, rural, and suburban) landscape. And it is important because of its influence on key tastemakers, whose writings and designs in the latter half of the 1800s defined and promoted the romantic landscape, the modern-day manifestation of which dominates our suburban landscape in places like Concord Estates.
13. See, for example, *Dedham Vale* (1828), which he considered perhaps his best, *Dedham Vale, Morning* (c. 1811), *Salisbury, Morning* (c. 1811), *Wivenhoe*

Park, Essex (1817), and *The Cornfield* (1826). For further discussion of the evolution in landscape painting, see Bazarov, *Landscape Painting;* Clark, *Landscape into Art;* and Crandell, *Nature Pictorialized.*

14. For a discussion of the manorial landscape and its evolution, see Cantor, *English Medieval Landscape;* and Hoskins, *Making of the English Landscape.*

15. Kassler, *Modern Gardens and the Landscape,* 11; for further detail on Pope's interpretation of genius loci, and its meaning as a garden design concept during the period, see Hunt, *Gardens and the Picturesque,* chapter 8.

16. *Sketches and Hints on Landscape Gardening* (1794), *The Theory and Practice of Landscape Gardening* (1803), and, *An Enquiry into the Changes of Taste in Landscape Gardening* (1806), which repeated, with little change, much of his previous published material.

17. Repton, *Art of Landscape Gardening,* 3. Repton quotes this stanza without citing his source.

18. K. Jackson, *Crabgrass Frontier,* 66.

19. Handlin, *American Home,* 47; see also 46.

20. F. Scott, *Art of Beautifying Suburban Home Grounds;* and Weidenmann, *Beautifying Country Homes: A Handbook of Landscape Gardening.*

21. F. Scott, *Art of Beautifying Suburban Home Grounds,* 11.

22. Ibid., 19.

23. Ibid., 107, 111.

CHAPTER 17. AMERICA'S LANDSCAPE ARCHITECT

Major sources, in addition to those listed for the previous chapter, include Olmsted, *Walks and Talks of an American Farmer in England;* Olmsted, Vaux, and Company, *Preliminary Report;* and Roper, *FLO.*

1. Olmsted and Kimball, *Forty Years of Landscape Architecture,* vol. 1, 61.

2. Olmsted, *Walks and Talks of an American Farmer in England,* 82–83.

3. Ibid., 90–91; this quote was taken from the first edition, second volume (New York: George P. Putnam and Company, 1852), and differs slightly from that in the 1859 edition, as Olmsted made minor edits in the later edition.

4. Ibid., 1852 ed., 154–55.

5. Ibid., 103.

6. Ibid., 1859 ed., 187–88.

7. Ibid., 62–64.

8. Olmsted, *Cotton Kingdom,* xvi.

9. Roper, *FLO,* 133.

10. Olmsted, *Cotton Kingdom,* xxvii, xxxvi.

11. Roper, *FLO,* 152.

12. Reps, *Making of Urban America,* 331.

13. Ibid., 330.

14. Roper, *FLO,* 135.

15. Ibid., 277.

16. K. Jackson, *Crabgrass Frontier,* 92.

17. Olmsted, Vaux, and Co., *Preliminary Report,* 26–27.

18. See ibid., 16; and K. Jackson, *Crabgrass Frontier*, 80.

19. Olmsted, Vaux and Co., *Preliminary Report*, 27–28; see also Roper, *FLO*, 323.

CHAPTER 18. CITY AS SUBURB

Primary sources for this chapter included Hagman, *Urban Planning and Land Development Control Law*; K. Jackson, *Crabgrass Frontier*; and M. Scott, *American City Planning*. I obtained additional insight from Professor Ken Pearlman, Department of City and Regional Planning, Ohio State University.

1. K. Jackson, *Crabgrass Frontier*, 190.
2. See, for example, Helphand, "Learning from Linksland."
3. K. Jackson, *Crabgrass Frontier*, 99.
4. Ibid., 188.
5. Ibid., 157.
6. Ibid., 159.
7. Hagman, *Urban Planning and Land Development Control Law*, 80–81.
8. 272 U.S. 365 (1926). In the two years following the Euclid decision, the high court decided two other zoning-related cases—*Zahn et al. v. Board of Public Works et al.*, 274 U.S. 325 (1927), and *Nectow v. City of Cambridge et al.*, 277 U.S. 183 (1928)—but has been mostly silent on the basic legality ever since. However, it has decided a number of cases that served to define the limits and proper procedures of zoning in response to issues that arose incrementally in the ensuing years; notable examples include the *Berman v. Parker* decision, which expanded the sphere of legitimate public and governmental concern to include aesthetics in 1954, and the *Village of Belle Terre v. Boraas*, 416 U.S. 1 (1973), where the court further expanded the police power to include the regulation of social lifestyle. Those decisions reflected the same general liberalization occurring simultaneously in environment-related decisions.
9. Ibid., at 369.
10. Ibid., at 388.
11. K. Jackson, *Crabgrass Frontier*, 193–94.
12. Ibid., 197–98.

Works Cited

Adams, Henry. *The Education of Henry Adams,* edited by Ernest Samuels. Boston: Houghton Mifflin, 1974.

Alder, Jonathan. *The Journal of Jonathan Alder,* edited by Johnda T. Davis. N.p., 1988.

Ambrose, Stephen E. *Undaunted Courage: Meriwether Lewis, Thomas Jefferson, and the Opening of the American West.* New York: Simon and Schuster, 1996.

Anderson, Frederick R. *NEPA in the Courts: A Legal Analysis of the National Environmental Policy Act.* Baltimore: Johns Hopkins University Press, 1973.

Atwater, Caleb. *History of the State of Ohio, Natural and Civil.* Cincinnati: Glezen and Shepard, 1838.

Bazarov, Konstantin. *Landscape Painting.* London: Octopus Books, 1981.

Beadle, J. H. *Western Wilds and the Men Who Redeem Them.* Cincinnati: Jones Brothers and Co., 1878.

Berquist, Goodwin, and Paul C. Bowers Jr. *The New Eden: James Kilbourne and the Development of Ohio.* Lanham: University Press of America, 1983.

Berry, Wendell. *The Unsettling of America: Culture and Agriculture.* San Francisco: Sierra Club Books, 1977.

Billing, John C. "Thomas Jefferson vs. Alexander Hamilton: Philosophies to the Origin of an American Spirit and Sense of Place." In *Proceedings of the Rural Planning and Development: Visions of the 21st Century Conference,* 166–77. Gainesville: University of Florida, 1991.

Boorstin, Daniel J. *The Americans: The Colonial Experience.* New York: Random House, 1958.

———. *The Americans: The National Experience.* New York: Vintage, 1965.

———. *The Americans: The Democratic Experience.* New York: Vintage, 1973.

Bromfield, Louis. *Louis Bromfield at Malabar: Writings on Farming and Country Life,* edited by Charles E. Little. Baltimore: Johns Hopkins University Press, 1988.

Brooks, Paul. *The House of Life: Rachel Carson at Work.* Boston: Houghton Mifflin, 1972.

———. *Speaking for Nature: How Literary Naturalists from Henry David Thoreau to Rachel Carson Have Shaped America.* Boston: Houghton Mifflin, 1980.

Cantor, Leonard, ed. *The English Medieval Landscape.* London: Croom Helm, 1982.

Cantor, Norman F. *Imagining the Law: Common Law and the Foundations of the American Legal System.* New York: HarperCollins, 1997.

Carson, Rachel. *Silent Spring.* New York: Houghton Mifflin, 1962.

Caudill, Harry M. *Night Comes to the Cumberlands: A Biography of a Depressed Area.* Boston: Little, Brown, and Co., 1963.

Chase, Alston. *Playing God in Yellowstone: The Destruction of America's First National Park.* Boston: Atlantic Monthly Press, 1986.

Chittenden, Hiram M. *The Yellowstone National Park.* Norman: University of Oklahoma Press, 1964.

Clark, Kenneth. *Landscape into Art.* Revised ed. New York: HarperCollins, 1979.

Clawson, Marion. *Man and Land in the United States.* Lincoln: University of Nebraska Press, 1964.

———. *America's Land and Its Uses.* Baltimore: Johns Hopkins University Press, for Resources for the Future, 1972.

———. *The Federal Lands Revisited.* Baltimore: Johns Hopkins University Press, for Resources for the Future, 1983.

Conzen, Michael P., ed. *The Making of the American Landscape.* Boston: Unwin Hyman, 1990.

Countryman, Edward. *Americans: A Collision of Histories.* New York: Hill and Wang, 1996.

Crandell, Gina. *Nature Pictorialized: "The View" in Landscape History.* Baltimore: Johns Hopkins University Press, 1993.

Cronon, William. *Changes in the Land: Indians, Colonists, and the Ecology of New England.* New York: Hill and Wang, 1983.

Cronon, William, George Miles, and Jay Gitlin, eds. *Under an Open Sky: Rethinking America's Western Past.* New York: W. W. Norton, 1992.

Day, Robert. " 'Sooners' or 'Goners,' They Were Hellbent on Grabbing Free Land." *Smithsonian* 20, no. 8 (1989): 192–206.

de Crèvecœur, Hector St. John. *Letters from an American Farmer.* New York: E. P. Dutton and Co., 1945.

Denig, Nancy Watkins. " 'On Values' Revisited: A Judeo-Christian Theology of Man and Nature." *Landscape Journal* 4, no. 2 (1985): 96–105.

de Tocqueville, Alexis. *Democracy in America,* edited by J. P. Mayer. New York: HarperPerennial, 1988.

DeVoto, Bernard. *The Course of Empire.* Boston: Houghton Mifflin, 1980.

Dickens, Charles. *Martin Chuzzlewit,* edited by Margaret Cardwell. London: Clarendon, 1982.

Divine, Robert A., et al. *America: Past and Present.* Vol. 1. 3rd ed. New York: HarperCollins, 1991.

Downing, Andrew Jackson. (A Facsimile Edition of) *A Treatise on the Theory and Practice of Landscape Gardening.* New York: Funk and Wagnalls, 1967.

Downs, Robert B. *Books That Changed America.* New York: Macmillan, 1970.

Ellis, Joseph J. *American Sphinx: The Character of Thomas Jefferson.* New York: Alfred A. Knopf, 1997.

Fabos, Julius Gy., Gordon T. Milde, and V. Michael Weinmayr. *Frederick Law Olmsted, Sr.: Founder of Landscape Architecture in America.* Amherst: University of Massachusetts Press, 1968.

Findley, Roger W., and Daniel A. Farber. *Environmental Law in a Nutshell.* 3rd ed. St. Paul: West Publishing, 1992.

Flint, Timothy. *The History and Geography of the Mississippi Valley.* Vol. 1. 3rd ed. Cincinnati: E. H. Flint, 1833.

Foster, Emily, ed. *The Ohio Frontier: An Anthology of Early Writings.* Lexington: University Press of Kentucky, 1996.

Fox, Stephen. *The American Conservation Movement: John Muir and His Legacy.* Madison: University of Wisconsin Press, 1981.

Gans, Herbert J. *The Levittowners: Ways of Life and Politics in a New Suburban Community.* New York: Vintage Books, 1967.

Giono, Jean. *The Man Who Planted Trees.* Chelsea: Chelsea Green, 1987.

Gist, Christopher. *Christopher Gist's Journals,* edited by William Darlington. 1893. Reprint, New York: Argonaut, 1966.

Goudie, Andrew. *The Human Impact on the Natural Environment.* 4th ed. Cambridge: MIT Press, 1994.

Hagman, Donald G. *Urban Planning and Land Development Control Law.* St. Paul: West Publishing, 1975.

Haines, Aubrey L. *The Yellowstone Story: A History of Our First National Park.* Vol. 1. Yellowstone National Park, Wyo.: Yellowstone Library and Museum Association, 1977.

Halpern, Daniel, ed. *On Nature: Nature, Landscape, and Natural History.* San Francisco: North Point, 1987.

Handlin, David P. *The American Home: Architecture and Society, 1815–1915.* Boston: Little, Brown, and Co., 1979.

Hanson, Victor Davis. *The Other Greeks: The Family Farm and the Agrarian Roots of Western Civilization.* New York: Free Press, 1995.

Hawthorne, Nathaniel. *The Scarlet Letter,* edited by Sculley Bradley, Richmond Croom Beatty, and E. Hudson Long. New York: W. W. Norton and Co., 1962.

Helphand, Kenneth I. "Learning from Linksland." *Landscape Journal* 14, no. 1 (1995): 74–86.

Hildreth, Samuel P. *Pioneer History: Being an Account of the First Examinations of the Ohio Valley and the Early Settlement of the Northwest Territory.* Cincinnati: H. W. Derby and Co., 1848.

Holmes, Oliver Wendell. *Collected Legal Papers.* New York: Peter Smith, 1952.

Hook, William John. "Ecological Christian Stewardship as a Paradigm for Christian Environmental Ethics." Ph.D. diss., Vanderbilt University, 1992.

Hoskins, W. G. *The Making of the English Landscape.* Harmondsworth, Middlesex, England: Penguin Books, 1970.

Howe, Henry. *Historical Collections of Ohio.* Cincinnati: E. Morgan and Co., 1857.

Howells, William Dean. *Suburban Sketches.* New York: Hurd and Houghton, 1871.

Hubbard, Henry Vincent, and Theodora Kimball. *An Introduction to the Study of Landscape Design.* New York: Macmillan, 1917.

Hughes, J. Donald. *Pan's Travail: Environmental Problems of the Ancient Greeks and Romans.* Baltimore: Johns Hopkins University Press, 1994.

Hunt, John Dixon. *Gardens and the Picturesque: Studies in the History of Landscape Architecture.* Cambridge: MIT Press, 1992.

Hurt, R. Douglas. *The Ohio Frontier: Crucible of the Old Northwest, 1720–1830.* Bloomington: Indiana University Press, 1996.

Jackson, John Brinckerhoff. *Landscapes: Selected Writings of J. B. Jackson,* edited by Ervin H. Zube. Amherst: University of Massachusetts Press, 1970.

———. *Discovering the Vernacular Landscape.* New Haven: Yale University Press, 1984.

———. *Figure in a Landscape: A Conversation with J. B. Jackson* (a film by Janet Mendelsohn and Claire Marino, producers). Los Angeles: Direct Cinema Limited, 1988.

Jackson, Kenneth T. *Crabgrass Frontier: The Suburbanization of the United States.* New York: Oxford University Press, 1985.

Jackson, Wes, Wendell Berry, and Bruce Coleman, eds. *Meeting the Expectations of the Land: Essay in Sustainable Agriculture and Stewardship.* San Francisco: North Point, 1984.

Jakle, John A. *Images of the Ohio Valley: A Historical Geography of Travel, 1740 to 1860.* New York: Oxford University Press, 1977.

James, Preston J., and Geoffrey J. Martin. *All Possible Worlds: A History of Geographical Ideas.* 2nd ed. New York: John Wiley and Sons, 1981.

Jefferson, Thomas. *Notes on the State of Virginia.* Boston: Lilly and Wait, 1832.

Johnson, Hildegard Binder. *Order upon the Land: The U.S. Rectangular Land Survey and the Upper Mississippi Country.* New York: Oxford University Press, 1976.

Jones, Robert Leslie. *A History of Agriculture in Ohio to 1880.* Kent, Ohio: Kent State University Press, 1983.

Kassler, Elizabeth B. *Modern Gardens and the Landscape.* Garden City, N.Y.: Doubleday and Co., 1984.

Knepper, George W. *Ohio and Its People.* Kent, Ohio: Kent State University Press, 1989.

Kolodny, Annette. *The Lay of the Land: Metaphor as Experience and History in American Life and Letters.* Chapel Hill: University of North Carolina Press, 1975.

———. *The Land before Her: Fantasy and Experience of the American Frontiers, 1630–1860.* Chapel Hill: University of North Carolina Press, 1984.

Koster, Donald N. *Transcendentalism in America*. New York: Twayne Publishers, 1975.

Lafferty, Michael B., ed. *Ohio's Natural Heritage*. Columbus: Ohio Academy of Science, 1979.

Lambert, Craig. "Safari on a City Street." *Harvard Magazine* 98, no. 3 (1996): 36–42.

Langford, Nathaniel P. "The Wonders of Yellowstone." *Scribner's Monthly* 2, no. 1 (May 1871): 1–17; continued in 2, no. 2 (June 1871): 113–28.

———. *Diary of the Washburn Expedition to the Yellowstone and Firehole Rivers in the Year 1870, by Nathaniel Pitt Langford*. St. Paul, Minn.: J. E. Haynes, 1905.

Lapping, Mark B. "Federal Rural Planning and Development Policy: An Interpretive History." In *Proceedings of the Rural Planning and Development: Visions of the 21st Century Conference*, 14–40. Gainesville: University of Florida, 1991.

Lee, Alfred E. *History of the City of Columbus, Capital of Ohio*. Vol. 1. New York: Munsell and Co., 1892.

Leopold, Aldo. *A Sand County Almanac, with Essays on Conservation from the Round River*. New York: Ballantine Books, 1966.

Lewis, Meriwether, and William Clark. *The Journals of Lewis and Clark*, edited by John Bakeless. New York: Penguin Books, 1964.

Little, Nathaniel. Personal Papers. Drawer No. 3, File PNF Lit., Document 73G121A. Worthington Historical Society, Worthington, Ohio.

Lowenthal, David. *George Perkins Marsh: Versatile Vermonter*. New York: Columbia University Press, 1958.

Marsh, George Perkins. *Man and Nature*, edited by David Lowenthal. Cambridge: Harvard University Press, Belknap Press, 1965.

Martin, Erik J. "A Voice for the Wilderness: Arthur H. Carhart." *Landscape Architecture* 76, no. 4 (1986): 70–75.

Marx, Leo. *The Machine in the Garden: Technology and the Pastoral Ideal in America*. New York: Oxford University Press, 1964.

Meinig, D. W., ed. *The Interpretation of Ordinary Landscapes: Geographical Essays*. New York: Oxford University Press, 1979.

———. *The Shaping of America: A Geographical Perspective on 500 Years of History*. Vol. 1: *Atlantic America, 1492–1800*. New Haven: Yale University Press, 1986.

———. *The Shaping of America: A Geographical Perspective on 500 Years of History*. Vol. 2: *Continental America, 1800–1867*. New Haven: Yale University Press, 1993.

Meisner, G., ed. *Landscape Architecture in Ohio: Designs for Livable Places*. Columbus: Ohio Chapter of the American Society of Landscape Architects, 1985.

Melville, Herman. *Moby-Dick*. New York: Macmillan, 1962.

Miller, Char. "The Greening of Gifford Pinchot." *Environmental History Review* 16, no. 3 (1992): 1–20.

Miller, Perry. *The American Transcendentalists, Their Prose and Poetry*. Garden City, N.Y.: Doubleday, 1957.

Mitchell, John Hanson. *Ceremonial Time: Fifteen Thousand Years on One Square Mile.* New York: Warner Books, 1984.

Muir, John. *The Yosemite.* San Francisco: Sierra Club Books, 1988.

———. *Our National Parks.* San Francisco: Sierra Club Books, 1991.

Nash, Roderick. *Wilderness and the American Mind.* 3rd ed. New Haven: Yale University Press, 1982.

———. *The Rights of Nature: A History of Environmental Ethics.* Madison: University of Wisconsin Press, 1989.

Newton, Norman. *Design on the Land: The Development of Landscape Architecture.* Cambridge: Harvard University Press, Belknap Press, 1971.

Nobles, Gregory H. *American Frontiers: Cultural Encounters and Continental Conquest.* New York: Hill and Wang, 1997.

Oelschlaeger, Max. *The Idea of Wilderness: From Prehistory to the Age of Ecology.* New Haven: Yale University Press, 1991.

Olmsted, Frederick Law. *Walks and Talks of an American Farmer in England in the Years 1850–51.* Columbus: Jos. H. Riley and Co., 1859.

———. *Forty Years of Landscape Architecture: Being the Professional Papers of Frederick Law Olmsted, Senior,* edited by Frederick Law Olmsted Jr. and Theodora Kimball. Vols. 1 and 2. New York: G. P. Putnam's Sons, 1922.

———. "The Yosemite Valley and the Mariposa Big Tree Grove." *Landscape Architecture* 43, no. 1 (1952): 12–25.

———. *The Cotton Kingdom: A Traveller's Observations on Cotton and Slavery in the American Slave States Based upon Three Former Volumes of Journeys and Investigations by the Same Author,* edited by Arthur M. Schlesinger. New York: Alfred A. Knopf, 1953.

———. *Civilizing American Cities: A Selection of Frederick Law Olmsted's Writings on City Landscapes,* edited by S. B. Sutton. Cambridge: MIT Press, 1971.

Olmsted, Vaux, and Company. *Preliminary Report upon the Proposed Suburban Village at Riverside, near Chicago.* New York: Sutton, Bowne, 1868.

Opie, John. *The Law of the Land: Two Hundred Years of American Farmland Policy.* Lincoln: University of Nebraska Press, 1987.

Palmer, Donald. *Looking at Philosophy: The Unbearable Heaviness of Philosophy Made Lighter.* 2nd ed. Mountain View: Mayfield Publishing, 1994.

Parkman, Francis. *The California and Oregon Trail: Being Sketches of Prairie and Rocky Mountain Life.* New York: Hurst and Co., 1849.

Peacefull, Leonard, ed. *A Geography of Ohio.* Kent, Ohio: Kent State University Press, 1996.

Petulla, Joseph M. *American Environmental History.* San Francisco: Boyd and Fraser, 1977.

Pinchot, Gifford. *Breaking New Ground.* New York: Harcourt, Brace, 1947.

Platt, Rutherford H. *Land Use and Society: Geography, Law, and Public Policy.* Washington, D.C.: Island Press, 1996.

Powell, John Wesley. *Report on the Lands of the Arid Region of the United States, with a More Detailed Account of the Lands of Utah,* edited by Wallace Stegner. Cambridge: Harvard University Press, Belknap Press, 1962.

Reisner, Marc. *Cadillac Desert: The American West and Its Disappearing Water.* New York: Viking Penguin, 1986.

Reps, John W. *The Making of Urban America: A History of City Planning in the United States.* Princeton: Princeton University Press, 1965.

Repton, Humphry. *An Inquiry into the Changes of Taste in Landscape Gardening.* London: J. Taylor, 1806.

———. *The Art of Landscape Gardening,* edited by John Nolen. Boston: Houghton Mifflin, 1907.

Roper, Laura Wood. *FLO: A Biography of Frederick Law Olmsted.* Baltimore: Johns Hopkins University Press, 1973.

Runte, Alfred. *National Parks: The American Experience.* Lincoln: University of Nebraska Press, 1979.

Sax, Joseph L. *Mountains without Handrails: Reflections on the National Parks.* Ann Arbor: University of Michigan Press, 1980.

Scott, Frank J. *The Art of Beautifying Suburban Home Grounds of Small Extent.* New York: American Book Exchange, 1881.

Scott, Mel. *American City Planning since 1890.* Berkeley and Los Angeles: University of California Press, 1969.

Simpson, John W. "The Emotional Landscape." *Landscape Architecture* 75, no. 3 (1985): 60–63, 108–9, 112–13.

———. "A Tale of Two Parks." *Landscape Architecture* 77, no. 3 (1987): 60–67.

Smith, Henry Nash. *Virgin Land: The American West as Symbol and Myth.* Cambridge: Harvard University Press, 1978.

Smith, James. *Scoouwa: James Smith's Indian Captivity Narrative.* Columbus: Ohio Historical Society, 1978.

Stegner, Wallace. *Beyond the Hundredth Meridian: John Wesley Powell and the Second Opening of the West.* Lincoln: University of Nebraska Press, 1954.

Steiner, Frederick. "The Evolution of Federal Agricultural Land Use Policy in the United States." *Journal of Rural Studies* 4, no. 4 (1988): 349–63.

Stilgoe, John R. *Common Landscape of America, 1580–1845.* New Haven: Yale University Press, 1982.

———. *Metropolitan Corridor: Railroads and the American Scene.* New Haven: Yale University Press, 1983.

———. *Borderland: Origins of the American Suburb, 1820–1939.* New Haven: Yale University Press, 1988.

Stone, Christopher D. *Should Trees Have Standing? Toward Legal Rights for Natural Objects.* Los Altos, Calif.: William Kaufmann, 1974.

Thoreau, Henry David. *Walden, or Life in the Woods, and On the Duty of Civil Disobedience.* New York: Macmillan, 1962.

———. *Thoreau's Vision: The Major Essays,* edited by Charles R. Anderson. Englewood Cliffs: Prentice-Hall, 1973.

Thrower, Norman J. W. *Original Land Survey and Land Subdivision: A Comparative Study of the Form and Effect of Contrasting Cadastral Surveys.* Chicago: Rand McNally and Co., 1966.

Tobey, George B., Jr. *A History of Landscape Architecture: The Relationship of People to Environment.* New York: American Elsevier Publishing, 1973.

Truettner, William H., ed. *The West as America: Reinterpreting Images of the Frontier, 1820–1920.* Washington, D.C.: Smithsonian Institution Press, 1991.

Tuan, Yi-Fu. *Space and Place: The Perspective of Experience.* Minneapolis: University of Minnesota Press, 1977.

Tunnard, Christopher. *The City of Man: A New Approach to the Recovery of Beauty in American Cities.* 2nd ed. New York: Charles Scribner's Sons, 1970.

Turner, Frederick Jackson. *The Frontier in American History.* Tucson: University of Arizona Press, 1986.

United States Department of the Interior. *Surface Mining and Our Environment.* Washington, D.C.: U.S. Government Printing Office, 1967.

United States Senate. *Report of Lieutenant Gustavus C. Doane upon the So-Called Yellowstone Expedition of 1870.* 41st Cong., 3rd sess., 1873, Exec. Doc. 51.

Van Doren, Charles. *A History of Knowledge: Past, Present, and Future.* New York: Ballantine Books, 1991.

Venable, W. H. *Footprints of the Pioneers in the Ohio Valley: A Centennial Sketch.* Cincinnati: Ohio Valley Press, 1888.

Viola, Herman J., and Carolyn Margolis, eds. *Seeds of Change.* Washington, D.C.: Smithsonian Institution Press, 1991.

Watts, May T. *Reading the Landscape: An Adventure in Ecology.* New York: Macmillan, 1957.

Webb, Walter Prescott. *The Great Plains.* Lincoln: University of Nebraska Press, 1959.

Weidenmann, Jacob. *Beautifying Country Homes, a Handbook of Landscape Gardening.* New York: Orange Judd, 1870.

West, Elliott. *The Way to the West: Essays on the Central Plains.* Albuquerque: University of New Mexico Press, 1995.

Whitman, Walt. *Walt Whitman: Complete Poetry and Selected Prose and Letters,* edited by Emory Holloway. London: Nonesuch, 1967.

Winks, Robin. "Dispelling the Myth." *National Parks* 70, nos. 7–8 (July–August 1996): 52–53.

Wirth, Conrad. *Parks, Politics, and the People.* Norman: University of Oklahoma Press, 1980.

Worster, Donald. *Nature's Economy: A History of Ecological Ideas.* Cambridge: Cambridge University Press, 1977.

———. *Rivers of Empire: Water, Aridity, and the Growth of the American West.* New York: Pantheon Books, 1985.

———. *The Wealth of Nature: Environmental History and the Ecological Imagination.* New York: Oxford University Press, 1993.

———. *An Unsettled Country: Changing Landscapes of the American West.* Albuquerque: University of New Mexico Press, 1994.

Zaslowsky, Dyan, and T. H. Watkins. *These American Lands: Parks, Wilderness, and the Public Lands.* Revised and expanded ed. Washington, D.C.: Island Press, 1994.

Zelinsky, Wilbur. *The Cultural Geography of the United States.* Revised ed. Englewood Cliffs, N.J.: Prentice-Hall, 1992.

Index

Moby-Dick (Melville), 139–140
Monroe, James, 60, 69, 79, 88
Moore, Charles, 167, 168
Moore, John, 13
Moran, Thomas, 122, 168, 169
Morrill Land Grant Act, 288, 363n5
Morton, Levi P., 191
Muir, Daniel, 151, 152, 153, 154, 157
Muir, John, 151–164, 201, 217; childhood and farm life of, 151–155; and
  early conservation movement, 178–
  180; early employment of, 155; and
  Emerson, 161–163; eye injury of, 159;
  forestry preservation philosophy of,
  189, 193–194; and Hetch Hetchy,
  196–198, 199–200; marriage and fatherhood of, 179; orchid sighting by,
  157–158; travels of, 157–158, 159–
  161; university education of, 156–157;
  writing career of, 163–164, 177–178;
  and Yosemite, 161–163, 177–178,
  182–185, 357n5
Mumford, Lewis, 150, 253
myths: Arcadian, 25, 107–108, 244, 281;
  Elzéard Bouffier, 3–5; Frontier Thesis,
  129–131; Great American Desert, 80–
  82, 88; Land of Gilpin, 92–96; of wilderness transformed into Garden of
  Eden, 25, 107–108, 349n21; Wild
  West, 97, 121–122

Napoleon I, 68, 69, 71
Nash, Roderick, 24–25, 201, 220
National Environmental Policy Act
  (NEPA), 227–233; Council of Environmental Quality (CEQ) created by, 229–
  230, 360n13; environmental impact
  statement (EIS) requirement of, 231–
  233, 361n15; history of, 227–229
national grid. *See* grid system
national parks: origin of concept of, 167,
  356n5; Yellowstone as, 165, 167–168,
  170–171, 357n12; Yosemite as, 183–
  185, 356n5. *See also* U.S. National
  Park Service
Native Americans: Euro-Americans captured and raised by, 11–14, 32; Euro-
  Americans' observations of, 19, 77;
  European diseases' effect on, 36, 99;
  fire use by, 31, 32, 172–173; government bureau studying, 128; in Great
  Plains, 86–87, 98–99; on humans as
  part of nature, 15; importance of horse
  to, 86–87; landscape values of, 14–15;
  and Northwest Territory land distribution, 55, 352n6; property/ownership
  conception of, 15; rationalizations for

displacing, 21–22; wilderness beliefs
  of, 22, 24; in Yellowstone, 172–173
natural resources: as commodities,
  348n12; inexhaustible supply of, 23
nature: control of, 41–42, 255, 342;
  humans as separate from, 6, 24, 244;
  Native American conception of, 14; in
  suburbia, 6, 255. *See also* wilderness
Nature (Emerson), 144
Nature's Economy (Worster), 141
Newlands Act, 128
New Orleans, 67
Niagara Falls, protection of, 181
*Night Comes to the Cumberlands: A Bi-
  ography of a Depressed Area* (Caudill),
  236, 237
Nixon, Richard, 227, 242
Noble, John W., 187
Northern Pacific Railroad (NPRR), and
  Yellowstone, 167, 169, 170, 173–174
Northwest Ordinance (1787), 17
Northwest Passage, expeditions to find,
  71–78
Northwest Territory: colonies' claiming
  lands of, 47; Euro-American settlement
  of, 17–18, 28–29; Land Ordinance of
  1785 applied to, 50; Native Americans
  of, and government land distribution,
  55, 352n6
Norton, Charles Eliot, 181, 302
Notes on the State of Virginia (Jefferson),
  59

Ohio: deforestation of, 30–31; Euro-
  American settlement of, 17–18, 28–29;
  Land Ordinance of 1785 applied to,
  49, 58; Native Americans of, and government land distribution, 55, 352n6
Ohio Company: Gist's, 32, 48; land speculation by, 53–54
Oklahoma Land Rush, 113–115
Olmsted, Denison, 292
Olmsted, Frederick Law, 288–313,
  364n5; Biltmore estate work by, 192,
  359n12; childhood of, 288–290; collaboration of Weidenmann with, 285;
  education of, 290–291, 292; farming
  career of, 293–294; landscape architecture career of, 303–313; Llewellyn
  Park, New Jersey plan of, 309–310;
  marriage of, 302; and Central Park,
  303–309, 356n5; and Niagara Falls,
  181; Riverside, Illinois, design of,
  309–312; sailing experience of, 291–
  292; sense of social justice of, 290,
  292, 296–297; walking tour of England by, 295–298; work in West by,

Illustrations:	Abigail Rorer
Compositor:	Prestige Typography
Text:	Sabon
Display:	Sabon
Printer and Binder:	Haddon Craftsmen, Inc.